BIFURCATION ANALYSIS OF FLUID FLOWS

A better understanding of the mechanisms leading a fluid system to exhibit turbulent behaviour is one of the grand challenges of the physical and mathematical sciences. Over the last few decades, numerical bifurcation methods have been extended and applied to several flow problems to identify critical conditions for fluid instabilities to occur. This book provides a state-of-the-art account of these numerical methods, with much attention to modern linear systems solvers and generalized eigenvalue problem solvers. These methods also have a broad applicability in industrial, environmental, and astrophysical flows. The book is a must-have reference for anyone working in scientific fields where fluid flow instabilities play a role. Exercises at the end of each chapter and Python code for the bifurcation analysis of canonical fluid flow problems provide practice material to get to grips with the methods and concepts presented in the book.

HENK A. DIJKSTRA is professor of dynamical oceanography at the Institute for Marine and Atmospheric Research Utrecht (IMAU) within the Department of Physics and Astronomy at Utrecht University, The Netherlands. He has been a member of the Dutch Royal Academy of Arts and Sciences (KNAW) since 2002. He received the Lewis Fry Richardson medal from the European Geosciences Union in 2005, he was elected a Fellow of the Society for Industrial and Applied Mathematics (SIAM) in 2009, and he was awarded an Advanced Grant from the European Research Council in 2021. He is (co-)author of several books, including *Nonlinear Physical Oceanography* (2005), *Dynamical Oceanography* (2008), *Nonlinear Climate Dynamics* (Cambridge University Press, 2013), and *Networks in Climate* (Cambridge University Press, 2019).

FRED W. WUBS is associate professor of numerical mathematics at the Bernoulli Institute for Mathematics, Computer Science and Artificial Intelligence within the Faculty of Science and Engineering of the University of Groningen, The Netherlands. He is the (co-)author of almost 70 publications in the areas of numerical treatment of (stochastic) partial differential equations, preconditioning of sparse linear systems, solution of large sparse eigenvalue problems, and high-performance computing.

BIFURCATION ANALYSIS OF FLUID FLOWS

FRED W. WUBS
University of Groningen

HENK A. DIJKSTRA
Utrecht University

CAMBRIDGE
UNIVERSITY PRESS

Shaftesbury Road, Cambridge CB2 8EA, United Kingdom

One Liberty Plaza, 20th Floor, New York, NY 10006, USA

477 Williamstown Road, Port Melbourne, VIC 3207, Australia

314–321, 3rd Floor, Plot 3, Splendor Forum, Jasola District Centre, New Delhi – 110025, India

103 Penang Road, #05–06/07, Visioncrest Commercial, Singapore 238467

Cambridge University Press is part of Cambridge University Press & Assessment, a department of the University of Cambridge.

We share the University's mission to contribute to society through the pursuit of education, learning and research at the highest international levels of excellence.

www.cambridge.org
Information on this title: www.cambridge.org/9781108495813

DOI: 10.1017/9781108863148

First published 2023

Printed in the United Kingdom by CPI Group Ltd, Croydon CR0 4YY

A catalogue record for this publication is available from the British Library

ISBN 978-1-108-49581-3 Hardback

Contents

Preface

Many fluid flows become unstable, leading, for example, to patterns of convection cells in a liquid layer heated from below, to propagating vortices for flows over a cylinder, or to droplet formation in capillary flows. While the physical processes in each of these cases can be completely different, on an abstract level similar phenomena do occur, such as a transition from steady to periodic behaviour. Dynamical systems theory and in particular bifurcation theory is an extremely powerful framework to understand the behaviour of fluid flow instabilities on this abstract level. Its concepts apply to many scientific fields, and hence its language provides an important communication tool between people of different research domains. Bifurcation theory also provides a systematic approach to assess sensitivity of the solutions of a mathematical model of a particular flow phenomenon to changes in parameters.

Often the application of bifurcation theory is thought to be limited to small-dimensional dynamical systems, that is, with models described by a few ordinary differential equations. However, over recent decades there has been a (relatively small) scientific community that has developed numerical techniques which enable the application of bifurcation theory to systems of partial differential equations. In this way, models of fluid flows can be analysed for bifurcation behaviour and the associated instability processes. Many in this community meet bi-annually at the International Symposium on Bifurcations and Instabilities in Fluid Dynamics; the ninth edition of this symposium was held in August 2022 in Groningen, the Netherlands.

This book developed from a course on 'Numerical Bifurcation Analysis of Large-Scale Systems' which we have taught in the national MasterMath framework in the Netherlands (https://elo.mastermath.nl) since 2015. The main aim of this course is to introduce the (mostly mathematics and physics) students to the numerical methods needed to apply bifurcation analysis to fluid flow models. The standard numerical methods used in many numerical bifurcation analysis codes, such as MatCont and AUTO, limit their application to relatively small-dimensional systems and hence these codes cannot be used for fluid flow problems. The numerical background needed to tackle fluid flow equations, where more sophisticated numerical methods are needed as well as examples of their application, is provided in the current book.

The book starts (in Chapter 1) with four canonical fluid flow problems which have been a focus of much research over the last century and describes the primary instabilities occurring. Hydrodynamic stability theory and bifurcation theory are briefly introduced in Chapter 2, where also an example system used throughout the book, the Ginzburg–Landau equation, is derived. The next two chapters provide basic theory on the partial differential equation systems of fluid flow (Chapter 3) as well as their discretization (Chapter 4). Chapter 5 then introduces the basic continuation methodology, describing how to determine branches of steady states (fixed points in dynamical systems theory) and time-periodic flows (limit cycles in dynamical systems theory) versus parameters of a fluid flow model. Chapters 6 and 9 focus on numerical solution techniques of the discrete equations, where in Chapter 6 it is assumed that the Jacobian matrix of the dynamical system is available and in Chapter 9 that only products of this matrix and a vector can be computed. Chapters 7 and 8 deal with iterative methods for solving large-dimensional linear systems of equations and for solving large-dimensional generalized eigenvalue problems. Finally, in Chapter 10, these techniques are applied to determine the primary bifurcations in the four canonical fluid flow problems of Chapter 1; the Python code to do so is available on GitHub.

Apart from students in mathematics and physics, the book will also be of interest to anyone working in scientific fields where hydrodynamic instabilities play a role. Someone with an interest in the fluid mechanics aspects, for example, may focus on Chapters 1, 4, 5, and 10. Another with an interest in the dynamical systems aspects might focus on Chapters 1, 2, 3, 5, and 10. Finally, a reader with an interest in the numerical methods could focus on Chapters 3, 4, 6, 7, 8, and 9. The numerical techniques presented here are not limited to traditional flows such as those presented in Chapter 1, but have a broad applicability in industrial, environmental, and astrophysical flows. Exercises at the end of each Chapter (except Chapter 10) provide many more details and practice material to absorb all the ideas presented in the book. Matlab codes are available online to assist with these exercises, and the solutions to most of the exercises are available on request. We hope that this material finds its way to the broader fluid mechanics community, in particular to the younger generation of applied mathematicians and fluid physicists.

Acknowledgements

The writing of this book has been a pleasure over the years due to our interaction with many students in the Netherlands and many colleagues in the fields of fluid dynamics and dynamical systems. First of all, we thank our colleagues at the Bernoulli Institute for Mathematics, Computer Science and Artificial Intelligence (Groningen) and the Institute for Marine and Atmospheric Research (Utrecht) for creating the environment in which this book could be written. We as authors have been able to generate externally funded joint projects over a period of more than 25 years and would like to thank all the PhD students and postdocs who have worked with us in these projects. In particular, we want to mention Erik Bernsen, Arjen Terwisscha van Scheltinga, Ena Tiesinga, Hakan Oksuzoglu, Matthijs den Toom, Arie de Niet, Jonas Thies, Jan Viebahn, Daniele Castellana, and more recently Erik Mulder and Sven Baars for all their efforts. We are also pleased to acknowledge our collaboration with and support from colleagues at the Netherlands eScience Center, in particular Inti Pelupessy, Jason Maasen, Ben van Werkhoven, and Marijn Verstraeten.

Sven Baars has made an important contribution to this book by generating all results in Chapter 10 and developing and maintaining the Python code FVM to make these calculations. For parts of the book we followed expositions in lecture notes of our former colleagues Eugen Botta and Henk van der Vorst. We learned much from these two giants and we thank them for their inspirational notes.

The generous support of the Netherlands Organization for Scientific Research (NWO) over the years is much appreciated. This support has been essential in enabling us to carry out research on the applications of dynamical systems theory to fluid flow problems. The support in supercomputing time on the machines at the Academic Computing Center in Amsterdam (SARA) through projects from the earlier National Computing Facilities Foundation (NCF) and more recently the Natural Science Branch of NWO is also much appreciated.

A substantial part of the book was written by one of the authors (Henk) during a three-month sabbatical at the Australian National University in 2019–20. He thanks Andy Hogg for a quiet and enjoyable time in Canberra and the Center for Climate Research at the Australian National University for their generous support.

Sarah Lambert and Matt Lloyd at Cambridge University Press have been very patient in waiting for the final version of the manuscript and very supportive during the writing process. We also thank Sarah Lambert for going through the text and providing useful comments on layout and style. Bret Workman is thanked for his great copy editing of the manuscript and Reshma Xavier (Integra) for carefully processing all corrections to the page proofs into the final version of the book.

Finally, we thank our spouses Tineke (Fred) and Julia (Henk) for allowing us to make time for writing this book during hours that also could have been spent with them.

1

Transitions in Fluid Flows

Order in the chaos:
the emergence of fluid flow patterns

One of the most fascinating phenomena in fluid dynamics is that flows can spontaneously reorganize their macro-scale behaviour when external conditions are changed. Such transitions are at the heart of pattern formation in fluids as intensively studied since the early twentieth century. The overall problem is the determination of the different (statistical) equilibrium flow patterns versus the applied forcing and external geometry of the flow. The complicating factor here is that, due to the non-linearity of the underlying processes, multiple flow patterns can exist under the same external conditions.

The study of transition phenomena in flows of liquids and gases is fundamental to many engineering processes, in particular those associated with mass and heat transfer, and hence of great practical interest. Transitions can lead to different momentum regimes (e.g., drag, torque) and heat transfer regimes. These transitions can suddenly (compared to changes in the forcing) lead to different operating conditions. Isothermal flows may undergo qualitative changes in separation behaviour and turbulence intensity; mean flows in turbulent buoyancy-driven convection can change their overall pattern; and plasma flows in a Tokamak reactor suddenly show strong oscillatory behaviour (Crawford and Knobloch, 1991; Dijkstra *et al.*, 2014).

Transitions in environmental flows, for example in weather and climate, are at the moment much studied in connection with climate change. For example, a reorganization of the large-scale Atlantic Ocean circulation may lead to a substantial change in the meridional heat transport affecting land temperatures over a large part of the globe (Rahmstorf, 2000). Critical conditions in such flows, at which they may undergo a large qualitative change, are associated with what is now often referred to as a 'tipping' point (Gladwell, 2000). Sub-components of the climate system with potential transition behaviour are indicated by 'tipping elements'

1

(Lenton *et al.*, 2008). From a practical point of view, one would like to determine these critical conditions and understand how to avoid undesirable transitions.

The fluid flow transitions are also interesting from a complex systems science point of view (Thurner *et al.*, 2018). Due to variations in forcing of the flow, the micro-scale fluid particles organize in particular ways so as to give different macroscopic patterns and hence are an example of emerging behaviour. Subsequent transitions when the forcing of the flow is increased lead to more complicated behaviour and eventually to turbulent flows. Understanding the (staged) transition to turbulence is one of the central problems of classical physics which is, despite much progress over the last decades, still unsolved (Eckert, 2019).

In this first chapter, we will describe canonical fluid dynamical systems which have been used to study flow transitions. The attractive aspects of these flows is that they are easily accessible experimentally and that there are mathematical equations accurately describing their behaviour. After presenting those equations in general form in Section 1.1, the following sections will deal with four typical flow configurations: the Lid-Driven Cavity flow (Section 1.2), the Taylor–Couette flow (Section 1.3), the Rayleigh–Bénard–Marangoni flow (Section 1.4), and the Differentially Heated Cavity flow (Section 1.5). For each case, we will shortly describe typical experimental results and present the specific mathematical equations describing the canonical experimental configuration.

1.1 General Fluid Dynamics Equations

The main set of equations in fluid dynamics describes the conservation of mass, momentum, and thermal energy, and contains an equation of state relating density, pressure, and temperature. We will consider here the case of a Newtonian fluid with constant heat capacity C_p, dynamic viscosity μ, and thermal conductivity k, which is relevant to understanding the results in many experiments.

Conservation of mass is expressed by the continuity equation,

$$\frac{\partial \rho}{\partial t} = \nabla \cdot (\rho \mathbf{u}), \tag{1.1}$$

where ρ is density and $\mathbf{u}$ the velocity vector. The Navier–Stokes equations describe the conservation of momentum. They are written as

$$\rho \left(\frac{\partial \mathbf{u}}{\partial t} + \mathbf{u} \cdot \nabla \mathbf{u} \right) = -\nabla p + \mu \nabla^2 \mathbf{u} + \frac{1}{3} \mu \nabla (\nabla \cdot \mathbf{u}) + \mathbf{f}, \tag{1.2}$$

where p is the pressure and $\mathbf{f}$ represents a body force. One has to interpret the maths shown in Equation (1.2) component-wise.

Thermal energy conservation is expressed as an equation for the temperature T given by

$$\rho C_p \left(\frac{\partial T}{\partial t} + \mathbf{u} \cdot \nabla T \right) = k\nabla^2 T + Q_T, \tag{1.3}$$

where Q_T is a heat source or heat sink.

These equations are supplemented by a general equation of state,

$$\rho = \rho(p, T);$$

for example, the ideal-gas law $p = \rho RT$, where R is the universal gas constant. This leads to six equations for the six unknowns: three components of the velocity, the temperature, the density, and the pressure. In addition to these main equations, there may be equations for so-called tracers, for example chemical components (e.g., salinity in the ocean) and moisture (e.g., in the atmosphere). These additional equations usually have the same form as Equation (1.3), that is, they are of the convection–diffusion type. Also, electromagnetic forces may be involved, for example for studying plasma flows, leading to the addition of Maxwell's equations. All equations have to be supplied with appropriate boundary conditions and an initial condition.

1.2 Lid-Driven Cavity Flow

Due to its simple geometry, the incompressible flow in lid-driven cavities plays an important role in fundamental fluid mechanics and serves often as a numerical benchmark problem. Batchelor (1956) already pointed out that lid-driven cavity flows exhibit almost all the phenomena that can possibly occur in incompressible flows: eddies, secondary flows, complex flow patterns, chaotic particle motions, and turbulence.

1.2.1 Experimental Results

A typical experimental configuration, used by, for example, Koseff and Street (1984), is shown in Fig. 1.1a. The liquid is contained in a three-dimensional cavity characterized by two aspect ratios: that of the length L to depth D, $A_x = L/D$; and that of the width B to depth, $A_y = B/D$. The density of the liquid is constant ρ, its constant kinematic viscosity is $\nu = \mu/\rho$, and the liquid is sheared at its top by a constant velocity U. To realize this, a belt-drive support structure was used in Koseff and Street (1984) in which the belt speed was constant up to 0.5 per cent.

A typical flow observed along the symmetry plane of the cavity (e.g., for $A_x = 1$ and $A_y = 3$ and for a Reynolds number $Re = UL/\nu = 3,300$ in Koseff and Street (1984)) is sketched in Fig. 1.1b. Overall, a one-cell flow structure is seen, with surface flow into the direction of the surface velocity, and smaller vortices appear in the corner regions at the bottom. The flow may not be stationary; for example,

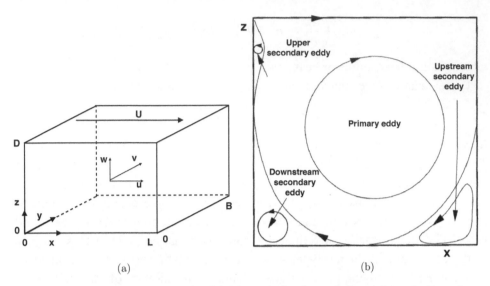

Figure 1.1 (a) Geometry of the lid-driven cavity flow of dimensions, L, B, and D, with a moving lid in the x-direction on the top having velocity U. (b) Typical flow observed in the lid-driven cavity (based on the results of Koseff and Street (1984)).

fluctuations in the vertical velocity were measured for $Re = 3,300$ (Koseff and Street, 1984), with a dominant frequency of about 0.01 Hz. In the streamwise (y-)direction, so-called Taylor–Görtler vortices were found and about eight pairs of vortices were found for $A_y = 3$. An example of this flow structure can be found in Freitas and colleagues (1985) and in Fig. 12 of Kuhlmann and Romanò (2019). The structure of the Taylor–Görtler vortices and the secondary corner vortices depend strongly on the geometry of the container.

For this same configuration ($A_x = 1, A_y = 3$), Aidun and colleagues (1991) found a stationary flow for values of Re up to $Re_c = 875 \pm 50$ and oscillatory flow for higher values of Re, in agreement with the results in Koseff and Street (1984) for $Re = 3,300$. For very large values of Re (e.g., Koseff and Street (1984) use $Re = 10,000$), irregular (turbulent) flows are found. An experimental regime diagram, mapping out the value of Re_c versus the parameters A_x and A_y and showing secondary flow types for higher values of Re, does not appear to be available in the literature.

This short summary of the experimental result on the lid-driven cavity flow already leads to intriguing fluid dynamical questions. For example, why is there a critical value of Re marking the transition from steady to oscillatory flows? Many other issues, for example the spatial structure of the corner flows and the dependence of the Taylor–Görtler vortices on A_y, have led to a number of fundamental theoretical studies (Kuhlmann and Romanò, 2019).

1.2.2 Governing Equations

Following the experimental setup used in Koseff and Street (1984), we consider the flow of a constant density Newtonian fluid in a three-dimensional cavity with length L, width B, and height D (Fig. 1.1a). The fluid motion has velocity vector $\mathbf{u} = (u, v, w)^T$ and is driven by a moving lid on the top with constant velocity U into the positive x-direction.

The governing equations for this problem are the incompressible Navier–Stokes equations given by

$$\frac{\partial u}{\partial t} + u\frac{\partial u}{\partial x} + v\frac{\partial u}{\partial y} + w\frac{\partial u}{\partial z} = -\frac{1}{\rho}\frac{\partial p}{\partial x} + \nu\nabla^2 u + f_x, \tag{1.4a}$$

$$\frac{\partial v}{\partial t} + u\frac{\partial v}{\partial x} + v\frac{\partial v}{\partial y} + w\frac{\partial v}{\partial z} = -\frac{1}{\rho}\frac{\partial p}{\partial y} + \nu\nabla^2 v + f_y, \tag{1.4b}$$

$$\frac{\partial w}{\partial t} + u\frac{\partial w}{\partial x} + v\frac{\partial w}{\partial y} + w\frac{\partial w}{\partial z} = -\frac{1}{\rho}\frac{\partial p}{\partial z} + \nu\nabla^2 w + f_z, \tag{1.4c}$$

$$\frac{\partial u}{\partial x} + \frac{\partial v}{\partial y} + \frac{\partial w}{\partial z} = 0, \tag{1.4d}$$

with $f_x = f_y = 0$ and $f_z = -g$, where g is the gravitational acceleration. *See example Ex. 1.1*

The boundary conditions are given by no-slip conditions on all the walls, formulated as

$$x = 0, L: \quad u = v = w = 0, \tag{1.5a}$$

$$y = 0, B: \quad u = v = w = 0, \tag{1.5b}$$

$$z = 0: \quad u = v = w = 0, \tag{1.5c}$$

$$z = D: \quad u = U, v = w = 0. \tag{1.5d}$$

As we will see in Chapter 10, for a given geometry and boundary conditions, the Reynolds number $Re = UL/\nu$ is determining the flow completely. There is no elementary non-trivial analytical solution known for the Lid-Driven Cavity flow.

Additional Material

- A relatively recent and extensive review of experimental results on the Lid-Driven Cavity flow can be found in Kuhlmann and Romanò (2019). They discuss also experimental results for extended configurations, for example where two walls of the cavity move with a (different) constant velocity.

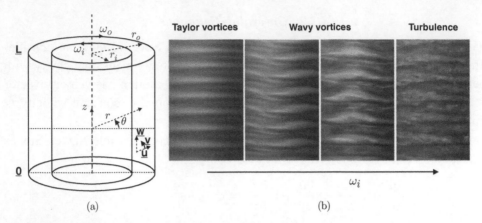

Figure 1.2 (a) Sketch of the Taylor–Couette flow configuration. (b) Observed pattern of Taylor vortices for four different values of the inner rotation rate ω_i, which increases from left to right (source: https://advlabs.aapt.org/wiki/Taylor-Couette_Flow), with $\omega_o = 0$.

1.3   Taylor–Couette Flow

The flow between concentric cylinders, or Taylor–Couette flow, is one of the canonical pattern formation flows, sometimes referred to as the 'Hydrogen Atom' or 'Drosophila' of fluid dynamics. It shows a staged transition to turbulence, for example as the rotation rate of the inner cylinder is increased, with a very rich behaviour.

1.3.1   Experimental Results

A typical experimental configuration (Fig. 1.2a) consists of two rotating cylinders enclosing a fluid. The inner cylinder of radius r_i rotates with angular velocity ω_i, and the outer one (radius r_o) with angular frequency ω_o. The cylinders have a finite length L, hence the geometry is characterized by the aspect ratio $\Gamma = L/d, d = r_o - r_i$, and the radius ratio $\eta = r_i/r_o$. The fluid has a constant density ρ and constant kinematic viscosity ν, and in this way the flow is characterized by the two Reynolds numbers

$$Re_i = \frac{r_i \omega_i d}{\nu} , \; Re_o = \frac{r_o \omega_o d}{\nu}. \qquad (1.6)$$

The first experiments were carried out by Taylor (1923) for the case $\omega_o = 0$, hence with only the inner cylinder rotating. For small values of ω_i a parallel flow exists, which can be determined analytically (see Subsection 1.3.2). As this flow was already observed experimentally by Couette (1890), it was named the Couette flow. Taylor observed that when ω_i exceeds a critical value, instability sets in and rows of cellular vortices develop. This is the so-called Taylor vortex flow, of which an example is shown in the 'Taylor vortices' panel of Fig. 1.2b. When ω_i is increased to higher values, the cell rows start to move in a wavy fashion after a transition to time dependence (two middle 'Wavy vortices' panels of Fig. 1.2b). For higher values of ω_i, the Taylor cells break down and a turbulent statistical equilibrium flow is established ('Turbulence' panel of Fig. 1.2b).

A theoretical analysis in the case of $\omega_o = 0$ and $\Gamma \to \infty$ (i.e., infinitely long cylinders) to explain the first transition from the Couette flow to the Taylor vortex flow was already done by Taylor (1923). The critical value of Re_i^c for $\eta = 0.8$ is given by $Re_i^c = 94.7$, and for $\eta = 0.9$ it is $Re_i^c = 131.6$ (Recktenwald *et al.*, 1993). Often, another parameter is used to characterize this critical condition for transition, the Taylor number Ta:

$$Ta = 4Re_i^2 \frac{1-\eta}{1+\eta},\tag{1.7}$$

with $Ta_c = 3,986$ for $\eta = 0.8$. The critical values agree well with experiments in very long cylinders (Chandrasekhar, 1961; Drazin and Reid, 2004; Chossat and Iooss, 2012). Snyder (1968) has given a semi-empirical equation for the critical condition from collected experimental data.

Detailed work on the Taylor vortex flows (which appear after the first transition) has been done by Mullin and coworkers (Mullin and Blohm, 2001; Mullin *et al.*, 2017). For $\eta = 0.833$, the regime diagram is given in Fig. 1.3, as based on Andereck and colleagues (1986). The complexity of the different flow regimes is striking, given the simplicity of the geometry and forcing, and there are still some unexplored regimes.

1.3.2 Governing Equations

For the configuration in Fig. 1.2a, the incompressible Navier–Stokes equations are conveniently written in cylindrical coordinates (r, θ, z) and the velocity vector (radial, azimuthal, and axial) as $\mathbf{u} = (u, v, w)^T$. The resulting equations are given by

$$\frac{\partial u}{\partial t} + u\frac{\partial u}{\partial r} + \frac{v}{r}\frac{\partial u}{\partial \theta} + w\frac{\partial u}{\partial z} - \frac{v^2}{r} =$$
$$\nu(\frac{1}{r}\frac{\partial}{\partial r}(r\frac{\partial u}{\partial r}) + \frac{1}{r^2}\frac{\partial^2 u}{\partial \theta^2} - \frac{u}{r^2} - \frac{2}{r^2}\frac{\partial v}{\partial \theta} + \frac{\partial^2 u}{\partial z^2}) - \frac{1}{\rho}\frac{\partial p}{\partial r} \quad -f_r,\tag{1.8a}$$

$$\frac{\partial v}{\partial t} + u\frac{\partial v}{\partial r} + \frac{v}{r}\frac{\partial v}{\partial \theta} + w\frac{\partial v}{\partial z} + \frac{uv}{r} =$$
$$\nu(\frac{1}{r}\frac{\partial}{\partial r}(r\frac{\partial v}{\partial r}) + \frac{1}{r^2}\frac{\partial^2 v}{\partial \theta^2} - \frac{v}{r^2} + \frac{2}{r^2}\frac{\partial u}{\partial \theta} + \frac{\partial^2 v}{\partial z^2}) - \frac{1}{\rho}\frac{1}{r}\frac{\partial p}{\partial \theta} \quad -f_\theta,\tag{1.8b}$$

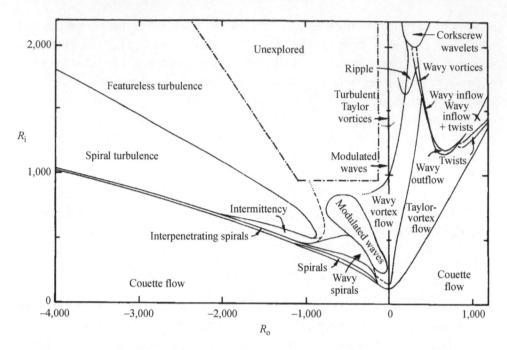

Figure 1.3 Regime diagram for the different flows as observed in Taylor–Couette flow experiments as from Andereck and colleagues (1986) for the case $\eta = 0.833$ and $\Gamma = 30$; here $R = Re$. (Published with permission from Cambridge University Press.)

$$\frac{\partial w}{\partial t} + u \frac{\partial w}{\partial r} + \frac{v}{r} \frac{\partial w}{\partial \theta} + w \frac{\partial w}{\partial z} \quad =$$

$$\nu \left(\frac{1}{r} \frac{\partial}{\partial r} \left(r \frac{\partial w}{\partial r} \right) + \frac{1}{r^2} \frac{\partial^2 w}{\partial \theta^2} + \frac{\partial^2 w}{\partial z^2} \right) - \frac{1}{\rho} \frac{\partial p}{\partial z} \quad - f_z, \qquad (1.8c)$$

$$\frac{1}{r} \frac{\partial (ru)}{\partial r} + \frac{1}{r} \frac{\partial v}{\partial \theta} + \frac{\partial w}{\partial z} = 0, \qquad (1.8d)$$

with $f_r = f_\theta = 0$ and $f_z = -g$.

The boundary conditions are given by

$$r = r_i: \quad u = w = 0, v = \omega_i r_i, \qquad (1.9a)$$

$$r = r_o: \quad u = w = 0, v = \omega_o r_o, \qquad (1.9b)$$

$$z = 0, L: \quad u = v = w = 0. \qquad (1.9c)$$

Periodic conditions apply in the θ direction, that is, $u(r, 0, z, t) = u(r, 2\pi, z, t)$ with similar conditions for the other velocity components and the pressure.

An analytical solution $(\bar{u}, \bar{v}, \bar{w}, \bar{p})$ of the equations, the Couette flow (Couette, 1890), exists for infinitely long cylinders ($L \to \infty$) and is given by

$$\bar{u} = \bar{w} = 0 \ , \ \bar{v} = ar + \frac{b}{r} \ , \ \bar{p} = -\rho g z + p_d(r), \tag{1.10a}$$

$$a = \frac{\omega_o r_o^2 - \omega_i r_i^2}{r_o^2 - r_i^2} \ , \ b = \frac{(\omega_i - \omega_o) r_o^2 r_i^2}{r_o^2 - r_i^2}, \tag{1.10b}$$

where $p_d(r) = \rho \int_{r_i}^{r} \bar{v}^2 / r \, dr$. There are also analytical solutions (in Bessel function series form) for $\omega_o = 0$ in finite-length containers (Wendl, 1999). *See example Ex. 1.2*

Additional Material

- An extensive review of experimental results on the Taylor–Couette flow can be found in Koschmieder (1993) and Grossmann and colleagues (2016). There is a very accessible Scholarpedia paper on the Taylor–Couette flow by Richard Lueptow, see www.scholarpedia.org/article/Taylor-Couette_flow, where also many other sources (videos, webpages) on the experimental results are provided.
- Since 1979, an international workshop, the ICTW (International Couette–Taylor Workshop), is organized bi-annually; see https://pof.tnw.utwente.nl/ictw/history.html.

1.4 Rayleigh–Bénard–Marangoni Flow

The Rayleigh–Bénard–Marangoni problem is another classic in fluid dynamics. A liquid layer heated from below shows a fascinating and rich set of flow patterns once a critical vertical temperature gradient is exceeded (Koschmieder, 1993).

1.4.1 Experimental Results

First experiments were carried out by Bénard (1901), with a circular container being filled with a viscous liquid such as silicone oil (Fig. 1.4a) with constant heat capacity C_p, dynamic viscosity μ, and thermal conductivity k. Air was situated above the upper surface of the liquid, and the temperature far from the air–liquid interface was nearly constant. This creates a temperature difference, $\Delta T = T_B - T_A$, between the bottom of the container and the surface of the liquid.

When the initially motionless liquid is heated from below, the liquid remains motionless below a critical value of ΔT, say ΔT_c. In this case, the heat transfer through the layer is only by heat conduction. When the temperature difference slightly exceeds ΔT_c, the liquid is set into motion and after a while the flow organizes itself into steady (often hexagonal) cellular patterns (Fig. 1.4b).

For the liquid, a linear equation of state,

$$\rho = \rho_0 (1 - \alpha_T (T - T_0)), \tag{1.11}$$

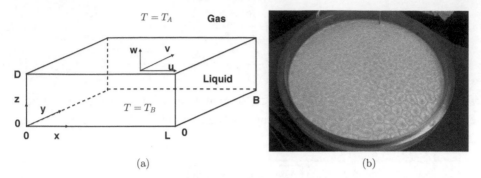

Figure 1.4 (a) Sketch of the model set-up and boundary conditions of the Rayleigh–Bénard–Marangoni problem. (b) Example of a flow pattern consisting of hexagons arising in a liquid heated from below (Matsson, 2008); see https://peer.asee.org/a-student-project-on-rayleigh-benard-convection.

is usually an adequate approximation, where α_T is the thermal compressibility coefficient, T_0 a reference temperature, and ρ_0 a reference density. In situations with an air–liquid interface, such as in Fig. 1.4a, the surface tension σ depends on the temperature. For many liquids, including water, a relation of the form

$$\sigma = \sigma_0(1 - \gamma_T(T - T_0)) \tag{1.12}$$

holds, where σ_0 is a reference surface tension. In this case, the surface tension decreases with increasing temperature, and γ_T is a constant. Surface tension gradients give rise to shear stresses (this is the Marangoni effect), with flows directed from low to high surface tension (Marangoni, 1865). When the upper surface is a rigid lid, as holds for many experiments performed, no Marangoni effects occur. *See example Ex. 1.3*

For experiments in rectangular containers, with length L, width B, and height D, the flow is characterized by the two aspect ratios $A_x = L/D$ and $A_y = B/D$, and four other parameters, namely the Rayleigh number Ra, the Prandtl number Pr, the Marangoni number Ma, and the Biot number Bi, defined by

$$Ra = \frac{\alpha_T g \Delta T D^3}{\nu \kappa}; \; Pr = \frac{\nu}{\kappa}; \; Ma = \frac{\sigma_0 \gamma_T \Delta T D}{\rho_0 \nu \kappa}; \; Bi = \frac{hD}{k},$$

where h is a (constant) surface heat transfer coefficient, $\nu = \mu/\rho_0$ is the kinematic viscosity, and $\kappa = k/(\rho_0 C_p)$ the thermal diffusivity.

For pure Rayleigh–Bénard convection, when the upper surface is a rigid wall, the parameters Bi and Ma do not appear; an experimentally determined regime diagram in the $Pr-Ra$ space is shown in Fig. 1.5a. The critical value of Ra in this case (with rigid bottom wall and $A_x, A_y \to \infty$) is $Ra_c = 1707.8$. In this case, roll cells appear above criticality; a sketch of such a flow pattern is shown in Fig. 1.5b. The Ra range

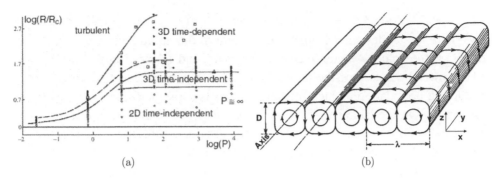

(a)   (b)

Figure 1.5 (a) Regime diagram in the $Pr - Ra$ space for pure Rayleigh–Bénard convection (Krishnamurti, 1973), where $P = Pr$ and $R = Ra$. (Published with permission from Cambridge University Press.) (b) Sketch of a pattern of roll cells in a rectangular container.

of steady roll cells is much larger for larger Pr liquids and disappears from liquids like mercury having a very small Pr. When Ra is increased, time dependence also occurs and finally the flow becomes turbulent.

For pure Bénard–Marangoni convection, we have $Ra = 0$ and the critical Marangoni number Ma_c depends on Bi; for $Bi = 0$ and again $A_x, A_y \to \infty$, its value is 79.6. There is no experimentally determined regime diagram of pure Bénard–Marangoni flows, since buoyancy effects are difficult to eliminate. It could, in principle, be done in space, but is practically difficult to carry out.

1.4.2 Governing Equations

In deriving the governing equations from the general equations in Section 1.1, often the Boussinesq approximation is made, which (in short) assumes that density variations are so small compared to the reference density ρ_0 that they only have to be taken into account in the volume (in this case buoyancy) force. The equations then become

$$\frac{\partial u}{\partial t} + u\frac{\partial u}{\partial x} + v\frac{\partial u}{\partial y} + w\frac{\partial u}{\partial z} = -\frac{1}{\rho_0}\frac{\partial p}{\partial x} + \nu\nabla^2 u + f_x, \qquad (1.13a)$$

$$\frac{\partial v}{\partial t} + u\frac{\partial v}{\partial x} + v\frac{\partial v}{\partial y} + w\frac{\partial v}{\partial z} = -\frac{1}{\rho_0}\frac{\partial p}{\partial y} + \nu\nabla^2 v + f_y, \qquad (1.13b)$$

$$\frac{\partial w}{\partial t} + u\frac{\partial w}{\partial x} + v\frac{\partial w}{\partial y} + w\frac{\partial w}{\partial z} = -\frac{1}{\rho_0}\frac{\partial p}{\partial z} + \nu\nabla^2 w + f_z, \qquad (1.13c)$$

$$\frac{\partial u}{\partial x} + \frac{\partial v}{\partial y} + \frac{\partial w}{\partial z} = 0, \qquad (1.13d)$$

$$\frac{\partial T}{\partial t} + u\frac{\partial T}{\partial x} + v\frac{\partial T}{\partial y} + w\frac{\partial T}{\partial z} = \kappa\nabla^2 T, \qquad (1.13e)$$

with $f_x = f_y = 0$ and $f_z = -\rho g/\rho_0 = -g(1 - \alpha_T(T - T_0))$. In these equations, (x, y, z) are the Cartesian coordinates of a point in the liquid layer, t denotes time, (u, v, w) is the velocity vector, p denotes pressure, and T is the temperature.

The lower boundary of the liquid (Fig. 1.4a) is considered to be a very good conducting boundary on which the temperature is constant T_B, and no-slip conditions apply. On the lateral walls (at $x = 0, L$ and $y = 0, B$), no-flux and no-slip conditions are prescribed. Let the non-deforming gas–liquid interface be located at $z = D$; then the general boundary conditions become

$$x = 0, L: \qquad\qquad u = v = w = \frac{\partial T}{\partial x} = 0, \qquad\qquad (1.14a)$$

$$y = 0, B: \qquad\qquad u = v = w = \frac{\partial T}{\partial y} = 0, \qquad\qquad (1.14b)$$

$$z = 0: \qquad\qquad T = T_B \; ; \; u = v = w = 0, \qquad\qquad (1.14c)$$

$$z = D: \quad \mu\frac{\partial u}{\partial z} = \frac{\partial \sigma}{\partial x} \; ; \; \mu\frac{\partial v}{\partial z} = \frac{\partial \sigma}{\partial y} \; ; \; w = 0 \; ; \; k\frac{\partial T}{\partial z} = h(T_A - T), \quad (1.14d)$$

where T_A is the temperature of the gas just above the interface. The first two equations in (1.14d) represent the Marangoni effect, and the last equation is the surface heat transfer condition. *See example Ex. 1.4*

For $\bar{u} = \bar{v} = \bar{w} = 0$, there is a steady state given by

$$\bar{T}(z) = T_B - \beta z; \; \beta = \frac{h(T_B - T_A)}{k + hD}, \qquad\qquad (1.15)$$

where β is the vertical temperature gradient over the layer. The corresponding pressure distribution is readily determined from (1.13), and if one chooses $T_0 = T_A$, this gives

$$\bar{p}(z) = p_0 + \rho_0 g\left((\alpha_T(T_B - T_A) - 1)z - \frac{\alpha_T\beta}{2}z^2\right). \qquad\qquad (1.16)$$

This motionless solution is characterized by only conductive heat transfer and is easily realized in laboratory experiments. Note that such a motionless solution exists for all values of the vertical temperature difference, $\Delta T = \beta D$.

If the upper surface is a rigid surface and kept at a constant temperature T_A, there are no Marangoni effects and the surface temperature is kept constant ($h \to \infty$ in (1.14)). In this case of pure Rayleigh–Bénard convection, the boundary conditions (1.14d) are replaced by

$$z = D: u = v = w = 0, T = T_A. \qquad\qquad (1.17)$$

The analytic solution for temperature (1.15) in this case still holds, but the limit $h \to \infty$ has to be taken for which $\beta \to (T_B - T_A)/D$.

Additional Material

- An extensive overview of experimental work on the Rayleigh–Bénard–Marangoni flows is given in Koschmieder (1993) and Getling (1998).
- The Bénard Centenary Review (Mutabazi *et al.*, 2010) contains many very interesting papers on Bénard's experiments and their extensions; also included is a great scientific biography of Henri Bénard by José Eduardo Wesfreid (pages 9–37).

1.5 Differentially Heated Cavity Flow

Because of its engineering importance, the flow in a cavity of which two opposite vertical sidewalls have different temperatures has been studied extensively. The main aim of these studies was to determine the heat transport from the heated wall towards the cooler wall.

1.5.1 Experimental Results

The first experiments with air in containers of different aspect ratios were performed by Eckert and Carlson (1961). As in the Rayleigh–Bénard problem, the flow is characterized by two aspect ratios $A_x = L/D$ and $A_y = B/D$, the Rayleigh number Ra, and the Prandtl number Pr. Eckert and Carlson (1961) used the Grashof number $Gr = Ra/Pr$ and showed that the flow is stationary up to a critical value (Gr_c) and unsteady thereafter.

Elder (1965) performed experiments in a container with $A_y >> A_x$ (such that the flow is approximately two-dimensional) while varying D/L over the range $[1, 60]$ for a liquid with $Pr = 10^3$. A sketch of the experimental configuration is shown in Fig. 1.6a. The value of Ra in Elder (1965) is based on the length L and given by

$$Ra = \frac{\alpha_T g \Delta T L^3}{\nu \kappa}. \tag{1.18}$$

For small values of Ra there is a wall-to-wall flow, the liquid descending along the cold wall and moving upward along the warm wall. Flow profiles for different values of Ra (Fig. 1.6b) show that secondary flows develop for higher Ra. The wavelength of these flows decreases with increasing values of Ra.

In an extensive series of experiments, Jannot and Mazeas (1973) determined (for the same configuration as in Elder (1965)) the Ra boundary between stationary and non-stationary regimes, namely the onset of time dependence in the flow. However, there does not appear to be a full experimental regime diagram of this flow. Variations of the configuration were introduced by Hart (1971), including

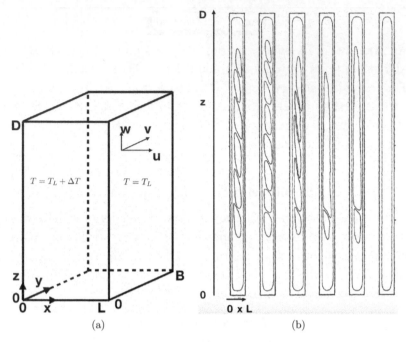

Figure 1.6 (a) Sketch of the experimental set-up used in Elder (1965). (b) Sketch of flows visualized (Elder, 1965) for different values of Ra (for $L = 2$ cm and $D = 38$ cm) with Ra increasing from right to left, showing the appearance of secondary flows with larger Ra. (Published with permission from Cambridge University Press.)

the effects of angular rotation (for their geophysical relevance), and by Imberger (1974), studying flows for very large A_x.

1.5.2 Governing Equations

The equations describing the Differentially Heated Cavity flow are the same as for the pure Rayleigh–Bénard problem (1.13) and are not repeated here; also, a linear equation of state is assumed to be adequate. The vertical sidewalls at $x = 0$ and $x = L$, which have a different temperature, are considered to be extremely good conducting boundaries on which the temperature is constant, and no-slip conditions apply. On the other walls (at $z = 0, D$ and $y = 0, B$) no-flux and no-slip conditions are prescribed. The boundary conditions then become (Fig. 1.6a) *See example Ex. 1.5*

$$x = 0: \quad u = v = w = T - (T_L + \Delta T) = 0, \tag{1.19a}$$

$$x = L: \quad u = v = w = T - T_L = 0, \tag{1.19b}$$

$$y = 0, B: \quad u = v = w = \frac{\partial T}{\partial y} = 0, \tag{1.19c}$$

$$z = 0, D: \quad u = v = w = \frac{\partial T}{\partial z} = 0. \tag{1.19d}$$

There is no elementary analytical solution for the differentially heated cavity problem, as even small temperature differences ΔT will cause a non-trivial flow in the cavity.

Additional Material

- An extensive, although already a bit dated, review on experimental work regarding the Differentially Heated Cavity flow can be found in Paolucci (1994). Variants of this flow, which also show very interesting behaviour, are laterally heated/cooled containers which are inclined with respect to gravity (Saury *et al.*, 2012).

1.6 Summary

- Fluid flows can undergo staged transitions when parameters, for example those related to the forcing of the flow, are varied. For large forcing (with respect to viscous and diffusive processes), turbulent flows result.
- We have focused on four classical flows: the Lid-Driven Cavity flow, the Taylor–Couette flow, the Rayleigh–Bénard–Marangoni flow, and the Differentially Heated Cavity flow. These flows have been studied extensively in laboratory experiments.
- For the Taylor–Couette flow and the Rayleigh–Bénard–Marangoni flow analytical parallel flow solutions exist. These become unstable at specific values of Reynolds and Rayleigh/Marangoni numbers, respectively, and non-parallel patterned flows (Taylor vortices and hexagonal/roll cell patterns) result.
- For even higher values of these parameters, a transition to time dependence occurs, resulting, for example, in wavy Taylor vortices in the Taylor–Couette flow.
- For the Lid-Driven Cavity flow and the Differentially Heated Cavity flow, no such parallel steady flow solutions exists, but the non-parallel steady flows also become unstable at specific values of the Reynolds and Rayleigh numbers, respectively.
- For each of the classical flows, the fluid dynamical equations and boundary conditions describe the dominant balances of mass, momentum, and heat in unprecedented detail.

1.7 Exercises

Exercise 1.1 *Consider the two-dimensional Lid-Driven Cavity flow in Cartesian coordinates* (x, z) *in a container of length L and height D. In this case, the stream function* ψ *can be introduced through*

$$u = \frac{\partial \psi}{\partial z} \; ; \; w = -\frac{\partial \psi}{\partial x}$$

We can, furthermore, non-dimensionalize the equations through scales L for length, U for velocity, and L/U for time.

a. Show that the governing equations can be written as

$$\frac{\partial \zeta}{\partial t} + \boldsymbol{u} \cdot \nabla \zeta = Re^{-1} \nabla^2 \zeta$$

$$\zeta = -\nabla^2 \psi,$$

where $\zeta = \partial w / \partial x - \partial u / \partial z$ *is the vertical component of the vorticity vector and* $Re = UL/\nu$.

b. Formulate the boundary conditions for ψ.

Exercise 1.2 *Consider the Taylor–Couette flow in the case* $\omega_i = \omega$ *and* $\omega_o = 0$. *A steady solution can be found of the form* $\bar{u} = \bar{w} = 0$ *and* $\bar{v} = \bar{v}(r, z)$.

a. Show that the governing equations for $\bar{v}$ *become*

$$\frac{1}{r}\frac{\partial}{\partial r}(r\frac{\partial \bar{v}}{\partial r}) + \frac{\partial^2 \bar{v}}{\partial z^2} - \frac{\bar{v}}{r^2} = 0$$

with boundary conditions $r = r_i: \bar{v} = \omega r_i; r = r_o: \bar{v} = 0;$ *and* $z = 0, L: \bar{v} = 0$.

b. Argue that when the vertical direction is unbounded, $L \to \infty$, *that* $\bar{v} = \bar{v}(r)$.

c. Determine $\bar{v}(r)$ *and* $\bar{p}(r, z)$ *and compare with (1.10).*

Exercise 1.3 *Consider a fluid particle in a motionless liquid under a vertical temperature gradient in the Rayleigh–Bénard–Marangoni experiment. Assume that the particle is moved upwards adiabatically.*

First, consider the case that the surface tension is constant and the particle is far from the air–liquid surface.

a. Sketch and determine the forces on the particle just after the perturbation.

b. Argue that the perturbation motion is amplified if some critical value of the Rayleigh number Ra is exceeded.

   Next, consider the case that gravity is absent, but surface tension is a monotonically decreasing function of temperature and the particle is moved to the surface.

c. Sketch and determine the forces on the particle just after the perturbation.

d. Argue that the perturbation motion is amplified if some critical value of the Marangoni Ma is exceeded.

Exercise 1.4 *Consider the pure Rayleigh–Bénard problem as in Section 1.4.2 and the case of an upper rigid surface with a fixed temperature T_A.*

a. When Ra $<$ Ra$_c$, the liquid is motionless. Determine the equations for the temperature field and the pressure field.

b. Argue that the temperature $\bar{T}$ and pressure $\bar{p}$ are only functions of z.

c. Use the boundary conditions to determine the basic state solution $(\bar{T}, \bar{p})$ and compare to the solution given in Section 1.4.2.

Exercise 1.5 *Consider the two-dimensional Differentially Heated Cavity flow in a Cartesian coordinate system (x, z). Make the equations non-dimensionless, by using scales $g\alpha_T \Delta T D^3/(\nu L)$ for velocity, D for length, and $\nu L/(g\alpha_T D^2 \Delta T)$ for time, and define a dimensionless temperature ϑ by $\vartheta = (T - T_L)/\Delta T$. Furthermore, introduce a stream function ψ such that*

$$u = \frac{\partial \psi}{\partial z} \; ; \; w = -\frac{\partial \psi}{\partial x}.$$

a. Show that the dimensionless equations are given by

$$GrA^2 \left(\frac{\partial \zeta}{\partial t} + u \frac{\partial \zeta}{\partial x} + w \frac{\partial \zeta}{\partial z} \right) = A\nabla^2 \zeta + \frac{\partial \theta}{\partial x},$$

$$\nabla^2 \psi = -\zeta,$$

$$GrPrA \left(\frac{\partial \vartheta}{\partial t} + u \frac{\partial \vartheta}{\partial x} + w \frac{\partial \vartheta}{\partial z} \right) = \nabla^2 \vartheta,$$

where ζ is the vertical component of the vorticity vector. The boundary conditions are

$$x = 0: \quad \psi = \frac{\partial \psi}{\partial x} = \vartheta = 0,$$

$$x = 1/A: \quad \psi = \frac{\partial \psi}{\partial x} = \vartheta - 1 = 0,$$

$$z = 0, 1: \qquad \psi = \frac{\partial \psi}{\partial z} = \frac{\partial \vartheta}{\partial z} = 0,$$

where $A = 1/A_x = D/L$ and $Pr = \nu/\kappa$. *Give an expression for the Grashof number Gr in this case.*

Next, we try to find asymptotic solutions in the limit $A \to 0$ of the form

$$\vartheta = \vartheta_0 + A\vartheta_1 + A^2\vartheta_2 \cdots \ ; \ \psi = \psi_0 + A\psi_1 + A^2\psi_2 \cdots$$

b. Show that

$$\psi \sim K_1 \left(\frac{z^4}{24} - \frac{z^3}{12} + \frac{z^2}{24} \right),$$

$$\vartheta \sim K_1 x + K_2 + K_1^2 GrPrA^2 \left(\frac{z^5}{120} - \frac{z^4}{48} + \frac{z^3}{72} \right)$$

with constants K_1 and K_2.

c. Describe a procedure for how the constants K_1 and K_2 can be determined.

2

Dynamical Systems Background

The need for a universal language:
from steady patterns to turbulence

Fluid transitions, such as those described in the previous chapter, are associated with the change in stability of flow patterns. Intuitively, an equilibrium flow (such as the Couette flow in the Taylor–Couette problem of Section 1.3) is stable when initial perturbations to the flow decay to zero in time. When critical values of parameters, such as the Reynolds numbers in the Taylor–Couette problem, are exceeded, perturbations on the flow will grow in time and eventually lead to a different equilibrium flow. In this chapter, we provide the background of this stability problem in the language of dynamical systems theory. It is written mostly as a refresher and to define the terminology which will be used in later chapters.

2.1 Stability of Fluid Flows

A general framework on the loss of stability of a fluid flow is given in Landau and Lifshitz (1987).

2.1.1 Stability Boundaries

We follow Joseph (1976) on stability boundaries, using the concept of asymptotic stability in the mean. Consider a constant-density (ρ) steady flow with an equilibrium velocity field $\bar{\mathbf{u}}$, referred to as the basic state, in a fixed flow domain $\mathcal{V}$. For a velocity perturbation $\tilde{\mathbf{u}}$ on $\bar{\mathbf{u}}$, the volume-averaged perturbation kinetic energy $\mathcal{E}$ is given by

$$\mathcal{E}(t) = \int_{\mathcal{V}} \frac{\rho}{2} \left\| \tilde{\mathbf{u}} \right\|^2 \, d^3 x. \tag{2.1}$$

19

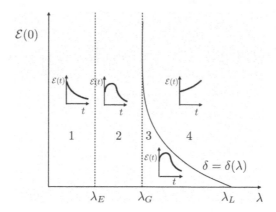

Figure 2.1 Plot of the different stability regimes, with 1: monotonic stability, 2: global stability, 3: conditional stability, and 4: instability. The control parameter is λ and the values of λ_E, λ_G, and λ_L are the energy, global, and linear stability boundaries, respectively. The curve $\delta(\lambda)$ bounds the region of conditional stability. Typical trajectories of the volume-averaged perturbation kinetic energy are sketched to illustrate the different behaviour in each domain.

For isothermal flows, such as the Lid-Driven Cavity flow and the Taylor–Couette flow, an equation for $\mathcal{E}$ can be obtained from the mechanical energy balance equation (Joseph, 1976).

The solution $\bar{\mathbf{u}}$ is said to be asymptotically stable in the mean if

$$\lim_{t \to \infty} \frac{\mathcal{E}(t)}{\mathcal{E}(0)} = 0, \qquad\qquad (2.2)$$

where $\mathcal{E}(0)$ is the initial value of $\mathcal{E}$ at $t = 0$. If there exists a positive constant δ such that (2.2) holds only when $\mathcal{E}(0) < \delta$, then the basic state is said to be conditionally stable. If $\delta \to \infty$, then the basic state is globally stable; and if (2.2) is satisfied and $d\mathcal{E}(t)/dt < 0$ holds for all $t > 0$, then the basic state is said to be monotonically stable. Note that this definition of stability does not a priori assume that the perturbations should be small compared to the basic state.

Let one of the parameters in a particular model be indicated by λ, for example, the Reynolds number in the Lid-Driven Cavity flow. Based on the concept of asymptotic stability in the mean, four regions can be distinguished (Fig. 2.1):

- In region 1, the basic state is monotonically stable; all perturbations, whatever their initial amplitude, have a monotonically decaying perturbation kinetic energy.
- In region 2, there may be perturbations which initially grow (not necessarily exponentially), but the perturbation kinetic energy eventually decays to zero for all initial amplitudes of the perturbations.

- Region 3 is a region of conditional instability. If the initial amplitude of the perturbations is small enough ($\mathcal{E}(0) < \delta(\lambda)$), the perturbation kinetic energy decays to zero, whereas if it is larger than some particular value $\delta(\lambda)$, the perturbation kinetic energy will increase. In the latter case, the perturbed state will evolve to a different state (than $\bar{\mathbf{u}}$), and the state $\bar{\mathbf{u}}$ is said to be (non-linearly) unstable to finite amplitude perturbations.
- In region 4, even infinitesimally small perturbations grow and the basic state is said to be (linearly) unstable.

From Fig. 2.1, stability boundaries have been defined (Joseph, 1976) according to the evolution of the perturbation kinetic energy $\mathcal{E}$. When $\lambda < \lambda_G$, then the basic state is globally stable and every perturbation decays to zero in time; λ_G is the global stability boundary and provides *sufficient conditions for stability*. If $\lambda < \lambda_E$, the basic state is monotonically stable; λ_E is called the energy stability boundary. If $\lambda_G < \lambda < \lambda_L$, then the basic state is conditionally stable: small amplitude disturbances decay, whereas excessively large perturbations grow. Beyond the linear stability boundary λ_L, infinitesimally small perturbations will grow, and this stability bound therefore provides *sufficient conditions for instability*. In summary, there are two cases of instability:

(i) Sub-critical instability: $\lambda_G < \lambda < \lambda_L$, the basic state is not globally stable.
(ii) Super-critical instability: $\lambda > \lambda_L$, the basic state is not linearly stable.

The linear stability boundary is obtained by linearizing the governing equations for the perturbations in their infinitesimally small amplitude. This linear stability problem leads to an eigenvalue problem which, except in some specific cases, also has to be solved numerically.

| Additional Material |

- Determination of the global and energy stability boundaries has to be done with the full non-linear equations. Use is made of variational principles, and many examples, also for the flows in Chapter 1, are provided in Joseph (1976) and Straughan (2004).

2.1.2 Linear Stability Boundary: An Example

As an example of a linear stability boundary, we present the famous (Rayleigh, 1916) result for pure Rayleigh–Bénard flow in a horizontally unbounded rectangular container ($A_x \to \infty, A_y \to \infty$) where the two horizontal solid walls are assumed to satisfy slip boundary conditions.

The governing equations (Section 1.4), for $Ma = 0, Bi \rightarrow \infty$, are non-dimensionalized using scales κ/D for velocity, D^2/κ for time, and D for length. Moreover, a dimensionless temperature ϑ is introduced through $T = (T_B - T_A)\vartheta + T_A$ and a dimensionless pressure P through $p = -g\rho_0 z + (\mu\kappa/D^2)P$, where the first term is the hydrostatic component. This leads to the non-dimensional problem

$$Pr^{-1}\left(\frac{\partial \mathbf{u}}{\partial t} + \mathbf{u} \cdot \nabla \mathbf{u}\right) = -\nabla P + \nabla^2 \mathbf{u} + Ra\ \vartheta\ \mathbf{e}_3, \qquad (2.3a)$$

$$\nabla \cdot \mathbf{u} = 0, \qquad (2.3b)$$

$$\frac{\partial \vartheta}{\partial t} + \mathbf{u} \cdot \nabla \vartheta = \nabla^2 \vartheta, \qquad (2.3c)$$

where $\mathbf{e}_3$ is the unit vector in vertical direction and with boundary conditions

$$z = 0: \vartheta = 1\ ;\ \frac{\partial u}{\partial z} = \frac{\partial v}{\partial z} = w = 0, \qquad (2.4a)$$

$$z = 1: \frac{\partial u}{\partial z} = \frac{\partial v}{\partial z} = w = \vartheta = 0. \qquad (2.4b)$$

In Equations (2.3)–(2.4), the two dimensionless parameters Pr (Prandtl) and Ra (Rayleigh) appear which are defined as

$$Ra = \frac{\alpha_T g(T_B - T_A)D^3}{\nu\kappa}\ ;\ Pr = \frac{\nu}{\kappa}.$$

The dimensionless motionless solution (the basic state) is given by

$$\bar{u} = \bar{v} = \bar{w} = 0\ ;\ \bar{\vartheta}(z) = 1 - z, \qquad (2.5a)$$

$$\bar{P}(z) = Ra(z - \frac{z^2}{2}), \qquad (2.5b)$$

being a solution for all values of Ra and Pr.

Infinitesimal perturbations on this basic state are assumed next, that is, $u = \bar{u} + \tilde{u}$ with similar expressions for the other variables, where the tilde indicates the perturbation quantities. Linearizing the equations around the background state (neglecting products of perturbation terms) leads to

$$Pr^{-1}\frac{\partial \tilde{\mathbf{u}}}{\partial t} = -\nabla \tilde{P} + \nabla^2 \tilde{\mathbf{u}} + Ra\ \tilde{\vartheta}\ \mathbf{e}_3, \qquad (2.6a)$$

$$\nabla \cdot \tilde{\mathbf{u}} = 0, \qquad (2.6b)$$

$$\frac{\partial \tilde{\vartheta}}{\partial t} - \tilde{w} = \nabla^2 \tilde{\vartheta} \qquad (2.6c)$$

with boundary conditions

$$z = 0, 1: \frac{\partial \tilde{u}}{\partial z} = \frac{\partial \tilde{v}}{\partial z} = \tilde{w} = \tilde{\vartheta} = 0. \qquad (2.7)$$

Next, a normal mode expansion is employed, that is, for $\tilde{w}$, *See example Ex. 2.1*

$$\tilde{w} = e^{\sigma t}H(\mathbf{x})W(z), \tag{2.8a}$$

$$H(\mathbf{x}) = \sum_{j=-N, j\neq 0}^{j=N} \mathbf{c}_j e^{i\mathbf{k}_j \cdot \mathbf{x}}, \tag{2.8b}$$

with similar expressions for the other quantities. Here $\mathbf{x} = (x, y, 0)$ and $\mathbf{k} = (k_x, k_y, 0)$ are the horizontal coordinate and wavenumber vector, respectively. The wavenumber vectors $\mathbf{k}_j$ differ only in orientation, $\mathbf{k}_{-j} = -\mathbf{k}_j$, $\mathbf{c}_{-j} = \mathbf{c}_j^*$ (where the $*$ indicates the complex conjugate) and $|\mathbf{k}_j| = k$. The function H represents possible two-dimensional space-filling patterns such as roll cells and hexagons.

Expressing $\tilde{P}, \tilde{u}, \tilde{v}$, and $\tilde{\vartheta}$ in terms of $\tilde{w}$ using the equations in (2.6) leads to the following problem (Getling, 1998):

$$\frac{\partial^2 H}{\partial x^2} + \frac{\partial^2 H}{\partial y^2} + k^2 H = 0, \tag{2.9a}$$

$$(D_z^2 - k^2 - \sigma)(D_z^2 - k^2 - \frac{\sigma}{Pr})(D_z^2 - k^2)W = -k^2 Ra\, W, \tag{2.9b}$$

where $D_z = d/dz$. The boundary conditions for W become

$$W(0) = W(1) = D_z^2 W(0) = D_z^2 W(1) = D_z^4 W(0) = D_z^4 W(1) = 0. \tag{2.10}$$

The eigenfunctions for W are given by

$$W_n(z) = \sin n\pi z, n = 1, 2, \ldots \tag{2.11}$$

and the eigenvalues σ (labelled by the vertical structure of the eigenfunctions) are given by

$$\sigma_n = -\frac{Pr+1}{2}(n^2\pi^2 + k^2) \pm \sqrt{(\frac{Pr-1}{2})^2(n^2\pi^2 + k^2)^2 + \frac{Ra Pr k^2}{n^2\pi^2 + k^2}}. \tag{2.12}$$

As $Ra \geq 0$, the eigenvalues are real and there is always a positive eigenvalue when

$$Ra > Ra_n(k) = \frac{(n^2\pi^2 + k^2)^3}{k^2}. \tag{2.13}$$

The so-called neutral curve for $n = 1$ ($\sigma_1 = 0$), showing $Ra_1(k)$ versus k, is plotted in Fig. 2.2. This curve has a minimum at (k_c, Ra_c) given by

$$k_c = \frac{\pi}{\sqrt{2}}\,;\; Ra_c = \frac{27}{4}\pi^4. \tag{2.14}$$

Hence, for $Ra > Ra_c$, there is a band of wavenumbers with $\sigma > 0$ and so these perturbations will grow exponentially. The value of Ra_c therefore provides sufficient conditions for instability and is the linear stability boundary. Note that multiple patterns (hexagons, rolls, represented by the function $H(x, y)$) become unstable under

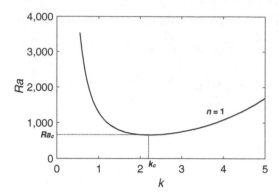

Figure 2.2 Neutral curve for pure Rayleigh–Bénard convection, showing $Ra_1(k)$ from (2.13) versus k (for $\sigma_1 = 0$) in a horizontally unbounded rectangular liquid layer with slip conditions at the top and bottom walls.

the same conditions, as the stability boundary depends only on the norm k of the wavenumber vector $\mathbf{k}$. *See example Ex. 2.2*

Additional Material

- Linear stability theory of fluid flows is standard material in textbooks on hydrodynamic stability theory, such as Chandrasekhar (1961) and Drazin and Reid (2004).
- Specific focus on the stability of the Rayleigh–Bénard–Marangoni flow and related flows can be found in Getling (1998) and Platten and Legros (1984).

2.2 Beyond Criticality: Weakly Non-linear Theory

When the control parameter λ is slightly above the linear stability boundary λ_c, such as $Ra > Ra_c$ in the Rayleigh–Bénard problem of the previous subsection, a band of wavenumbers destabilizes the basic flow as the associated spatial patterns grow exponentially in time. At some later time, the linear theory is no longer valid because the amplitude of the non-linear terms can no longer be neglected. In this section, the so-called weakly non-linear theory is presented where non-linear effects are taken into account in the regime

$$\frac{|\lambda - \lambda_c|}{|\lambda_c|} < \epsilon, \tag{2.15}$$

where $\epsilon \ll 1$.

When at least one of the horizontal dimensions is unbounded, and hence the importance of boundary conditions in this direction on the finite amplitude flows is neglected, the weakly non-linear analysis leads to a Ginzburg–Landau equation. When the effects of horizontal boundary conditions cannot be neglected, the weakly non-linear approach will lead to a set of amplitude equations. We will describe both approaches in what follows for a general flow problem for the state vector $\mathbf{\Phi}$ (e.g., $\mathbf{u}, p$) with governing equations

$$\mathcal{M}\frac{\partial \mathbf{\Phi}}{\partial t} + \mathcal{G}(\mathbf{\Phi}) = \mathbf{f}, \tag{2.16}$$

with appropriate boundary conditions on the boundary of the flow domain. Here $\mathcal{M}$ is a linear operator, $\mathcal{G}$ is a non-linear operator, and $\mathbf{f}$ represents the forcing. Note that the operator $\mathcal{M}$ does not need to the identify, as in the incompressible general fluid dynamics equations there is an equation without a time derivative (i.e., the continuity equation).

2.2.1 Ginzburg–Landau Equations

Consider first the case where one of the horizontal dimensions of the problem is unbounded, which allows the existence of traveling wave solutions of the linear stability problem. The background state $\bar{\mathbf{\Phi}}$ is assumed to be steady and satisfies

$$\mathcal{G}(\bar{\mathbf{\Phi}}) = \mathbf{f}. \tag{2.17}$$

We next now consider perturbations $\boldsymbol{\phi} = \mathbf{\Phi} - \bar{\mathbf{\Phi}}$, and the system of equations for the perturbations $\boldsymbol{\phi}$ can be written in general as

$$(\mathcal{M}\frac{\partial}{\partial t} + \mathcal{L})\boldsymbol{\phi} + \mathcal{N}(\boldsymbol{\phi})\boldsymbol{\phi} = 0 \tag{2.18}$$

with appropriate boundary conditions. Here $\mathcal{L}$ is a linear operator and $\mathcal{N}$ a non-linear operator which we assume to represent a quadratic non-linearity (as in the governing equations in Section 1.1). The linear stability problem (where non-linear interactions of the perturbations are neglected) is formulated as

$$(\mathcal{M}\frac{\partial}{\partial t} + \mathcal{L})\tilde{\boldsymbol{\phi}} = 0 \tag{2.19}$$

for infinitesimally small perturbations $\tilde{\boldsymbol{\phi}}$.

Assume that the x-direction is unbounded such that traveling wave solutions of (2.19) exist with wavenumber k and complex growth factor σ. For convenience, we will consider the two-dimensional Cartesian case (coordinates (x, z)), with

$$\tilde{\boldsymbol{\phi}}(x, z, t) = \hat{\boldsymbol{\phi}}(z)e^{ikx+\sigma t} + c.c. \tag{2.20}$$

where *c.c.* indicates complex conjugate. Substitution of (2.20) into (2.19) gives a boundary value problem for the eigenpair $(\sigma, \hat{\boldsymbol{\phi}})$, that is,

$$(\hat{\mathcal{M}}(k)\sigma + \hat{\mathcal{L}}(k))\hat{\boldsymbol{\phi}} = 0 \tag{2.21}$$

with appropriate boundary conditions. Here the operators $\hat{\mathcal{M}}$ and $\hat{\mathcal{L}}$ are the Fourier transforms of the original operators in the x-direction.

The eigenvalue is written as $\sigma = \sigma_R + i\sigma_I$ and considered as a function of the wavenumber k and the control parameter λ. The neutral curve $\sigma_R(k, \lambda) = 0$ provides sufficient conditions for instability, and in many applications, the neutral curve has a minimum at (k_c, λ_c) (see Fig. 2.2) at which

$$\sigma_R(k_c, \lambda_c) = 0 \; ; \; \frac{\partial \sigma_R}{\partial k}(k_c, \lambda_c) = 0 \; ; \; \frac{\partial^2 \sigma_R}{\partial k^2}(k_c, \lambda_c) < 0. \tag{2.22}$$

In what follows we will use $\omega_c = \sigma_I(k_c)$, and also assume that the mode $k = 0$ is damped.

Assume now conditions just above criticality, that is,

$$\lambda = \lambda_c + m\epsilon^2, \tag{2.23}$$

where $\epsilon \ll 1$ and $m = \mathcal{O}(1)$. As the neutral curve can be approximated by a parabola $\lambda - \lambda_c \sim (k - k_c)^2$, this implies that $|k - k_c| = \mathcal{O}(\epsilon)$. The unstable traveling waves are hence limited to a narrow band around k_c which can be interpreted as a wave packet with central wavenumber k_c. This wave packet evolves on a time scale which is large compared to typical wave periods $2\pi/\omega_c$ and is characterized by scales

$$T = \epsilon^2 t \; ; \; X = \epsilon(x - c_g t), \tag{2.24}$$

where $c_g = \partial \sigma_I / \partial k$ is the group velocity. The long spatial variable X is a slowly moving coordinate, traveling with the group velocity of the growing wave packet.

The scaling leads to transformations for $\phi(x, X(x, t), z, t, T(t))$ as

$$\frac{\partial}{\partial t} \;\; \rightarrow \;\; \frac{\partial}{\partial t} - \epsilon c_g \frac{\partial}{\partial X} + \epsilon^2 \frac{\partial}{\partial T}, \tag{2.25a}$$

$$\frac{\partial}{\partial x} \;\; \rightarrow \;\; \frac{\partial}{\partial x} + \epsilon \frac{\partial}{\partial X}. \tag{2.25b}$$

The final amplitude of the perturbations will be small compared to that of the background state (for λ close to λ_c), so the solution vector is expanded in terms of the small parameter ϵ and Fourier modes of the marginally stable wave $E = \exp(i(k_c x + \omega_c t))$, that is,

$$\phi = \epsilon \Phi^{(11)} E + \epsilon^2 (\Phi^{(02)} + \Phi^{(12)} E + \Phi^{(22)} E^2) + \epsilon^3 \Phi^{(13)} E + \ldots + c.c. \tag{2.26}$$

where the $\Phi^{(ij)} = \Phi^{(ij)}(X, z, T)$ depend on the slow time scale and long spatial scale.

Substitution of (2.26) into (2.18) and collecting terms of the same order (in ϵ and E) gives at $\mathcal{O}(\epsilon E)$ the linear stability problem

$$(i\omega_c \hat{\mathcal{M}}(k_c) + \hat{\mathcal{L}}(k_c))\mathbf{\Phi}^{(11)} = 0. \tag{2.27}$$

As the left-hand side does not operate on the large scales of X and T, we can write $\mathbf{\Phi}^{(11)} = A(X, T)\mathbf{\Psi}$, where $\mathbf{\Psi} = \hat{\phi}(z)$ is the eigenvector at $k = k_c$ from (2.20) and (2.20). The weakly non-linear analysis eventually leads to an equation for the scalar (but complex) amplitude $A(X, T)$. At $\mathcal{O}(\epsilon^2 E)$, the equations are

$$(i\omega_c \hat{\mathcal{M}}(k_c) + \hat{\mathcal{L}}(k_c))\mathbf{\Phi}^{(12)} = -(i\omega_c \hat{\mathcal{M}}_k(k_c) + \hat{\mathcal{L}}_k(k_c) - c_g\hat{\mathcal{M}}(k_c))\frac{\partial \mathbf{\Phi}^{(11)}}{\partial X}, \tag{2.28}$$

where the subscript k indicates differentiation to k. At $\mathcal{O}(\epsilon^2)$ and $\mathcal{O}(\epsilon^2 E^2)$, one finds

$$\hat{\mathcal{L}}(0)\mathbf{\Phi}^{(02)} = -2\mathcal{R}(\mathcal{N}(\mathbf{\Phi}^{(11)})\mathbf{\Phi}^{(11)*}), \tag{2.29a}$$

$$(2i\omega_c \hat{\mathcal{M}}(2ik_c) + \hat{\mathcal{L}}(2ik_c))\mathbf{\Phi}^{(22)} = -\mathcal{N}(\mathbf{\Phi}^{(11)})\mathbf{\Phi}^{(11)}, \tag{2.29b}$$

where $\mathcal{R}$ indicates real part and $*$ again complex conjugate. Using the relation $\mathbf{\Phi}^{(11)} = A(X, T)\mathbf{\Psi}$ in the equations of (2.28) and (2.29) leads to

$$\mathbf{\Phi}^{(12)} = \frac{\partial A}{\partial X}\mathbf{\Psi}^{(12)} \; ; \; \mathbf{\Phi}^{(02)} = |A|^2\mathbf{\Psi}^{(02)} \; ; \; \mathbf{\Phi}^{(22)} = A^2\mathbf{\Psi}^{(22)}, \tag{2.30}$$

where the vectors $\mathbf{\Psi}^{(12)}$, $\mathbf{\Psi}^{(02)}$, and $\mathbf{\Psi}^{(22)}$ satisfy

$$(i\omega_c \hat{\mathcal{M}}(k_c) + \hat{\mathcal{L}}(k_c))\mathbf{\Psi}^{(12)} = -(i\omega_c \hat{\mathcal{M}}_k(k_c) + \hat{\mathcal{L}}_k(k_c)$$
$$-c_g\hat{\mathcal{M}}(k_c))\mathbf{\Psi}, \tag{2.31a}$$

$$\hat{\mathcal{L}}(0)\mathbf{\Psi}^{(02)} = -2\mathcal{R}(\mathcal{N}(\mathbf{\Psi})\mathbf{\Psi}^*), \tag{2.31b}$$

$$(2i\omega_c \hat{\mathcal{M}}(2ik_c) + \hat{\mathcal{L}}(2ik_c))\mathbf{\Psi}^{(22)} = -\mathcal{N}(\mathbf{\Psi})\mathbf{\Psi} \tag{2.31c}$$

and these equations are complemented with the appropriate boundary conditions at each order of the expansion.

Differentiation of the eigenvalue problem (2.21) to k and use of the group velocity at criticality gives $\mathbf{\Psi}^{(12)}$ from (2.31a) as

$$\mathbf{\Psi}^{(12)} = -i\frac{\partial \mathbf{\Psi}}{\partial k} \tag{2.32}$$

evaluated at criticality. The left-hand sides of (2.31b,c) are non-singular and hence can be solved for $\mathbf{\Psi}^{(02)}$ and $\mathbf{\Psi}^{(22)}$. At $\mathcal{O}(\epsilon^3 E^2)$ a singular problem is obtained for $\mathbf{\Psi}^{(13)}$, that is,

$$(i\omega_c \hat{\mathcal{M}}(k_c) + \hat{\mathcal{L}}(k_c))\mathbf{\Psi}^{(13)} = -(\mathcal{M}(k_c)\mathbf{\Psi}\frac{\partial A}{\partial T} + m\Gamma A + \Sigma\frac{\partial^2 A}{\partial X^2} + \Lambda A|A|^2), \tag{2.33}$$

where

$$\mathbf{\Gamma} = (\hat{\mathcal{L}}_\lambda(k_c) - i\omega_c\hat{\mathcal{M}}_\lambda(k_c))\mathbf{\Psi}, \tag{2.34a}$$

$$\mathbf{\Sigma} = \frac{1}{2}(i\omega_c\hat{\mathcal{M}}_{kk}(k_c) - \hat{\mathcal{L}}_{kk}(k_c) - 2c_g\hat{\mathcal{M}}_k(k_c))\mathbf{\Psi} \tag{2.34b}$$

$$+ i(i\omega_c\hat{\mathcal{M}}_k(k_c) - c_g\hat{\mathcal{M}}(k_c) - \hat{\mathcal{L}}_k(k_c))\mathbf{\Psi}_k,$$

$$\mathbf{\Lambda} = \mathcal{N}(\mathbf{\Psi})\mathbf{\Psi}^{(02)} + \mathcal{N}(\mathbf{\Psi}^{(02)})\mathbf{\Psi} + \mathcal{N}(\mathbf{\Psi}^{(22)})\mathbf{\Psi}^* + \mathcal{N}(\mathbf{\Psi}^*)\mathbf{\Psi}^{(22)}. \tag{2.34c}$$

See example Ex. 2.3

In general, the right-hand side of (2.33) is not contained in the range of the linear operator on the left-hand side. Since the kernel of the operator $i\omega_c\hat{\mathcal{M}}(k_c) + \hat{\mathcal{L}}(k_c)$ has dimension 1, it is spanned by one vector, here indicated by $\mathbf{\Omega}$; this implies that

$$\mathbf{\Omega}^H(i\omega_c\hat{\mathcal{M}}(k_c) + \hat{\mathcal{L}}(k_c))\mathbf{W} = 0, \tag{2.35}$$

where $\mathbf{W}$ is the right-hand side of (2.33) and the superscript H indicates Hermitian transposed. The resulting amplitude equation derived from (2.33) and (2.35) is the Ginzburg–Landau equation

$$\frac{\partial A}{\partial T} = \gamma_1 A + \gamma_2 \frac{\partial^2 A}{\partial X^2} - \gamma_3 A|A|^2, \tag{2.36}$$

where

$$\gamma_1 = m\frac{\mathbf{\Omega}^H\mathbf{\Gamma}}{\mathbf{\Omega}^H\hat{\mathcal{M}}(k_c)\mathbf{\Psi}}, \tag{2.37a}$$

$$\gamma_2 = \frac{\mathbf{\Omega}^H\mathbf{\Sigma}}{\mathbf{\Omega}^H\hat{\mathcal{M}}(k_c)\mathbf{\Psi}}, \tag{2.37b}$$

$$\gamma_3 = -\frac{\mathbf{\Omega}^H\mathbf{\Lambda}}{\mathbf{\Omega}^H\hat{\mathcal{M}}(k_c)\mathbf{\Psi}}. \tag{2.37c}$$

In the remainder of the book, we will use the Ginzburg–Landau equation (2.36) as a one-dimensional partial differential equation for the complex amplitude $A(X, T)$ to illustrate bifurcation behaviour of typical fluid flows. As we will see, the behaviour of the solutions of this equation is very rich.

2.2.2 Amplitude Equations

In case the geometry of the problem is such that no traveling wave solutions exist (e.g., a bounded geometry in all directions), we can still obtain a reduced model near criticality through a Galerkin-type projection using the eigenfunctions of the linear operator.

Suppose that the original problem (2.16) is discretized (see Chapter 4) using a spectral, finite difference or finite element method. The set of discretized non-linear differential equations is rewritten in the general form,

$$M\frac{d\Phi}{dt} + L\Phi + N(\Phi, \Phi) = \mathbf{f}, \tag{2.38}$$

to explicitly show the linear part L, the non-linear part N and the forcing $\mathbf{f}$. Let $\bar{\Phi}$ be the solution to the steady problem, that is,

$$L\bar{\Phi} + N(\bar{\Phi}, \bar{\Phi}) = \mathbf{f}. \tag{2.39}$$

The solution to (2.38) is now decomposed into this steady state and a remainder time-dependent part,

$$\Phi = \bar{\Phi} + \phi. \tag{2.40}$$

After substitution into (2.38), the linearized flow ϕ is governed by

$$M\frac{d\phi}{dt} + J\phi = 0, \tag{2.41}$$

where the total Jacobian J is defined as

$$J = L + N(\bar{\Phi}, \cdot) + N(\cdot, \bar{\Phi}). \tag{2.42}$$

The linear operators have an eigenvector decomposition,

$$\Lambda^H J R = \Sigma \; ; \; \Lambda^H M R = I. \tag{2.43}$$

Here R and Λ denote the right- and left-hand eigenspaces of the linear operator J, I is the identity, and the diagonal matrix Σ contains the corresponding eigenvalues, that is,

$$\begin{aligned}
R &= \begin{pmatrix} \mathbf{r}_1 & \mathbf{r}_2 & \cdots & \mathbf{r}_r \end{pmatrix}, & \text{(2.44a)} \\
\Lambda &= \begin{pmatrix} \mathbf{l}_1 & \mathbf{l}_2 & \cdots & \mathbf{l}_r \end{pmatrix}, & \text{(2.44b)} \\
\Sigma &= \text{diag}(\sigma_1 & \sigma_2 & \cdots & \sigma_r), & \text{(2.44c)}
\end{aligned}$$

where $r = \text{rank}(\Sigma) \leq d$, with d being the number of degrees of freedom of the dynamical system; the preceding inequality is due to the singular nature of M. Relation (2.43) states that Λ and R are a bi-orthogonal set of eigenvectors, and this property will be used in the Galerkin projection that follows.

With the use of this eigenbasis, the perturbation ϕ is expanded in n right-hand eigenvectors:

$$\phi = R_n a = \sum_{j=1}^{n} \mathbf{r}_j a_j(t). \tag{2.45}$$

The matrix R_n denotes the n-dimensional subspace of R of suitably chosen right-hand vectors, and Λ_n is its adjoint subspace.

Substitution of (2.40) into (2.38) and using (2.39) and (2.45) yields

$$MR_n \frac{da}{dt} + JR_n a + N(R_n a, R_n a) = 0. \tag{2.46}$$

Projection onto the left-hand eigenbasis Λ_n and the use of the bi-orthogonality relation (2.43) results in the set of coupled amplitude equations,

$$\frac{da}{dt} - Sa + n(a, a) = \mathbf{0}. \tag{2.47}$$

The operators in the projected system are defined as

$$S \;=\; \Lambda_n^H J R_n, \tag{2.48a}$$

$$n(a, a) \;=\; \Lambda_n^H N(R_n a, R_n a). \tag{2.48b}$$

In terms of the individual components, the evolution of amplitudes $a_j(t), j = 1, \ldots, n$ is governed by

$$\frac{da_j}{dt} - \sum_{k=1}^{n} b_{jk} a_{jk} + \sum_{k=1}^{n} \sum_{l=1}^{n} c_{jkl} a_k a_l = 0 \quad , \quad j = 1, \ldots, n. \tag{2.49}$$

The coefficients in the projected system are defined as

$$b_{jk} \;=\; \mathbf{l}_j^H J \mathbf{r}_k, \tag{2.50a}$$

$$c_{jkl} \;=\; \mathbf{l}_j^H N(\mathbf{r}_k, \mathbf{r}_l). \tag{2.50b}$$

The main frustration with these amplitude equation models is the non-correspondence in dynamical behaviour between the reduced model and the full model when the order of truncation of the reduced model is changed (Van der Vaart *et al.*, 2002). An enormously rich behaviour may be found in many reduced models, which has fascinated researchers so much that critically examining the relation between the full and reduced models is often omitted. An example hereof is the famous Lorenz model (Lorenz, 1963), where the dynamics bear little resemblance to that of the underlying full model of Rayleigh–Bénard convection (Curry *et al.*, 1984). Hence, while this approach is fairly standard, the domain in parameter space where a close correspondence exists between the dynamical behaviour contained in (2.49), for a chosen value of n, and the full model is a priori unclear.

| Additional Material |

- Amplitude equations and the Ginzburg–Landau equation play an important role in theories in the broad research area of pattern formation; see
  Rabinovich *et al.* (2000) and Hoyle (2006).

2.3 Dynamical Systems

The stability bounds do indicate where (in parameter space) flows become sensitive to perturbations, but do not give any answer on what new patterns can arise. Dynamical systems theory, and in particular bifurcation theory, is a systematic approach to determine which equilibrium flow patterns are possible near a stability boundary. Although dynamical systems theory can in general be formulated for infinite-dimensional systems, such as the Navier–Stokes equations, we present here the necessary concepts for finite-dimensional systems. The reason is that in a numerical approach, eventually (often high-dimensional) finite-dimensional dynamical systems are handled.

2.3.1 Continuous versus Discrete Systems

A general first-order system of ordinary differential equations (ODEs) can be written as the continuous time dynamical system

$$\frac{d\mathbf{x}}{dt} = \mathbf{f}(\mathbf{x}, \lambda, t), \tag{2.51}$$

where $\mathbf{x}$ is the state vector in the state space $\mathbb{R}^d$, $\mathbf{f}$ is a smooth (sufficiently differentiable) vector field, λ is a real parameter, and t denotes time. The number d is referred to as the dimension or the number of degrees of freedom of the dynamical system. When the vector field $\mathbf{f}$ does not depend explicitly on time, that is,

$$\frac{d\mathbf{x}}{dt} = \mathbf{f}(\mathbf{x}, \lambda), \tag{2.52}$$

the dynamical system is called autonomous; otherwise, it is called non-autonomous. A trajectory of the dynamical system, starting, for example, at $\mathbf{x}_0$, is a curve $\mathbf{x}(t)$ satisfying (2.51). Hence, at each point the vector field $\mathbf{f}$ is tangent to the curve and $\mathbf{x}(t)$ is a solution of the equations of (2.51).

When time is discrete, with counter k, we obtain a discrete dynamical system of the form

$$\mathbf{x}_{k+1} = \mathbf{f}_k(\mathbf{x}_k, \lambda). \tag{2.53}$$

A continuous time dynamical system can be analyzed as a discrete time dynamical system through a so-called Poincaré map. To define a Poincaré map, a hypersurface Σ^+ in the state space $\mathbb{R}^d$, for example a line segment in two-dimensional state space or a plane in three-dimensional state space, is chosen such that each trajectory is not tangent to it for all time t, that is, when

$$\mathbf{n} \cdot \mathbf{f} \neq 0. \tag{2.54}$$

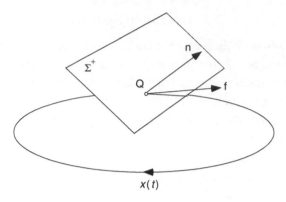

Figure 2.3 Sketch of a Poincaré section Σ^+. A periodic orbit is sketched which intersects the Poincaré section at the point Q. The vector **f** is the tangent to the trajectory at Q and **n** is the outward normal to Σ^+.

Here, **n** is the normal to the hypersurface (Fig. 2.3) and **f** the right-hand side of (2.51); this hypersurface is called a Poincaré section. Let a trajectory intersect a Poincaré section at successive intersections indicated by $\{\mathbf{x}_1, \mathbf{x}_2, \mathbf{x}_3, \ldots\}$; then the Poincaré map $\mathcal{P}: \Sigma^+ \to \Sigma^+$ is defined as

$$\mathbf{x}_{k+1} = \mathcal{P}\mathbf{x}_k. \tag{2.55}$$

2.3.2 Stability Theory of Fixed Points

A solution $\bar{\mathbf{x}} \in \mathbb{R}^d$ of an autonomous continuous dynamical system at a parameter value λ is a fixed point if

$$\mathbf{f}(\bar{\mathbf{x}}, \lambda) = 0, \tag{2.56}$$

and hence any trajectory with initial conditions on the fixed point will remain there forever. In this section, we are interested in the transient behaviour of small perturbations on such a fixed point.

In the analysis of the linear stability of a particular fixed point $\bar{\mathbf{x}}$, small perturbations **y** are assumed to be present, that is,

$$\mathbf{x} = \bar{\mathbf{x}} + \mathbf{y}, \tag{2.57}$$

and linearization of (2.52) around $\bar{\mathbf{x}}$ gives

$$\frac{d\mathbf{y}}{dt} = J(\bar{\mathbf{x}}, \lambda)\mathbf{y}, \tag{2.58}$$

where J is the Jacobian matrix given by

$$J = \begin{pmatrix} \frac{\partial f_1}{\partial x_1} & \cdots & \frac{\partial f_1}{\partial x_d} \\ \cdots & \cdots & \cdots \\ \frac{\partial f_d}{\partial x_1} & \cdots & \frac{\partial f_d}{\partial x_d} \end{pmatrix}. \tag{2.59}$$

The solution of (2.58) with initial condition $\mathbf{y}_0$ is given by

$$\mathbf{y}(t) = e^{Jt}\mathbf{y}_0, \tag{2.60}$$

and hence the time behaviour depends on the eigenvalues of the Jacobian matrix J. The corresponding eigenvectors are usually referred to as the normal modes.

If J is decomposed as $J = U\Sigma U^{-1}$, where Σ contains the eigenvalues of J, e^{Jt} is given by

$$e^{Jt} = \sum_{k=0}^{\infty} \frac{1}{k!}(Jt)^k = U\left[\sum_{k=0}^{\infty} \frac{1}{k!}(\Sigma t)^k\right]U^{-1} = Ue^{\Sigma t}U^{-1}. \tag{2.61}$$

If all eigenvalues of J, say $\sigma_1, \ldots, \sigma_d$ have negative real parts, that is, $\mathcal{R}(\sigma_j) < 0$ for all j, then the fixed point is linearly stable. For this case, indeed all trajectories of (2.58) will approach $\mathbf{y} = 0$ asymptotically, that is, for $t \to \infty$. When at least one of the eigenvalues σ_k has a positive real part, $\mathcal{R}(\sigma_k) > 0$, the fixed point is said to be unstable.

2.4 Bifurcation Theory of Fixed Points

Bifurcation theory addresses changes in the qualitative behaviour of a dynamical system as one or several of its parameters vary. If J in (2.59) has no purely imaginary eigenvalues, $\bar{\mathbf{x}}$ is called a hyperbolic fixed point. Near such a fixed point, the local solution structure of the linearized system is the same as that of the non-linear system. This is a consequence of the so-called Hartman–Grobman theorem (Guckenheimer and Holmes, 1990). When qualitative changes occur in the fixed-point solutions of the dynamical system, such as the changes in type or number of solutions, the dynamical system is said to have undergone a bifurcation. This can only occur at non-hyperbolic fixed points. In the state-parameter space formed by $(\mathbf{x}, \lambda)$, locations at which bifurcations occur are called bifurcation points. A bifurcation that needs at least k parameters to occur is called a co-dimension-k bifurcation.

The center manifold theorem (Guckenheimer and Holmes, 1990) implies that it is possible to (locally) reduce the dynamics. Typically, taking $\bar{\mathbf{x}} = 0$ for simplicity, one has

$$\frac{d\mathbf{x}}{dt} = L\mathbf{x} + N(\mathbf{x}), \tag{2.62}$$

where N, which depends on the parameter λ, has a Taylor expansion starting with at least quadratic terms, $\mathbf{x} \in \mathbb{R}^m$ and L has m eigenvalues with zero real part. Having reduced the system (2.51) into the system (2.62), it is possible to find a change of coordinates so that the system becomes 'as simple as possible'. The resulting vector field thus obtained is called the normal form. This procedure is an extension of the reduction to Jordan form for matrices to the non-linear case. Normal form

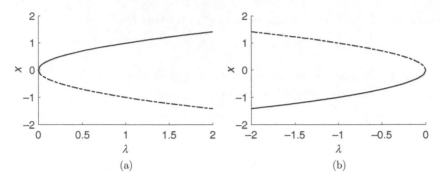

Figure 2.4 Super-critical (a) and sub-critical (b) saddle-node bifurcation. The solid (dash-dotted) branches indicate stable (unstable) solutions.

theory (Guckenheimer and Holmes, 1990) provides a way to classify the different kind of bifurcations that may occur with only knowledge of the eigenvalues that lie on the imaginary axis, that is, those of L.

In the case $m = 1$, there are three important normal forms:

1. *Saddle-node bifurcation*: this corresponds to the case where the system (2.62), when reduced to its normal form, is

$$\frac{dx}{dt} = \lambda \pm x^2. \tag{2.63}$$

The sign characterizes super-criticality ($\lambda - x^2$) or sub-criticality ($\lambda + x^2$). In the super-critical case, it is straightforward to check that the branch of solutions $x = \sqrt{\lambda}$ is linearly stable and the branch $x = -\sqrt{\lambda}$ is unstable (see Fig. 2.4).

2. *Trans-critical bifurcation*: in this case the normal form is given by

$$\frac{dx}{dt} = \lambda x \pm x^2. \tag{2.64}$$

In both sub-critical and super-critical cases, there is an exchange of stability from stable to unstable fixed points and vice versa as the parameter λ is varied through the bifurcation at $\lambda = 0$ (see Fig. 2.5).

3. *Pitchfork bifurcation*: the normal form is

$$\frac{dx}{dt} = \lambda x \pm x^3. \tag{2.65}$$

In the super-critical situation ($dx/dt = \lambda x - x^3$), there is a transfer of stability from the symmetric solution $x = 0$ to the pair of conjugated solutions $x = \pm\sqrt{\lambda}$

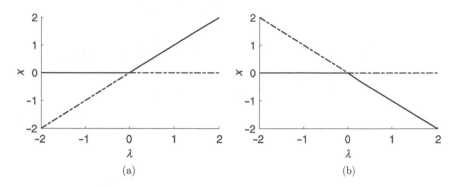

Figure 2.5 Super-critical (a) and sub-critical (b) trans-critical bifurcation.

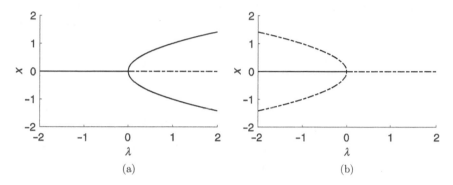

Figure 2.6 Super-critical (a) and sub-critical (b) pitchfork bifurcation.

(Fig. 2.6a). The system remains in a neighbourhood of the equilibrium so that one observes a soft or non-catastrophic loss of stability. In the sub-critical case ($dx/dt = \lambda x + x^3$), the situation is very different, as can be seen in Fig. 2.6b. The domain of attraction of the fixed point (the set of initial conditions which end up at the fixed point for infinite time) is bounded by the unstable fixed points and shrinks as the parameter λ approaches zero. The system is thus pushed out from the neighbourhood of the now unstable fixed point leading to a *sharp* or catastrophic loss of stability. Decreasing again the parameter to negative values will not necessarily return the system to the previously stable fixed point, since it may have already left its domain of attraction.

Whereas in cases 1–3, the number of fixed points changed as the parameter was varied, it is also possible that a steady solution becomes unstable to time-periodic disturbances. This so-called *Hopf bifurcation* occurs only in dynamical systems with $m > 1$ and corresponds to the special case of a simple conjugate pair of pure imaginary eigenvalues $\sigma = \pm i\omega$ crossing the imaginary axis, leading to the normal form (for $m = 2$)

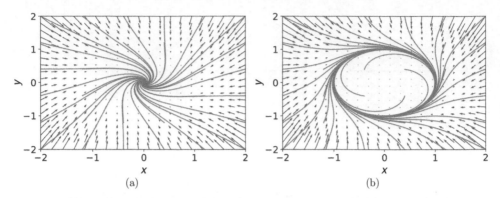

Figure 2.7 Phase portraits for a super-critical Hopf bifurcation (at $\lambda = 0$) where the arrows indicate the slope dy/dx at each location. (a) For $\lambda = -1 < 0$, there is only one stable fixed point. (b) A stable limit cycle $x^2 + y^2 = 1$ appears for $\lambda = 1 > 0$.

$$\frac{dx}{dt} = \lambda x - \omega y \pm x(x^2 + y^2), \tag{2.66a}$$

$$\frac{dy}{dt} = \omega x + \lambda y \pm y(x^2 + y^2). \tag{2.66b}$$

Phase portraits of the super-critical Hopf bifurcation ($-$ sign in (2.66)) show that for $\lambda < 0$, the fixed point ($x = 0, y = 0$) is stable (Fig. 2.7a). For $\lambda > 0$, a periodic orbit (with period $2\pi/\omega$) appears (Fig. 2.7b), which is called a *limit cycle* (an isolated periodic orbit).

The normal form (2.66) can also be written in polar coordinates $x = r\cos\theta, y = r\sin\theta$ as

$$\frac{dr}{dt} = \lambda r \pm r^3, \tag{2.67a}$$

$$\frac{d\theta}{dt} = \omega. \tag{2.67b}$$

Similar to the pitchfork bifurcation case, the sign determines whether the Hopf bifurcation is super-critical or sub-critical. *See example Ex. 2.4*

Example 2.1 *Consider the following particular case of the Ginzburg–Landau equation*

$$\frac{\partial A}{\partial t} = (\alpha + i\beta)A + \gamma_2 \frac{\partial^2 A}{\partial x^2} - \gamma_3 A|A|^2 \tag{2.68}$$

with real $\alpha, \beta, \gamma_2 > 0$ and $\gamma_3 > 0$. Assume first that $\gamma_2 = 0$. To find an x-independent solution, which is called the Stokes wave, we substitute $A = \rho \exp(i\theta)$ into (2.68), where ρ and θ are both real. After cancelling the common factor $\exp(i\theta)$, one obtains the equation

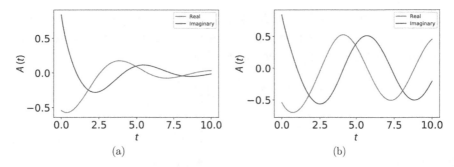

Figure 2.8 (a) Trajectories of (2.68) for $\gamma_2 = 0$, $\gamma_3 = 1$, $\beta = 1$ and values of (a) $\alpha = -0.25$ and (b) $\alpha = 0.25$.

$$\frac{d\rho}{dt} + i\rho\frac{d\theta}{dt} = i\rho\beta + \rho(\alpha - \gamma_3\rho^2). \qquad (2.69)$$

Collecting real and imaginary parts, we obtain two equations:

$$\frac{d\rho}{dt} = \rho(\alpha - \gamma_3\rho^2), \qquad (2.70a)$$

$$\frac{d\theta}{dt} = \beta. \qquad (2.70b)$$

For $\alpha < 0$, the steady state $\bar{\rho} = 0$ is stable (Fig.2.8a). Clearly $\bar{\rho} = 0$ is unstable if $\alpha > 0$, but then stable states $\bar{\rho} = \sqrt{\alpha/\gamma_3}$ exist. For $\rho \neq 0$, θ is a linear function of t, that is, $\theta = c + \beta t$. So, the solution is

$$A = \sqrt{\alpha/\gamma_3} \exp(ic + i\beta t). \qquad (2.71)$$

This periodic solution with period $2\pi/\beta$ is also a limit cycle for $\alpha > 0$ (Fig. 2.8b). Now, if we choose $\gamma_2 > 0$, then this has a stabilizing effect, so α needs to be larger than a certain positive value for the periodic solution to exist.

Additional Material

- The text in these sections is meant as a very short recap of dynamical systems theory. Other books are much more suitable to learn about this theory. Introductions with many examples can be found in Strogatz (1994), Verhulst (2000), and Perko (2013). More advanced texts are Guckenheimer and Holmes (1990) and Kuznetsov (1995).

2.5 Bifurcation Theory of Limit Cycles

In this section, we provide a brief overview of bifurcations of periodic orbits of autonomous dynamical systems when a single parameter is varied.

Assume that one has a limit cycle, say indicated by γ, of the original system (2.52) for a parameter λ that we omit in the notations for simplicity, and whose corresponding solution is $\bar{\mathbf{x}}(t) = \bar{\mathbf{x}}(t + p)$, where p is the period of the orbit. We consider an infinitesimal perturbation $\mathbf{y}(t)$ of γ, that is, we let $\mathbf{x}(t) = \bar{\mathbf{x}}(t) + \mathbf{y}(t)$ in (2.52), and neglecting quadratic terms, one then obtains

$$\dot{\mathbf{y}} = J(\bar{\mathbf{x}}(t))\mathbf{y}, \qquad (2.72)$$

and $J(\bar{\mathbf{x}}(t))$ is a p-periodic matrix.

It can be shown (Guckenheimer and Holmes, 1990) that, using the fundamental solution matrix Y of the system (2.72), it follows that

$$Y(t + p) = \Phi Y(t). \qquad (2.73)$$

With $Y(0) = I$, it follows that $\Phi = Y(T)$. The matrix Φ is called the mon-odromy matrix, and its eigenvalues $\rho_1, \ldots, \rho_d$ are called the Floquet multipliers. The monodromy matrix is not uniquely determined by the solutions of (2.72), but its eigenvalues are. Since the perturbation $\mathbf{y}(t) = \bar{\mathbf{x}}(t + \epsilon) - \bar{\mathbf{x}}(t)$, ϵ small, is p-periodic, it immediately implies that Φ has an eigenvalue $\rho_1 = +1$, that is, perturbations along γ neither diverge nor converge. The linear stability of γ is thus determined by the remaining $d - 1$ eigenvalues.

Let Σ^+ be a (fixed) local cross section of dimension $d - 1$ (see Fig. 2.3) of the limit cycle γ such that the periodic orbit is not tangent to this hypersurface, and denote $\mathbf{x}^\star$ the intersection of Σ^+ with γ. There is a nice geometrical interpretation of the monodromy matrix in terms of the Poincaré map defined as $\mathcal{P}(\mathbf{x}) = \phi_\tau(\mathbf{x})$, where $\mathbf{x}$ is assumed to be in a neighbourhood of $\mathbf{x}^\star$, and τ is the time taken for the orbit $\phi_t(\mathbf{x})$ to first return to Σ^+ (as $\mathbf{x}$ approaches $\mathbf{x}^\star$, τ will tend to p). After a change of basis such that the matrix Φ has a column $(0, \cdots 0, 1)^T$ corresponding to the unit eigenvalue, the remaining block $(d-1) \times (d-1)$ matrix corresponds to the linearized Poincaré map. These remarks show that the bifurcations of limit cycles are related to the behaviour of a discrete dynamical system (from the Poincaré map) $\mathbf{x}_{n+1} = \mathcal{P}\mathbf{x}_n$.

As an example, consider the system of equations for the supercritical Hopf bifurcation, in polar coordinates given by (2.67) as

$$\frac{dr}{dt} = \lambda r - r^3,$$
$$\frac{d\theta}{dt} = \omega.$$

At $\lambda = 0$, a Hopf bifurcation occurs and, for $\lambda > 0$, a periodic orbit having a period $p = 2\pi/\omega$ exists and is described by

$$r = \sqrt{\lambda} \, ; \, \theta = \omega t. \qquad (2.75)$$

To determine the stability of the periodic orbit coming from (2.66) for $\lambda > 0$ and $\omega \neq 0$, we write the solution as

$$\bar{x}(t) = \sqrt{\lambda} \, \cos \omega t,$$
$$\bar{y}(t) = \sqrt{\lambda} \, \sin \omega t.$$

The Jacobian matrix at the periodic orbit can be obtained from (2.66) and is given by

$$J(\bar{\mathbf{x}}(t), \lambda) = \begin{pmatrix} \lambda - 3x^2(t) - y^2(t) & -\omega - 2x(t)y(t) \\ \omega - 2x(t)y(t) & \lambda - 3y^2(t) - x^2(t) \end{pmatrix}. \tag{2.76}$$

Next, to determine the monodromy matrix, the system $d\mathbf{y}_j/dt = J(\bar{\mathbf{x}}(t), \lambda)\mathbf{y}_j$ has to be solved for $j = 1, 2$ with $y_1(0) = (1, 0)$ and $y_2(0) = (0, 1)$ as initial conditions. This has to be done numerically, and the monodromy matrix Φ is found from

$$\Phi = (y_1(\frac{2\pi}{\omega}), y_2(\frac{2\pi}{\omega})). \tag{2.77}$$

The Floquet multipliers are determined as the two eigenvalues of the matrix Φ. The first one ($\rho_1 = 1$) is unity, and the second one determines the stability of the periodic orbit, and in this case it is within the unit circle ($| \rho_2 |< 1$) for $\lambda > 0$; so the periodic orbit is stable.

The value of ρ_2 can be analytically determined by first defining a Poincaré section (for certain θ_0) as

$$\Sigma^+ = \{(r, \theta) \in \mathbb{R} \times [0, 2\pi) \mid \theta = \theta_0\}. \tag{2.78}$$

In this case, the normal in polar coordinates is given by $\mathbf{n} = (0, 1)$ and (on the periodic orbit) $\mathbf{f} = (0, \omega)$ such that Σ^+ is a one-sided Poincaré section if $\omega \neq 0$. If we choose $\theta_0 = \pi/2$, then in Cartesian coordinates, the Poincaré section is parallel to the y-axis. For example, one could take the interval $y \in (0, 1]$ as a Poincaré section.

The Poincaré map can in this case be explicitly computed because explicit solutions exist of the trajectories for all initial conditions (r_0, θ_0) at $t = t_0$. Using the indefinite integral,

$$\int \frac{dx}{\alpha_1 x^3 + \alpha_2 x} = \frac{1}{2\alpha_2} \ln | \frac{x^2}{\alpha_1 x^2 + \alpha_2} |, \tag{2.79}$$

the closed-form solution $(r(t), \theta(t))$ is

$$r(t; t_0) = \left[(\frac{1}{r_0^2} - \frac{1}{\lambda})e^{-2\lambda(t - t_0)} + \frac{1}{\lambda} \right]^{-\frac{1}{2}},$$
$$\theta(t; t_0) = \omega(t - t_0) + \theta_0.$$

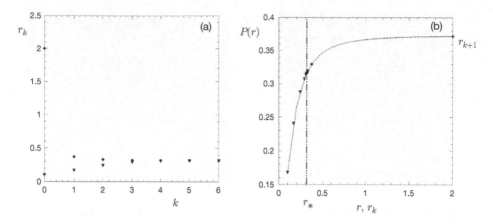

Figure 2.9 (a) Intersections r_k of the two trajectories for $\lambda = 0.1$ and $\omega = 1.0$ with the one-sided Poincaré section $r > 0, \theta_0 = \pi/2$. Subsequent intersections are labelled with k; the diamonds and triangles represent the intersections of the trajectories starting in $(0,2)$ and $(0,0.1)$, respectively. (b) Plot of the Poincaré map $\mathcal{P}(r)$ as in (2.81) together with the intersections r_k as in (a), but replotted as r_{k+1} versus r_k. The fixed point of the Poincaré map $\mathcal{P}(r)$ is at $r_* = \sqrt{\lambda} = 0.3162$.

A trajectory with initial conditions at (r_0, θ_0) intersects Σ^+ at times $t_k = t_0 + 2k\pi/\omega$. As the time difference between subsequent intersections (needed for the Poincaré map) is $2\pi/\omega$, this gives

$$r_{k+1} = \mathcal{P}(r_k) = \left[(\frac{1}{r_k^2} - \frac{1}{\lambda})e^{\frac{-4\pi\lambda}{\omega}} + \frac{1}{\lambda} \right]^{-\frac{1}{2}}. \tag{2.81}$$

Fixed points of the Poincaré map $\mathcal{P}(r)$ are defined by $\mathcal{P}(r_*) = r_*$, and a short calculation gives that $r_* = \sqrt{\lambda}$. The Poincaré map $\mathcal{P}$ is plotted as the dotted curve in Fig. 2.9b for the case $\lambda = 0.1$, $\omega = 1$. The intersections r_k of the two trajectories with the Poincaré section defined by $r > 0, \theta_0 = \pi/2$ are plotted versus k (which monitors the subsequent intersections) in Fig. 2.9a. They are replotted in Fig. 2.9b as r_{k+1} versus r_k and indeed move along the Poincaré map; with increasing k, the fixed point r_* is reached. *See example Ex. 2.5*

To determine the stability of the periodic orbits, the stability of the fixed point of the Poincaré map is considered. The Jacobian of the Poincaré map will indicate whether intersections of trajectories drift away (if positive) or are attracted to (if negative) the fixed point of the Poincaré map. For the periodic orbit the stability can be determined from

$$\frac{d\mathcal{P}}{dr}(r = \sqrt{\lambda}) = \frac{d}{dr}\left[(\frac{1}{r^2} - \frac{1}{\lambda})e^{\frac{-4\pi\lambda}{\omega}} + \frac{1}{\lambda} \right]^{-\frac{1}{2}}(r = \sqrt{\lambda}) = e^{\frac{-4\pi\lambda}{\omega}}. \tag{2.82}$$

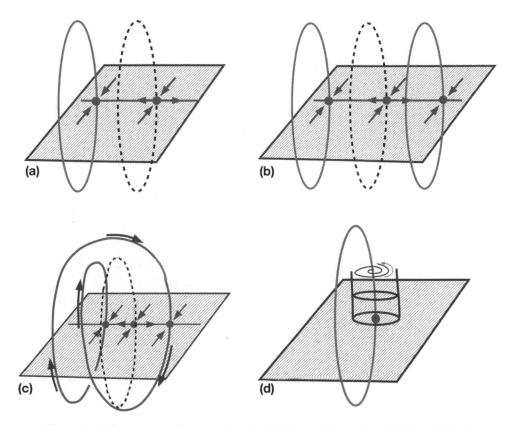

Figure 2.10 State space view associated with (a) a saddle-node and (b) a pitchfork bifurcation of a periodic orbit, (c) a period-doubling or flip bifurcation of a periodic orbit, and (d) a Neimark–Sacker or torus bifurcation. The original periodic orbit is shown as the dashed curve in (a), (b), and (c).

The norm of the right-hand side of this equation is smaller than unity for $\lambda > 0$ and hence the periodic orbit is stable.

The bifurcation theory for fixed points of the iterative map with an eigenvalue having unit norm is completely analogous to the bifurcation theory for fixed points of continuous systems with an eigenvalue on the imaginary axis. Periodic orbits become unstable when Floquet multipliers ρ_i cross the unit circle as the parameter λ is changed.

There are three important cases (Kuznetsov, 1995):

1. A real Floquet multiplier is crossing the unit circle $\rho(\bar{\lambda}) = 1$ at $\lambda = \bar{\lambda}$, which gives a saddle-node bifurcation of a periodic orbit, also called a cyclic-fold bifurcation (Fig. 2.10a). This situation can be shown to be topologically equivalent to the one-dimensional discrete dynamical system

$$x_{n+1} = \mathcal{P}(x_n), \text{ with } \mathcal{P}(x) = \lambda + x \pm x^2. \tag{2.83}$$

Consider the super-critical case $\mathcal{P}(x) = \lambda + x - x^2$ and assume that $\bar{\lambda} = 0$ for simplicity. As λ becomes positive, two fixed points $x_1^{\star}$ and $x_2^{\star}$ of the iterative map (2.83) appear which are solutions of $\mathcal{P}(x) = x$. These two fixed points correspond to the appearance of two new families of periodic orbits. One family is stable ($\mathcal{P}'(x_1^{\star}) < 1$), while the other is unstable ($\mathcal{P}'(x_2^{\star}) > 1$). Like in the case of fixed points, particular constraints (such as symmetry) may lead to trans-critical or pitchfork bifurcations (see Fig. 2.10b).

2. A real Floquet multiplier is crossing the unit circle at $\lambda = \bar{\lambda}$ with $\rho(\bar{\lambda}) = -1$. This situation is called flip or period-doubling bifurcation and has no equivalent for fixed points. The system is topologically equivalent to

$$x_{n+1} = \mathcal{P}(x_n), \text{ with } \mathcal{P}(x) = -(1 + \lambda)x \pm x^3. \tag{2.84}$$

This situation corresponds to the pitchfork case for the second iterate $\mathcal{P}^2$ map. Again consider (with $\bar{\lambda} = 0$) the super-critical case $\mathcal{P}(x) = -(1 + \lambda)x + x^3$. As λ becomes positive, two fixed points of the second iterate $\mathcal{P}^2$ appear which are not fixed points of the first iterate. This means that another stable peri-odic orbit of period $2p$ arises, whereas the original periodic orbit γ becomes unstable (Fig. 2.10c). The corresponding trajectories alternate from one side of γ to the other along the direction of the eigenvector associated with the Floquet multiplier $\rho = -1$.

3. A pair of complex conjugate Floquet multipliers ρ crosses the unit circle at $\lambda = \bar{\lambda}$ such that $|\rho(\bar{\lambda})| = |e^{i\varphi}| = 1$. This bifurcation is called a Neimark–Sacker or torus bifurcation (Fig. 2.10d). If one assumes after reduction on a two-dimensional invariant manifold that $d\rho(\lambda)/d\lambda \neq 0$ at $\lambda = \bar{\lambda}$, then there is a change of coordinates such that the Poincaré map takes the following form in polar coordinates (r, θ):

$$\begin{aligned} \mathcal{P}_r(r, \theta) &= r + c(\lambda - \bar{\lambda})r + ar^3, \\ \mathcal{P}_\theta(r, \theta) &= \theta + \varphi + br^2, \end{aligned} \tag{2.85}$$

where a, b, and c are parameters. Provided $a \neq 0$, this normal form indicates that a closed curve generically bifurcates from the fixed point; this closed curve corresponds to a two-dimensional invariant torus.

2.6 Summary

> • Based on the concept of asymptotic stability in the mean, fluid flows are either monotonically stable, globally stable, conditionally stable, or unstable.

- The linear stability boundary provides sufficient conditions for instability and it can, in general, be computed by solving an eigenvalue problem. The energy stability boundary can be determined by solving a non-linear optimization problem (not discussed here).
- The non-linear behaviour of the flow near the linear stability boundary can be studied using weakly non-linear theory. If the geometry is unbounded in one direction, Ginzburg–Landau theory can be used. If not, the theory of amplitude equations can be used.
- Dynamical systems theory for finite-dimensional autonomous systems provides a framework to understand transition behaviour in fluid flows when parameters are changed. When only one parameter is involved, only four types of bifurcations occur: saddle-node, trans-critical, pitchfork, and Hopf bifurcations. The first three bifurcations lead to multi-stable systems, and the Hopf bifurcation leads to oscillatory behaviour.
- Periodic orbits in autonomous dynamical systems, such as those arising at Hopf bifurcations, can become unstable when a parameter is varied. Apart from saddle-node, trans-critical, and pitchfork bifurcations, period doubling and torus bifurcations also can occur.

2.7 Exercises

Exercise 2.1 *For the Rayleigh–Bénard convection problem, the so-called principle of exchange of stability holds, which states that the eigenvalue σ is real at criticality.*

a. Multiply (2.6a) with $\tilde{\mathbf{u}}$ and (2.6c) with $\tilde{\vartheta}$ and integrate the result over the flow domain, substitute solutions proportional to $e^{\sigma t}$, and derive an equation for the eigenvalue $\sigma = \sigma_R + i\sigma_I$.

b. Prove that $\sigma_I = 0$.

Exercise 2.2 *Consider the pure Bénard–Marangoni problem in a horizontally unbounded configuration, with $Ra = 0$, as discussed in Section 1.3.*

a. Show that the background solution for $Bi \to \infty$ is the same as that for the Rayleigh–Bénard problem.

b. Determine in this case, the critical value of Ma associated with the linear stability boundary as a function of k.

Exercise 2.3 *A crucial step in the derivation of the Ginzburg–Landau equation (2.36) is the application of the Fredholm alternative to the singular non–self-adjoint system at $\mathcal{O}(\epsilon^3 E)$ and in particular the determination of the vector $\boldsymbol{\Omega}$ in (2.37). Numerically, this can be done by first computing a Singular Value Decomposition (SVD) of the linear operator as follows:*

$$i\omega_c \hat{M}(k_c) + \hat{L}(k_c) = WSV^H,$$

where $S = diag(s_1, \ldots, s_d)$ is the diagonal matrix of singular values with $s_d = 0$. The matrices W and V are orthonormal, and the superscript H indicates the Hermitian transpose.

Define $C = \sigma_d \hat{M} + \hat{L}$, where σ_d is the eigenvalue near criticality.

a. Show that $\boldsymbol{\Omega}$ can be determined efficiently from the problem

$$CC^H \boldsymbol{\Omega} = 0.$$

b. Show that in the Ginzburg–Landau equation, the coefficients $\gamma_i, i = 1, 2$ can be determined from

$$\gamma_1 = m\frac{\partial \sigma}{\partial \lambda} \; ; \; \gamma_2 = -\frac{1}{2}\frac{\partial^2 \sigma}{\partial k^2}.$$

c. How can the group velocity c_g be efficiently computed using the SVD given at the beginning of this exercise?

Exercise 2.4 *Consider the following system of equations:*

$$\frac{dA}{dt} = AB - A,$$
$$\frac{dB}{dt} = -A^2 - B + \gamma,$$

for real functions $A(t)$ and $B(t)$ with $t \in [0, \infty)$; $\gamma \geq 0$ is a real number.

a. Determine the fixed points $(\bar{A}, \bar{B})$ of these equations with γ as a control parameter.

b. Determine the linear stability of the fixed points $(\bar{A}, \bar{B})$. What special phenomenon occurs at $\gamma = 9/8$ on the branches for which $\bar{A} \neq 0$?

Exercise 2.5 *With a scaling $\tau = \rho_* T$, $\chi = \eta X$, and $a(\chi, \tau) = a_\infty^{-1} A(X, T) exp(-im\omega_\lambda T)$, with*

$$\rho_* = m(\sigma_R)_\lambda \; ; \; \eta = \sqrt{-\frac{2m(\sigma_R)_\lambda}{(\sigma_R)_{kk}}} \; ; \; a_\infty = \sqrt{m\frac{(\sigma_R)_\lambda}{\mathcal{R}(\gamma_3)}},$$

the Ginzburg–Landau (2.36) can be transformed into

$$\frac{\partial a}{\partial \tau} = a + (1 + i\alpha_1)\frac{\partial^2 a}{\partial \chi^2} - (1 + i\alpha_2)a|a|^2.$$

a. Show that the Ginzburg–Landau equation has only bounded solutions when $\mathcal{R}(\gamma_3) > 0$.

b. Determine the coefficients α_1 and α_2.

c. Determine the Stokes wave solution $a_S(\tau)$ of the Ginzburg–Landau equation.

3

Well-Posed Problems

*Understanding the problem
is the key to its solution*

Building reliable and efficient software for bifurcation analysis of flow problems is not an easy task. In this chapter, we will discuss the issue of under which conditions a problem defined by partial differential equations (PDEs) is well-posed. This essentially will boil down to the question of which initial and boundary conditions are required for these equations to have robust solutions. Furthermore, we will introduce here the notion of projections which plays such an important role in constructing finite element methods and in iterative schemes to solve linear equations and eigenvalue problems.

3.1 From Governing Equations to Numerical Solution

When numerical computations fail, this can have different causes. By means of an example (Fig. 3.1) we will discuss these causes and show how this leads to the mathematical concept of a well-posed problem.

3.1.1 Error Types

Consider a typical fluid flow, such as in one of the experiments shown in Chapter 1. By providing the equations to model these flows, several assumptions were made, such as a constant viscosity and thermally insulated walls. Through these assumptions, we have introduced *modeling errors*. Assume for illustration that the problem is defined by a linear partial differential equation

$$Lu = f, \qquad (3.1)$$

where u represents the state of the unknown flow and f is the forcing. The question is how these modeling errors affect the solution, in particular whether they are very

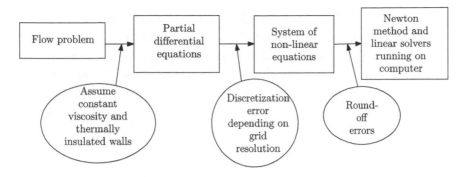

Figure 3.1 Steps to solve a general fluid dynamical problem on a computer.

sensitive to them or not. Of course, one hopes that small modeling errors will lead to small errors in the solution. One way to analyze this is by adjusting the forcing f such that the real flow, say U, satisfies the equation, say

$$LU = f_a. \tag{3.2}$$

In general, the adjustment $f_a - f$ will be small, but we hope that it will also give a small difference between the real solution U and the computed solution u. This is the case if our equations are stable,[1] that is, small changes in the forcing f, say δf, lead to only small changes in the solution u, say δu. So, we like to have some bounded positive K such that $||\delta u|| \leq K||\delta f||$ for arbitrary small δf.

In the next step of the process, we are going to approximate the equations on a grid which has a typical mesh width h; this will lead to an *approximation error*. In general, the approximation leads to a system of equations $L_h u_h = f_h$. Similar to the previous step, we want stability of this system in order to be able to show that the difference $u - u_h$ is small. Suppose we adapt the right-hand side f_h such that the solution u of (3.1) satisfies $L_h u = f_{h,a}$; then the difference $u - u_h$ is the solution of $L_h(u - u_h) = f_{h,a} - f_h$ and we can bound the difference if this discrete problem is stable, that is, if there exists a bounded positive K_h such that $||\delta u_h|| \leq K_h||\delta f_h||$ for arbitrary small δf_h.

The final step in the process is solving the system of equations $L_h u_h = f_h$. On the computer we will not be able to compute the solution exactly, because it uses a floating point system, which means that the set of real numbers is approximated by a finite set of numbers. This will introduce *round-off errors* during the processing of the algorithm. It is important that such round-off errors do not accumulate in such a way that one cannot find a reasonable approximation to the solution of the system. Hence, we want to be able to bound the error in the solution of the linear system in terms of the unit round, that is, the smallest positive number ε such that the floating-point representation of $1 + \varepsilon$ is different from that of 1; for standard precision, $\varepsilon = 10^{-16}$. This is governed by the stability of the numerical method used to solve the system, say Gaussian Elimination with pivoting. So, in fact we will find a solution $u_{h,\varepsilon}$ contaminated by round-off errors which is an exact solution of a perturbed system $L_{h,\varepsilon} u_{h,\varepsilon} = f_{h,\varepsilon}$ and u_h is a solution of the adjusted system $L_{h,\varepsilon} u_h = f_{h,\varepsilon,a}$. And hence we have $L_{h,\varepsilon}(u_{h,\varepsilon} - u_h) = f_{h,\varepsilon,a} - f_{h,\varepsilon}$. The difference will be small if the solution method is stable so there exists a bounded positive K_ε such that $\|\delta u_{h,\varepsilon}\| \leq K_\varepsilon \|\delta f_{h,\varepsilon}\|$.

Next, we can look to the total difference:

$$\||U - u_{h,\varepsilon}\| \leq \|U - u + u - u_h + u_h - u_{h,\varepsilon}\| \leq \|U - u\| + \|u - u_h\|$$
$$+ \|u_h - u_{h,\varepsilon}\|$$
$$\leq K\|f_a - f\| + K_h\|f_{h,a} - f_h\| + K_\varepsilon\|f_{h,\varepsilon,a} - f_{h,\varepsilon}\|.$$

So, we see that the outcome of the numerical simulation $u_{h,\varepsilon}$ is a good approximation of the reality U if all the sub-problems are stable and the model, discretization, and round-off errors are all small enough.

Be aware of the fact that one can construct an unstable approximation of a stable equation and one can use an unstable method to solve a stable linear system.

3.1.2 Workflow

An overview of the different steps to compute a solution to a PDE is given in Fig. 3.2. If a computation fails, for example by giving unrealistic values in the solution, this can be caused by numerical instability or by an ill-posed problem. Usually this is caused by imposing wrong boundary conditions. Therefore, in the following we will study well-posedness of equations for a few well-known classes of PDEs (the *inspection* step in Fig. 3.2). The result of this step is that we have a well-posed Initial Boundary Value problem (IBVP), a Boundary Value problem (BVP), or an Eigenvalue problem (EVP). Continuous EVPs occur naturally if we want to study the stability of steady-state solutions of a PDE.

A PDE is defined in every point in the domain, that is, in an infinite number of points. A computer can only solve finite problems, and therefore in the next chapter we will discuss how one can transform a continuous problem into a discrete finite problem (the *discretization* step in Fig. 3.2). For the IBVP this happens usually in two steps. First, we discretize in space, leading to a system of ordinary differential equations (SODEs), and next we discretize in time, leading to a set of recurrence equations which is explicit or implicit depending on the time integration method used. The latter leads to a non-linear system of equations (NLS) to be solved at every time step. For the BVP and EVP, we just discretize in space and find an NLS and algebraic eigenvalue problem (AEVP).

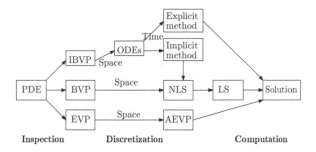

Figure 3.2 Sequence of steps to solve a PDE problem on a computer.

> **Additional Material**
>
> • A lot of material on partial differential equations can be found in textbooks such as Grossmann and colleagues (2007), Mattheij and colleagues (2005), Morton and Mayers (2005), and Gockenbach (2002).

The next step is to solve the NLS, referred to as the *computation* step in Fig. 3.2. The NLS is solved by a sequence of linear problems (LS) in a Newton-type method, each of which can be solved by a direct method (Chapter 6) or by an iterative method (Chapters 7 and 8). For the AEVP, which is in fact a special NLS, special methods have been developed (Chapters 7 and 8). At first sight it might seem strange that both iterative solvers for linear systems and AEVP solvers are in the same chapter. However, we will see that they have a common starting point.

In this way, one can view the computation of the solution of a PDE on a computer as the result of a recursion of models. Similarly to the modelling treated in the previous section, each model in this recursion should be a stable model and an approximation of the previous in order to let the computed solution converge to that of the problem we want to solve. *Eventually, the goal is to obtain a sufficiently accurate solution in as short a time as possible subject to constraints on memory.* Hence, it is important to know what the consequences of certain choices in the modelling process are. For instance, for constructing the mathematical model, it is recommended to find the simplest model that describes the phenomenon and exploit symmetry in the problem (e.g., this may lead to a self-adjoint problem). In the discretized model, try to keep possible symmetry in (e.g., self-adjointness leads to a symmetric discrete problem), and try to keep certain physical/mathematical properties in (positivity of solution, conservation of mass, monotony). Finally, for choosing the solution method, exploit the sparsity and symmetry (symmetric matrices need less storage and computer time).

3.1.3 Stability in the Sense of Hadamard

A PDE gives a relation between the partial derivatives of a function u of the n independent variables $x_1, \ldots, x_n$. The *order* of the PDE is that of the highest-order derivative occurring in it, for example,

$$F(x_1, \ldots, x_n, u, u_{x_1}, \ldots, u_{x_n}) = 0 \quad \text{with} \quad u_{x_i} = \frac{\partial u}{\partial x_i} \qquad (3.3)$$

is a PDE of first order. The PDE is called *quasi-linear* if F is linear in its highest-order derivatives. The PDE is *linear* if F is linear in all its arguments, except for the x_i's. *See example Ex. 3.1*

Usually a PDE is considered only on a part of $\mathbb{R}^n$. Then, for the uniqueness of the solution we need initial and boundary conditions. These conditions cannot be taken arbitrarily. Moreover, they should be such that the solution is stable with respect to perturbations in these conditions, which leads to the following definition.

Definition 3.1 *A PDE plus initial and boundary conditions is said to be well-posed in the sense of Hadamard if it has a unique solution and is stable with respect to perturbations in the data (that is, coefficients, forcing).*

While describing the classification of PDEs in the next sections, we will also indicate which initial and boundary conditions make a problem well-posed. Beforehand we could say that boundary and initial conditions should be such that all integration constants should be fixed and none of them should be over-specified. Note that well-posedness can already be an issue for linear systems.

There exist problems which are not stable in the sense of Hadamard, often called *inverse problems*. This occurs if perturbations grow to arbitrary size in finite time. An example of this is the backward heat equation. In general, these are hard to deal with in the presence of rounding errors. Special measures have to be taken, such as adapting the equations such that rounding errors are kept at bay, called *regularization*.

3.2 First-Order PDEs

In the subsequent sections we will distinguish various classes of first-order PDEs, going from scalar one-dimensional equation types, via vector one-dimensional equations to multi-dimensional systems of equations.

3.2.1 First-Order Scalar PDEs

Consider the initial value problem of the first order consisting of a PDE for scalar unknown u

$$\frac{\partial u}{\partial t} + a\frac{\partial u}{\partial x} = 0, \qquad a > 0, \tag{3.4}$$

where a is real. This equation is called the *transport equation*, or *first-order wave equation*. It is easy to see that $u(x, t) = f(x - at)$ is the general solution of this equation, which means that the solution is constant for lines $x - at = $ constant. These lines are called characteristics. In Fig. 3.3 we have drawn these characteristics in the (x, t)-plane for several cases. If we prescribe the value of u on some point on a characteristic, then the solution is known on the whole of the characteristic. So, if we prescribe u on a curve as depicted in Fig. 3.3a, then the solution in the whole plane is determined. However, care should be taken in choosing this curve. In Fig. 3.3b, for example, the curve crosses some of the characteristics twice. In this way, conflicting values might be prescribed for the solution on these characteristics, which is undesirable.

In numerical simulations we can in general only deal with bounded domains. Examples of such domains are depicted in Figs. 3.3c and 3.3d. The thick border-lines indicate where we could prescribe u. Figure 3.3c fits the case where we want to integrate forward in time, while Fig. 3.3d is apt for integrating backwards in time. Note that in the case a switches sign and becomes negative, the characteristics will have negative tangents in the (x, t)-plane. Hence in that case, one should prescribe a condition at the right in Fig. 3.3c, and at the left in Fig. 3.3d to make the solution unique.

If a is not constant, the characteristics will not be straight anymore; but u will still be constant along a characteristic (at least for homogeneous equations, that is, if there is no forcing). Moreover, it is the value of a at the boundary which determines whether a boundary condition needs to be imposed (Mattheij *et.al.*, 2005, Section 2.1). *See example Ex. 3.2*

3.2.2 Systems of First-Order Equations

Consider the system

$$\frac{\partial \mathbf{u}}{\partial t} + J\frac{\partial \mathbf{u}}{\partial x} = 0, \tag{3.5}$$

where J is a constant real $n \times n$ matrix and $\mathbf{u}$ an n-dimensional vector function depending on x and t. In order to better understand the behaviour of the solutions of this equation, we will try to write it in its most simple way, called a 'normal form'. In this case, we want it to be as close as possible to a set of independent scalar transport equations, which we have already analyzed in the previous section. For this analysis, we restrict to the case that J is diagonalizable, hence there exists a non-singular matrix Q such that

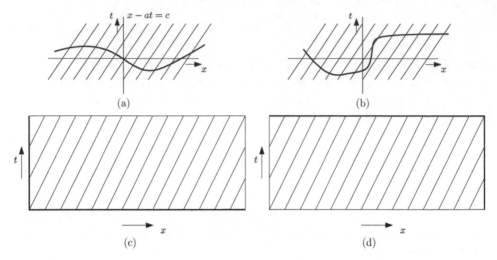

Figure 3.3 Characteristics: (a) prescribing the solution on the curve determines
solution in whole plane, (b) part of characteristics cross curves twice, which leads
to overspecification, and (c, d) thick lines indicate where to prescribe the solution.

$$Q^{-1}JQ = \Lambda = \begin{bmatrix} \lambda_1 & & & \\ & \lambda_2 & & \\ & & \ddots & \\ & & & \lambda_n \end{bmatrix}.$$

The system is called *hyperbolic* if all eigenvalues λ_i of J are real. In that case
we can, using $\mathbf{w} = Q^{-1}\mathbf{u}$, transform the system (3.5) into n decoupled scalar
(transport) equations

$$\frac{\partial w_i}{\partial t} + \lambda_i \frac{\partial w_i}{\partial x} = 0, \qquad i = 1, 2, \dots, n.$$

Here, every λ_i defines a characteristic direction along which the corresponding
w_i is constant. As in the scalar case, the sign of λ_i determines where a boundary condition must be prescribed. However, in general we cannot prescribe the
quantity we need at the boundary directly; but the imposed boundary together
with the quantity that is propagated along the characteristic towards that boundary from within the domain must specify the desired quantity (Mattheij *et al.*,
2005, section 2.2). Note that non-hyperbolic systems do exist; an important
example being the Cauchy–Riemann equations (3.9), as discussed in what follows.
See example Ex. 3.3.a–c

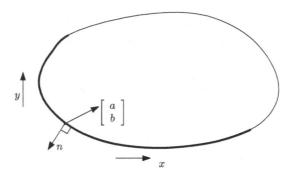

Figure 3.4 The fat part of the domain boundary indicates where we have to pre-scribe u if we want to integrate forward in time. Also, an initial condition over the whole domain is needed.

3.2.3 Higher Space Dimensions

In two-dimensional space (Fig. 3.4), the first-order wave equation is

$$\frac{\partial u}{\partial t} = -\left(a\frac{\partial u}{\partial x} + b\frac{\partial u}{\partial y} \right). \tag{3.6}$$

The right-hand side may also be written as $-((a,b)^T, \nabla u)$ and hence it is a deriva-tive in the direction of $(a,b)^T$. If we want to integrate forward in time, then we clearly need an initial condition. On the boundary we need to consider the gener-alization of the one-dimensional (1D) case. Observe that for a or b is zero we are in the 1D case. The criteria we used there are converted in the more general form where we prescribe u in a point on the boundary if the inner product $((a,b)^T, \mathbf{n}) < 0$ in that point, where $\mathbf{n}$ is the outward pointing normal. Or otherwise stated, if the vector $(a,b)^T$ points into the domain, we need to prescribe u. We have to consider each point of the boundary to get the correct boundary conditions to make the problem well posed.

3.3 Scalar Second-Order PDEs

In this section, we will discuss different types of scalar second-order PDEs, their 'normal forms', and the conditions for well-posedness.

3.3.1 Classification

Consider the following second-order PDE for the scalar u:

$$F\left(x_1,\ldots,x_n, u, \frac{\partial u}{\partial x_1} \cdots \frac{\partial u}{\partial x_n} \right) + \sum_{i,j=1}^{n} \frac{\partial}{\partial x_i}\left(J_{ij}\frac{\partial u}{\partial x_j} \right) = 0, \tag{3.7}$$

where the J_{ij} may be functions of x_i, $i = 1, \ldots, n$. As the highest-order derivatives determine the type of the equation, we have to study the matrix J given by

$$
J = \begin{bmatrix} J_{11} & \cdots & J_{1n} \\ \vdots & \ddots & \vdots \\ J_{n1} & \cdots & J_{nn} \end{bmatrix}.
$$

For sufficiently smooth u, this matrix can always be chosen symmetric, because if not, we can permute freely the order of the partial derivatives and the symmetric matrix is simply $(J + J^T)/2$. We say that the equation is *elliptic* if all eigenvalues of J have the same sign; *hyperbolic* if all eigenvalues of J have the same sign, except one which has an opposite sign; and *parabolic* if all eigenvalues of J have the same sign, except one which is zero. These are the three classes we will consider in this book. Of course, one can easily imagine other combinations of eigenvalues, but the ones mentioned here are the most relevant. If J depends on x, then the type of the PDE can depend on x as well. For example $u_{xx} = xu_{yy}$ is an equation which is elliptic in the region $x < 0$, hyperbolic in the region $x > 0$, and parabolic on the line $x = 0$ (Mattheij *et al.*, 2005, Section 2.4). *See example Ex. 3.4.a–b*

Example 3.1 (Ginzburg–Landau equation) *Re-write Equation (2.36) into the shape (3.7), assuming that γ_2 is constant:*

$$
-\frac{\partial A}{\partial t} + \gamma_1 A - \gamma_3 A |A|^2 + \gamma_2 \frac{\partial^2 A}{\partial x^2} = 0.
$$

So, the matrix J is

$$
J = \begin{bmatrix} 0 & 0 \\ 0 & \gamma_2 \end{bmatrix},
$$

and the problem is parabolic if $\gamma_2 \neq 0$.

By a coordinate transformation one can diagonalize J and even make all non-zero diagonal elements of magnitude one. In this way, we find for each class one normal form, that is, a simplest form showing the properties of the class. First, note that the value of the eigenvalues is not important in the type of the PDE; therefore, second-order PDEs are equivalent to a form in which J is diagonal with elements either 0, -1, or 1 on the diagonal. The consequence of this is that when we know which boundary and/or initial conditions make the normal form well posed, then we know it also is well posed for all elements in its class.

We show the process of creating a normal form for a linear problem. Given the problem

$$
\left(\frac{\partial}{\partial x_1} \cdots \frac{\partial}{\partial x_n} \right) J \left(\frac{\partial}{\partial x_1} \cdots \frac{\partial}{\partial x_n} \right)^T u = f,
$$

apply a coordinate transformation

$$\left(\frac{\partial}{\partial x_1} \cdots \frac{\partial}{\partial x_n}\right)^T = S \left(\frac{\partial}{\partial y_1} \cdots \frac{\partial}{\partial y_n}\right)^T$$

such that $S^T J S$ is diagonal and all diagonal elements have magnitude one or zero. This is a so-called *congruence transformation*, and it is known that such a trans-formation does not change the signs of the eigenvalues. It is not hard to find such an S. Since J is real and symmetric, we know that there exists an orthogonal matrix Q that will diagonalize J, hence $Q^T J Q = D$. Next, we pre- and post-multiply by the same non-singular diagonal matrix $\hat{D} = \sqrt{|D|}^{-1}$ such that we get the desired diagonal matrix. Thus, $S = Q\hat{D}$ (Mattheij *et al.*, 2005, Section 2.4).

The highest-order classes of PDEs introduced in the previous section have the following normal forms:

$$\frac{\partial^2 u}{\partial y_1^2} + \cdots + \frac{\partial^2 u}{\partial y_n^2} \quad \text{(elliptic)},$$

$$\frac{\partial^2 u}{\partial y_n^2} - \left(\frac{\partial^2 u}{\partial y_1^2} + \cdots + \frac{\partial^2 u}{\partial y_{n-1}^2}\right) \quad \text{(hyperbolic)},$$

$$\frac{\partial u}{\partial y_n} - \left(\frac{\partial^2 u}{\partial y_1^2} + \cdots + \frac{\partial^2 u}{\partial y_{n-1}^2}\right) \quad \text{(parabolic)}.$$

Examples in two-dimensional PDEs of the respective types just mentioned are

$$\text{Poisson equation:} \quad \frac{\partial^2 u}{\partial x^2} + \frac{\partial^2 u}{\partial y^2} = f, \tag{3.8a}$$

$$\text{wave equation:} \quad \frac{\partial^2 u}{\partial t^2} - \frac{\partial^2 u}{\partial x^2} = 0, \tag{3.8b}$$

$$\text{heat equation:} \quad \frac{\partial u}{\partial t} - \frac{\partial^2 u}{\partial x^2} = 0. \tag{3.8c}$$

If in the Poisson equation the right-hand side is zero, then it is called a *Laplace equation*. The Poisson and Laplace equations are well posed if we prescribe along the whole boundary u or u_n (recall that $u_n = (\nabla u, \mathbf{n})$ where $\mathbf{n}$ is the outward-pointing normal on the boundary), or a combination of these $u + \alpha u_n$, with $\alpha > 0$, which are called *Dirichlet*, *Neumann*, and *Robin* condition, respectively. This makes elliptic equations to BVPs (Section 3.5.3). For the heat equation we need an initial condition next to the boundary conditions in space, which are the same as those for the elliptic case. The wave equation even needs two initial conditions. So hyperbolic and parabolic equations are IBVPs.

If one tries to characterize the equations, one could say that for an elliptic equation one can prescribe, on each part of the boundary, one type of boundary condition. It does not matter which one, but we only prescribe one of these. However, this is different for the hyperbolic and parabolic equations. For the hyperbolic equation, one can both go backward and forward in time (simply replace t by $-t$ to see that the same equation occurs), where at the associated time boundary we must prescribe two conditions as stated earlier. On the space boundaries we must prescribe exactly one condition like in the elliptic case.[2] For the parabolic case the t direction has completely different derivatives from those in the space directions. If we replace here t by $-t$, the time derivative will change sign. This means you go backward in time when going forward in the new variable.

A problem which is ill posed is, for instance, the backward heat equation $u_t = -u_{xx}$ which is integrated from $t = 0$ to some time $T > 0$ using an initial condition and boundary conditions. This problem is unstable irrespective of the initial/boundary conditions. In this case we should integrate backwards in time in order to have a stable solution. For some other examples see Mattheij *et al.* (2005, Section 2.6).

In the previous discussion, we have classified scalar second-order PDEs and mentioned in general terms which boundary conditions ought to be used. To be more precise on the kind of boundary conditions needed to get a well-posed problem, we need some more theory which will be developed in the next sections. As the reader will notice, these sections are quite theoretical. However, since many numerical methods are based on them, we cannot skip these sections without missing the goal of this book to acquaint you with the mathematical theory of the numerical methods used to solve bifurcation problems.

In general, a second-order PDE can also be transformed to a system of first-order PDEs. For instance, by introducing the unknowns

$$u_1 = \frac{\partial u}{\partial x}, \quad u_2 = \frac{\partial u}{\partial y},$$

the Laplace equation $\Delta u = 0$ can be written as

$$\frac{\partial u_1}{\partial x} + \frac{\partial u_2}{\partial y} = 0, \quad \frac{\partial u_1}{\partial y} - \frac{\partial u_2}{\partial x} = 0. \tag{3.9}$$

These are the so-called Cauchy–Riemann equations which are of the form (3.5) with matrix

$$J = \begin{bmatrix} 0 & -1 \\ 1 & 0 \end{bmatrix}.$$

The eigenvalues of J are $\pm i$, and hence this is not a hyperbolic system of PDEs but it is elliptic (just as its corresponding second-order representation). *See example Ex. 3.3.h*

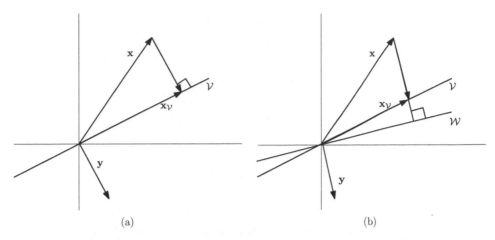

Figure 3.5 Projections: (a) orthogonal projection: $\mathbf{y} = x_{\mathcal{V}} - x$ perpendicular to $\mathcal{V}$;
(b) oblique projection: $\mathbf{y} = x_{\mathcal{V}} - x$ perpendicular to $\mathcal{W}$, fails if $\mathcal{W} \perp \mathcal{V}$.

So, for many problems there are multiple formulations. Analytically there is no essential difference between them, but numerically there is. For instance, if the original equations are the Cauchy–Riemann equations, then it is, on the one hand, attractive to solve the associated Laplace equation because many fast Poisson solvers are available. However, we have to perform numerical differentiation to obtain our final result. It is known that this operation amplifies round-off errors (Burden and Faires, 2001, section 4.1). So, with the original formulation we can obtain a higher accuracy.

3.4 Projections

Projections are a central notion in numerical mathematics. Projection is used in a variety of circumstances, of which we will mention a few in the following subsections. Via orthogonal and oblique (skew) projections we will get to the Galerkin and Petrov–Galerkin projections which play a prominent role in (i) the weak form of a PDE, (ii) the discretization that is derived from it, and (iii) the solution of linear and eigenvalue problems.

3.4.1 Projection Types

Consider first the (orthogonal) projection problem, as sketched in Fig. 3.5a. Let $\mathcal{V}$ be an m-dimensional linear sub-space of $\mathbb{R}^n$, and $\mathbf{x}$ a vector in $\mathbb{R}^n$ with $\mathbf{x} \notin \mathcal{V}$. The orthogonal projection of $\mathbf{x}$ on $\mathcal{V}$ is finding a vector $\mathbf{x}_{\mathcal{V}} \in \mathcal{V}$ such that there exists a $\mathbf{y} \perp \mathcal{V}$ with $\mathbf{x} + \mathbf{y} = \mathbf{x}_{\mathcal{V}}$.

To find a solution to this problem we need a basis for $\mathcal{V}$ and the standard inner product. In the standard inner product we can re-write the problem to finding a $\mathbf{x}_\mathcal{V} \in \mathcal{V}$ such that $(\mathbf{v}, \mathbf{x} - \mathbf{x}_\mathcal{V}) = 0$ for all $\mathbf{v} \in \mathcal{V}$. Now, suppose that the columns of the matrix V span the space $\mathcal{V}$. Then it holds that for arbitrary $\hat{\mathbf{y}}$ in $\mathbb{R}^m$, the vector $V\hat{\mathbf{y}}$ is in $\mathcal{V}$. (Similarly there exists a $\hat{\mathbf{y}}$ for every $\mathbf{y} \in \mathcal{V}$ such that $\mathbf{y} = V\hat{\mathbf{y}}$.) With this, the inner product form of our problem can be written as follows: find $\hat{\mathbf{x}}_\mathcal{V} \in \mathbb{R}^m$ such that $(V\hat{\mathbf{y}}, \mathbf{x} - V\hat{\mathbf{x}}_\mathcal{V}) = 0$ for all $\hat{\mathbf{y}} \in \mathbb{R}^m$. The expression can be re-written as $\hat{\mathbf{y}}^T(V^TV\hat{\mathbf{x}}_\mathcal{V} - V^T\mathbf{x}) = 0$. Since $\hat{\mathbf{y}}$ is arbitrary in $\mathbb{R}^m$, this is equivalent to $V^TV\hat{\mathbf{x}}_\mathcal{V} = V^T\mathbf{x}$. Solving this equation gives us the unique solution $\hat{\mathbf{x}}_\mathcal{V}$, and from this we can compute the projection $\mathbf{x}_\mathcal{V} = V\hat{\mathbf{x}}_\mathcal{V}$.

Next, the oblique/skew projection problem is considered (Fig. 3.5b). Let $\mathcal{V}$ be an m-dimensional linear sub-space of $\mathbb{R}^n$. Furthermore, let $\mathbf{x}$ be a vector in $\mathbb{R}^n$ and $\mathbf{x} \notin \mathcal{V}$; see Fig. 3.5b. Compute the skew projection $\mathbf{x}_\mathcal{V} \in \mathcal{V}$ of $\mathbf{x}$ such that $\mathbf{y} \equiv \mathbf{x}_\mathcal{V} - \mathbf{x}$ is perpendicular to $\mathcal{W}$, where $\mathcal{W}$ has the same dimension as $\mathcal{V}$. Note that for $\mathcal{W} = \mathcal{V}$, this will lead again to an orthogonal projection.

The solution follows along the same line as for the orthogonal projection, but now we need also a basis for $\mathcal{W}$, which we assume to be the columns of the matrix W. With the inner product we have to find a $\hat{\mathbf{x}}_\mathcal{V}$ in $\mathbb{R}^m$ such that $(W\hat{\mathbf{z}}, \mathbf{x} - V\hat{\mathbf{x}}_\mathcal{V}) = 0$ for all $\hat{\mathbf{z}} \in \mathbb{R}^m$, which leads straightforwardly to $W^TV\hat{\mathbf{x}}_\mathcal{V} = W^T\mathbf{x}$. This will give a unique solution if W^TV is non-singular. As a special case assume that $W = AV$; hence, $W = AV$ is generated from V by applying a matrix A to it. Then V^TA^TV should be non-singular. This is, for instance, the case if A is a positive definite matrix.

It can be shown by the definition of inner products that $(\mathbf{x}, A\mathbf{y})$ with A a symmetric positive definite matrix defines an inner product; it is sometimes indicated by $(\mathbf{x}, \mathbf{y})_A$. We will call this a non-standard inner product. Orthogonal and skew projections can straightforwardly be generalized to these inner products. If $(\mathbf{x}, A\mathbf{y}) = 0$, then one says that $\mathbf{x}$ is A-orthogonal to $\mathbf{y}$.

Example 3.2 *One can also define an inner product for functions, and an example is $(u, v) = \int_0^1 u(x)v(x)dx$. Moreover, one can define linear sub-spaces, for example, the linear space consisting of all quadratic polynomials on the interval [0,1]. Projection can also be defined in such spaces. For instance, one can project $\exp(x)$ on the space of all quadratic polynomials on the interval [0,1]. To start with, we need again a basis for the linear sub-space, which we can choose, $[1, x, x^2]$, and call it $V(x)$. Now, as above, we have to find $\hat{\mathbf{a}}_\mathcal{V} = [a, b, c]^T$ such that for all elements $\hat{\mathbf{a}} \in \mathbb{R}^3$ we have See example Ex. 3.5*

$$(V(x)\hat{\mathbf{a}}, V(x)\hat{\mathbf{a}}_\mathcal{V} - \exp(x)) = 0.$$

In the following we use inner products on matrices which we define for general $V(x)$ with m components and $W(x)$ with n components by

$$(V(x), W(x)) \equiv \begin{bmatrix} (v_1, w_1) & \cdots & (v_1, w_n) \\ \vdots & \ddots & \vdots \\ (v_m, w_1) & \cdots & (v_m, w_n) \end{bmatrix}.$$

Now, since $\hat{\mathbf{a}}$ *is arbitrary it should also hold that*

$$(V(x), (V(x)\hat{\mathbf{a}}_{\mathcal{V}} - \exp(x))) = 0,$$

which can be simplified to

$$(V(x), V(x))\hat{\mathbf{a}}_{\mathcal{V}} = (V(x), \exp(x)). \tag{3.10}$$

Using the above definition of this inner product, this equation has a 3×3 *matrix on the left and a vector with three components on the right. See example Ex. 3.6*

3.4.2 Projection and Approximation

There is a relation between projection and approximation. The projection of $\exp(x)$ in the previous paragraph yields an approximation of this function within the space spanned by $[1, x, x^2]$. This approximation will become more accurate if we extend the basis with x^3, x^4 etc. In fact, the same system will follow from the associated least squares approach, that is,

$$\hat{\mathbf{a}}_{\mathcal{V}} = \operatorname{argmin}_{\hat{\mathbf{a}}} ||V(x)\hat{\mathbf{a}} - \exp(x)||,$$

where $||u||^2 = (u, u)$.

Another example is trying to find an approximate solution for $A\mathbf{x} = \mathbf{b}$ with A a symmetric positive definite matrix. For that we can use the A-orthogonal projection problem to approximate x in the sub-space $\mathcal{V}$, which re-writes the equation for the orthogonal problem to $V^T A V \hat{\mathbf{x}}_{\mathcal{V}} = V^T A\mathbf{x}$. Though we do not know $\mathbf{x}$, we do know $A\mathbf{x}$ which is $\mathbf{b}$. Hence, we find the approximation of $\mathbf{x}$ in $\mathcal{V}$ to be $V\hat{\mathbf{x}}_{\mathcal{V}}$, where $\hat{\mathbf{x}}_{\mathcal{V}}$ follows from $V^T A V \hat{\mathbf{x}}_{\mathcal{V}} = V^T \mathbf{b}$. This is one of the pillars of the conjugate gradients (CG) method to solve $A\mathbf{x} = \mathbf{b}$. Another one is the step-by-step extension of $\mathcal{V}$ such that better and better approximations are found.

In general, every system of non-linear equations can be cast in a root-finding problem, so, for instance, in the finite-dimensional case we have an $\mathbf{r}(\mathbf{x})$: $\mathbb{R}^n \to \mathbb{R}^n$, for which we want to find the $\mathbf{x}$ for which $\mathbf{r} = 0$. Another way of stating this is that we want to find $\mathbf{x}$ in $\mathbb{R}^n$ such that $\mathbf{r}(\mathbf{x})$ is perpendicular to every element in $\mathbb{R}^n$. Now n may be very big (or infinite, as in (3.30)), and in order to make this problem easier to handle we restrict $\mathbf{x}$ to a sub-space $\mathcal{V}$ of $\mathbb{R}^n$ called the *search space*. However, by this restriction we cannot expect that $\mathbf{r}(\mathbf{x})$ will become equal to zero, hence we require that $\mathbf{r}(\mathbf{x})$ will be perpendicular to a space $\mathcal{W}$, called the *test space*, which has the same dimension as $\mathcal{V}$ in order to have as many unknowns as equations. Now it is easy to show that if the columns of V and W form a basis of $\mathcal{V}$

and $\mathcal{W}$, respectively, we obtain the projected problem $W^T \mathbf{r}(V\hat{\mathbf{x}}_V) = 0$ of reduced size. The , following. We have a *Galerkin projection/approximation* if $\mathcal{W} = \mathcal{V}$ and a *Petrov–Galerkin projection/approximation* if $\mathcal{W} \neq \mathcal{V}$.

Example 3.3 *Consider the eigenvalue problem $A\mathbf{x} = \lambda\mathbf{x}$, where A is of order n. Here* $\mathbf{r}(\mathbf{x}, \lambda) = A\mathbf{x} - \lambda\mathbf{x}$. *Hence, the projected problem is* $W^T(AV\hat{\mathbf{x}}_V - \lambda V\hat{\mathbf{x}}_V) = 0$ *or* $W^T AV\hat{\mathbf{x}}_V = \lambda W^T V\hat{\mathbf{x}}_V$.

Example 3.4 *Consider the equation $u'' = f(x)$ on [0,1] with $u \in C_0^2$, where $C_0^2 = \{u \in C^2 | u(0) = u(1) = 0\}$. Now we define $r(u) = u'' - f(x)$. In inner product form we can write the problem, for example, as follows: find $u \in C_0^2$ such that $(w, r(u)) = 0$ for all w in test space C^0. Now, suppose $\mathcal{V}$ is a finite-dimensional sub-space of C_0^2 with basis $\{\phi_1(x), \ldots, \phi_N(x)\}$ and $\mathcal{W}$ of C^0 with basis $\{\psi_1(x), \ldots, \psi_N(x)\}$. Then our problem transforms into the projected problem: find $a_1, \ldots, a_N$ such that $(\psi_i, r(\sum_{j=1}^N a_j\phi_j)) = 0$ for $i = 1, \ldots, N$. Due to the bi-linearity of the inner product, we can re-write this as $\sum_{j=1}^N (\psi_i, \phi_j'')a_j = (\psi_i, f)$ for $i = 1, \ldots, N$. Note that this is just a linear system. See example Ex. 3.7*

Suppose A is an $m \times n$ matrix of full rank with $m > n$. If we minimize in the standard inner product $\|\mathbf{b} - A\mathbf{x}\|^2 = (\mathbf{b} - A\mathbf{x}, \mathbf{b} - A\mathbf{x})$, then this will lead to a projection. Minimizing can be done by differentiating with respect to each element of $\mathbf{x}$ consecutively and setting that derivative to zero. Using the product rule for multiplication, this leads for the differentiation with respect to x_i to $(A\mathbf{e}_i, \mathbf{b} - A\mathbf{x}) + (\mathbf{b} - A\mathbf{x}, A\mathbf{e}_i) = 0$. Since this equation holds for all i, it can be re-written as $A^T(\mathbf{b} - A\mathbf{x}) = 0$. One can interpret this as $\mathbf{x}$ being in the search space $\mathbb{R}^m$ and tested against a test space generated by the columns of A. Alternatively, one can also see this as trying to find a vector in the search space spanned by the columns of A tested against the same space, which will lead to an orthogonal projection of $\mathbf{b}$ on the space spanned by the columns of A.

Additional Material

- A much more in-depth discussion on the application of projection methods in iterative schemes to solve linear systems of equations and eigenvalue problems can be found in Saad (2003).

3.5 Boundary Conditions and Well-Posedness

In this section, we dig a bit deeper into the theory of PDEs, which is needed for addressing the well-posedness of problems. In fact, we will apply the projection approach of the previous section to PDEs. We did this already in Example 3.4

for homogeneous Dirichlet boundary conditions. Here we will apply it to the case with more general boundary conditions. For this case, we need to define operators including boundary conditions, which we will do in the following section. After introducing those, we can apply the projection approach straightforwardly.

3.5.1 Linear Operators on Function Spaces

Suppose we have defined a PDE on a domain Ω with boundary Γ and a standard inner product $(u, v)_\Omega \equiv \int_\Omega uv \, d\Omega$. Likewise we can define a standard inner product on any part Γ_1 of the boundary by $(u, v)_{\Gamma_1} \equiv \int_{\Gamma_1} uv \, d\Gamma$. In Example 3.8, it is shown that one can express a boundary integral in a domain integral using Dirac delta functions. In the following we will use the term operator; an *operator* maps a function to another function, for example the derivative operator $\mathcal{D}$ defined by $\mathcal{D}u(x) = du(x)/dx$, the integral operator $\mathcal{O}$ by $\mathcal{O}u(x) = \int_0^x u(t)dt$, the identity operator $\mathcal{I}$ by $\mathcal{I}u(x) = u(x)$, and the squaring operator $\mathcal{S}$ by $\mathcal{S}(u(x)) = u(x)^2$. Apart from the last one, these are all linear operators. *See example Ex. 3.8*

Using the inner product, we can define the adjoint of an operator. Likewise one can define operators that live on the boundary of a domain. We will denote a linear operator on the domain by $\mathcal{L}$ and one on the boundary Γ_2 by $\mathcal{B}$. On $\Gamma_1 \equiv \Gamma/\Gamma_2$ we assume the functions to be zero. Together, these define a linear operator $\mathcal{A}$ implicitly by the requirement that

$$(v, \mathcal{A}u)_\Omega = (v, \mathcal{L}u)_\Omega + (v, \mathcal{B}u)_{\Gamma_2} \tag{3.11}$$

for all sufficient smooth u and v, all being zero at Γ_1. The *adjoint operator* of an operator $\mathcal{A}$ is the operator $\mathcal{A}^*$ defined by $(\mathcal{A}^*v, u)_\Omega = (v, \mathcal{A}u)_\Omega$ for all u and v. A *self-adjoint operator* is an operator for which $\mathcal{A} = \mathcal{A}^*$. For later use, we also mention that a *skew-adjoint operator* is an operator for which $\mathcal{A}^* = -\mathcal{A}$. *See example Ex. 3.9*

For linear second-order elliptic PDEs on a d-dimensional space, any self-adjoint operator can be written as the following pair of a domain and boundary operator:

$$\mathcal{L}u = -\sum_{i,j=1}^d \frac{\partial}{\partial x_i}\left(a_{ij}\frac{\partial u}{\partial x_j}\right) + qu = -\nabla \cdot (A\nabla u) + qu, \tag{3.12a}$$

$$\mathcal{B}u = (A\nabla u, \mathbf{n}) + su \text{ on } \Gamma_2, \tag{3.12b}$$

on a space where $u = 0$ on Γ_1. Here all the coefficients may depend on x, and the positive definite A (positive definite) is as defined in Section 3.3.1. In fact, the operators occurring in the Poisson and wave equation are self-adjoint, where for the wave equation, one must take a domain Ω in the (x, t)-space.

3.5.2 Weak Form of PDEs

Now we are ready to apply the projection approach to PDEs with more general boundary conditions than in Example 3.4. Suppose we have a problem $\mathcal{L}u = f$ on Ω and $\mathcal{B}u = g$ on Γ_2 and $u = 0$ on the remaining part of the boundary Γ_1. Together, $\mathcal{L}$ and $\mathcal{B}$ define a linear operator $\mathcal{A}$ as in (3.11). Consistent with this definition of $\mathcal{A}$, we define f by

$$(v, \mathrm{f})_\Omega = (v, f)_\Omega + (v, g)_{\Gamma_2}. \tag{3.13}$$

Next, we can define the residual $r(u) = \mathcal{A}u - \mathrm{f}$ and apply the Galerkin projection approach leading to the problem formulation:

$$\text{find } u \in \mathcal{V} \text{ such that } (v, \mathcal{A}u - \mathrm{f})_\Omega = 0 \text{ for all } v \in \mathcal{V}. \tag{3.14}$$

Suppose that u_{ex} is the exact solution of $r(u) = 0$; then, we can replace f by $\mathcal{A}u_{ex}$ and the problem turns into

$$\text{find } u \in \mathcal{V} \text{ such that } (v, \mathcal{A}(u - u_{ex}))_\Omega = 0 \text{ for all } v \in \mathcal{V}. \tag{3.15}$$

So u will be the oblique projection of u_{ex} on $\mathcal{V}$, where $\mathcal{W} = \mathcal{A}\mathcal{V}$. This turns over in a projection in an inner product based on $\mathcal{A}$, if the latter is self-adjoint and positive definite. In this case, a minimization is attached to the problem:

$$u = \mathrm{argmin}_{w \in \mathcal{V}} \|w - u_{ex}\|_{\mathcal{A}}, \tag{3.16}$$

where the norm is induced by the associated inner product $(u, v)_\mathcal{A} = (u, \mathcal{A}v)_\Omega$. *So, u is the least-squares approximation of the exact solution in $\mathcal{V}$ in the $\mathcal{A}$-inner product or, otherwise stated, the $\mathcal{A}$-orthogonal projection of u_{ex} on $\mathcal{V}$.* We can also write this minimization as

$$u = \mathrm{argmin}_{w \in \mathcal{V}} J(w), \text{ where } J(w) \equiv \frac{1}{2}(w, w)_\mathcal{A} - (w, \mathrm{f})_\Omega. \tag{3.17}$$

In applications, $J(w)$ represents an energy. The solution minimizes the energy subject to certain forcings, which is known as Dirichlet's principle. $J(w)$ is called a *functional*, which is a 'function' where you enter a function and which gives back a number. Formula (3.14) is also called the *variational form* or *weak form* of the PDE problem and if there exists an associated minimization form (3.17) also that one may be called like that.

We can also follow the opposite route by defining a general form of $J(w)$ and derive the associated PDEs and boundary conditions by minimization. For example, let

$$J(w) = \int_0^1 \left[\tfrac{1}{2}p \left(\frac{dw}{dx} \right)^2 + \tfrac{1}{2}qw^2 - fv \right] dx + \left(\tfrac{1}{2}sw(1)^2 - gw(1) \right). \tag{3.18}$$

Here p, q, and f are functions belonging to $C[0, 1]$ with $p > 0$ and $q \geq 0$ and s and g are constants with $s \geq 0$.

Now, $J(w)$ has a stationary point for $w = u$ if, for $w = u + \varepsilon v$, it holds that

$$\frac{d}{d\varepsilon}J(u + \varepsilon v)\bigg|_{\varepsilon=0} = 0.$$

See example Ex. 3.10

This leads to the following: find u such that

$$\int_0^1 \left[p\frac{dv}{dx}\frac{du}{dx} + vqu - vf \right] dx + v(1)(su(1) - g) = 0 \qquad (3.19)$$

for all v. So, the stationary solution has to satisfy this equation (and $u(0) = 0$) for all v with $v(0) = 0$.

Now we perform a partial integration on (3.19) and use the fact that $v(0) = u(0) = 0$. Then for the stationary point we have to find u such that

$$-\int_0^1 v\left[\frac{d}{dx}\left(p\frac{d}{dx}u \right) - qu + f \right] dx + v(1)\left[p\frac{du}{dx}(1) + su(1) - g \right] = 0$$

for all v. Hence, since v satisfies $v(0) = 0$, but is arbitrary on the rest of the domain, we have that the parts between brackets must be zero. Hence, we find

$$-\frac{d}{dx}\left(p\frac{d}{dx}u \right) + qu = f \qquad \text{in } (0, 1), \qquad (3.20a)$$

$$p\frac{du}{dx} + su = g \qquad \text{for } x = 1, \qquad (3.20b)$$

$$u = 0 \qquad \text{for } x = 0. \qquad (3.20c)$$

Note that we started off with only a boundary condition at $x = 0$, but that the minimization gives us also a boundary condition (3.20b) at $x = 1$. We call the latter boundary condition a *natural boundary condition*. The general name for a PDE that is derived from a minimization, here (3.20a), is *Euler–Lagrange differential equation*. For more on the *Calculus of Variations* and its applications, we refer to Gelfand and Fomin (2000).

Note that we could have taken a shorter, more abstract, route by identifying from the functional (3.18) and its general form (3.17) that in this case

$$(v, u)_A = \int_0^1 \left[p\frac{dv}{dx}\frac{du}{dx} + qvu \right] dx + sv(1)u(1),$$

$$(v, f)_\Omega = \int_0^1 fv\,dx + gv(1),$$

which is equivalent to (3.14). The latter can be written as follows: find u such that

$$(v, u)_A = (v, f)_\Omega$$

for all v, which is the same as (3.19) if we substitute the expressions. After that, the same route can be followed, so using partial integration, the equations of (3.20) emerge.

The expression $(v, u)_A$ is a *bi-linear form*, that is, it is linear in both its arguments, and we will also denote it as $a(v, u)$; moreover, $(v, f)_\Omega$ is a linear form and will also be denoted by $F(v)$. Using these definitions, we can also write (3.14) as

$$a(v, u) = F(v). \tag{3.21}$$

Note that the preceding bi-linear form is symmetric, that is, $a(v, u) = a(u, v)$. A symmetric $a(\cdot, \cdot)$ is equivalent to a self-adjoint operator $\mathcal{A}$. In this terminology, $J(w) = \frac{1}{2}a(w, w) - F(w)$. We call (3.14) the *weak form* of the equation. Note that not all PDE problems can be cast in a minimization form, but usually they can be put into a weak form. *See example Ex. 3.11*

Note that the functional (3.18) we started from has only first derivatives in it but that the resulting equation (3.20a) contains second-order derivatives. Hence, the equation requires more smoothness than the functional. Also note that the smoothness of u is determined by that of f. If it admits a solution of the equations of (3.20), then the same solution will be found from the minimization process. However, if it does not admit a solution of the equations of (3.20) but still does for the functional, then we say that we have found a *weak solution* of the equations. In general, the more derivatives of f are continuous, the more are those of u.

For the form with only first derivatives we can use a set of functions for which the function itself and its derivative are square integrable on [0,1] and that satisfy the so-called *essential boundary condition* $u(0) = 0$. Recall that a square integrable function v satisfies $||v|| < \infty$. This leads naturally to the definition of the *Sobolev norm*,

$$||v||_{H^1(0,1)} \equiv \sqrt{||v||^2 + ||\frac{dv}{dx}||^2}, \tag{3.22}$$

and defines the function space $H^1(0, 1)$ by requiring that $||v||_{H^1(0,1)} < \infty$.

Example 3.5 *Analogy Continuous and Discrete Case*
Consider

$$J(w) = \frac{1}{2}\left(\frac{dw}{dx}, \frac{dw}{dx}\right) - (w, f)$$

with $w(0) = 0$ on [0,1].
We split the interval into two equal parts and consider the discretization

$$\begin{bmatrix} \frac{dw}{dx}\left(\frac{1}{4}\right) \\ \frac{dw}{dx}\left(\frac{3}{4}\right) \end{bmatrix} \approx 2\begin{bmatrix} -1 & 1 & 0 \\ 0 & -1 & 1 \end{bmatrix}\begin{bmatrix} w(0) \\ w\left(\frac{1}{2}\right) \\ w(1) \end{bmatrix} \equiv D\mathbf{w},$$

where $\mathbf{w} = [w(0), w\left(\frac{1}{2}\right), w(1)]^T$. *The discrete analogue of* $J(w)$ *is found by approximating the continuous integration in the inner product by the midpoint rule* $(\int_a^b f(x)dx \approx (b - a)f((a+b)/2)$ *for the two intervals*

$$\hat{J}(\mathbf{w}) = \frac{1}{2}\frac{1}{2}(D\mathbf{w}, D\mathbf{w})_h - \frac{1}{2}(\mathbf{w}, \hat{\mathbf{f}})_h,$$

where $(w, f) \approx \frac{1}{2}\frac{1}{2}\left[\left(w(0) + w\left(\frac{1}{2}\right)\right)f\left(\frac{1}{4}\right) + \left(w\left(\frac{1}{2}\right) + w(1)\right)f\left(\frac{3}{4}\right)\right] = \frac{1}{2}\left(\mathbf{w}, \frac{1}{2}\left[f\left(\frac{1}{4}\right),\right.\right.$ $\left.\left.\left(f\left(\frac{1}{4}\right) + f\left(\frac{3}{4}\right), f\left(\frac{3}{4}\right)\right)\right]^T\right) \equiv \frac{1}{2}(\mathbf{w}, \hat{\mathbf{f}})_h$. *The subscript* h *denotes the standard inner product.*

We have to minimize $\hat{J}(\mathbf{w})$ *subject to the condition* $w(0) = 0$ *and get, analogously to the continuous case,*

$$(D\mathbf{v}, D\mathbf{u})_h = (\mathbf{v}, \hat{\mathbf{f}})_h \tag{3.23}$$

which should hold for arbitrary $\mathbf{v}$ *with* $v(0) = 0$. *Now* $(D\mathbf{w}, D\mathbf{u})_h = (\mathbf{w}, D^T D\mathbf{u})_h$, *where*

$$D^T D = 4 \begin{bmatrix} 1 & -1 & \\ -1 & 2 & -1 \\ & -1 & 1 \end{bmatrix}.$$

Hence, (3.23) is equivalent to

$$v(0)\left[u(0) - u\left(\frac{1}{2}\right)\right] - u(0)v\left(\frac{1}{2}\right) + \left(\begin{bmatrix} v\left(\frac{1}{2}\right) \\ v(1) \end{bmatrix}, \begin{bmatrix} 2 & -1 \\ -1 & 1 \end{bmatrix}\begin{bmatrix} u\left(\frac{1}{2}\right) \\ u(1) \end{bmatrix}\right)_h = \frac{1}{4}(\mathbf{v}, \hat{\mathbf{f}})_h.$$

Now we enter the conditions $v(0) = u(0) = 0$ *and obtain*

$$\left(\begin{bmatrix} v\left(\frac{1}{2}\right) \\ v(1) \end{bmatrix}, \begin{bmatrix} 2 & -1 \\ -1 & 1 \end{bmatrix}\begin{bmatrix} u\left(\frac{1}{2}\right) \\ u(1) \end{bmatrix}\right) = \frac{1}{4}\left(\left[0, v\left(\frac{1}{2}\right), v(1)\right]^T, \hat{f}\right)_h$$

for arbitrary $v\left(\frac{1}{2}\right)$ *and* $v(1)$. *This finally leads to*

$$\begin{bmatrix} 2 & -1 \\ -1 & 1 \end{bmatrix}\begin{bmatrix} u\left(\frac{1}{2}\right) \\ u(1) \end{bmatrix} = 1/4 \begin{bmatrix} f\left(\frac{1}{2}\right) \\ \frac{1}{2}f\left(\frac{1}{4}\right) \end{bmatrix}.$$

The preceding example is relevant for finite difference and finite volume discretization of Laplace operators. It gives us a means to derive a stable discretization on quite general grids (see Section 4.3.2 from Chapter 4).

3.5.3 Lax–Milgram Theorem and Poincaré's Inequality

The Lax–Milgram theorem is a generalization for the weak form (3.21) of the statement that a positive definite matrix, not necessarily symmetric, is invertible, and as a consequence the problem $Ax = b$ has a unique solution for any right-hand side b. The theorem reads as follows.

Theorem 3.1 (Lax–Milgram theorem) *Let V be a Hilbert space where the bi-linear form $a(v, u)$ satisfies the coercivity condition:*

$$a(v, v) \geq c||v||_V^2 \qquad (3.24)$$

with $c > 0$ for all $v \in V$. Moreover, there should exist a K and M such that for all $u, v \in V$

$$|a(v, u)| \leq M||v||_V||u||_V, \qquad (3.25a)$$

$$|F(v)| \leq K||v||_V. \qquad (3.25b)$$

Then the problem: find $u \in V$ such that $a(v, u) = F(v)$ for all $v \in V$ has a unique solution. See example Ex. 3.12

Recall that V is a Hilbert space if it has a inner product and that the limit of any sequence in the space is also an element of that space. The space $H^1[0, 1]$, introduced in the previous section, is an example of a Hilbert space.

At this point we will not go into the details of this theorem; for that, see Quarteroni and colleagues (2007) Section 12.4.4. The most important condition is the *coercivity condition* which is the generalization of the positive-definiteness condition for matrices.

The boundedness condition (3.25a) is a generalization of the Cauchy–Schwarz inequality to the bi-linear form. In (3.25b) we could define K by $K_F = \sup_{v \in V} |F(v)|/||v||_V$, which reminds us of matrix norms induced by vector norms.

The boundedness of F and the coercivity of the bi-linear form leads to the following stability statements.

Theorem 3.2 (Stability) *The Lax–Milgram theorem leads to the following stability results: (i) $||u||_V \leq K_F/c$, and (ii) $||u||_V \leq ||Au||/c$.*

Proof The proof of (i) is left as an exercise. *See example Ex. 3.13*
For (ii) we note that $a(u, u) = (Au, u)$, hence the coercivity condition can also be written as

$$c||u||_V^2 \leq (Au, u) \leq ||Au||||u|| \leq ||Au||||u||_V.$$

Dividing by $c||u||_V$ leads to the result.   ∎

This theorem also entails that the solution is stable in the sense of Hadamard (see Definition 3.1) with respect to perturbations in the linear form, as can be shown in what follows. Suppose we perturb the linear form by a linear form $\delta(v)$ which is bounded by $K_\delta||v||_V$. This makes also $F(v) + \delta(v)$ bounded. Now we have two problems to solve,

$$a(v, u) = F(v) \text{ and } a(v, \tilde{u}) = F(v) + \delta(v),$$

and we like to bound the difference $||\tilde{u} - u||_V$ by K_δ. We can subtract the two equations and find $a(v, \tilde{u} - u) = \delta(v)$. This problem also satisfies the conditions of the Lax–Milgram theorem, and then part (i) from Th. 3.2 gives $||\tilde{u} - u||_V \leq K_\delta/c$ which shows the stability. Note that by the definition of the linear form (3.13) $\delta(v)$ can mean a perturbation on both the domain and the boundary.

An important role in proving coercivity is played by the Poincaré inequality, which we will discuss here for the one-dimensional case. We consider here the interval $[0, L]$ instead of $[0, 1]$, since L plays an important role in the bound.

Theorem 3.3 (Poincaré inequality) *Let* $(f, g) \equiv \int_0^L f(x)g(x)dx$ *and* $||\cdot||$ *the induced norm. Then for any* $u \in H^1$ *it holds that* $||u - u(y)|| \leq L||du/dx||$, *for any (fixed)* y *on the interval* $[0,L]$.

Proof Obviously we have

$$u(x) - u(y) = \int_y^x \frac{du}{dx}(s)ds.$$

Hence for any x and y

$$|u(x) - u(y)| = |\int_y^x \frac{du}{dx}(s)ds| \leq \int_0^{l.} |\frac{du}{dx}(s)|ds$$

$$= \left(1, |\frac{du}{dx}|\right) \leq ||1|| \, ||\frac{du}{dx}|| = \sqrt{L}||\frac{du}{dx}||.$$

Now it holds that $\int_0^L (u(x) - u(y))^2 dx = ||u - u(y)||^2$, so

$$||u - u(y)||^2 \leq || \left(\sqrt{L}||\frac{du}{dx}||\right) ||^2 = L||\frac{du}{dx}||^2||1||^2$$

$$= L^2||\frac{du}{dx}||^2,$$

from which the assertion follows. ∎

A few special cases are given in the corollary.

Corollary 3.1

1. $\max(||u - \max_x(u)||, ||u - \min_x(u)||) \leq L||\frac{du}{dx}||$.
2. *If* $u(0)$ *or* $u(L)$ *is zero, then* $||u|| \leq L||\frac{du}{dx}||$.

The Poincaré inequality is essential in proving coercivity, given certain boundary conditions.

Example 3.6 *Suppose we have $a(u, v) = \int_0^1 du/dx \, dv/dx \, dx$, where $u, v \in H_0^1$ and the subscript 0 indicates here that the functions in the space are zero at $x = 0$. Then we have to prove that $a(u, u) \geq c\|u\|_{H_0^1}^2$. This proceeds as follows:*

$$a(u, u) = \|\frac{du}{dx}\|^2 = \frac{1}{2}\|\frac{du}{dx}\|^2 + \frac{1}{2}\|\frac{du}{dx}\|^2 \geq \frac{1}{2L^2}\|u\|^2 + \frac{1}{2}\|\frac{du}{dx}\|^2$$

$$\geq \min\left(\frac{1}{2L^2}, \frac{1}{2}\right) \|u\|_{H_0^1}^2,$$

which proves the desired property with $c = \min\left(\frac{1}{2L^2}, \frac{1}{2}\right)$.

Note that we could have split up $\|du/dx\|^2$ also in other positive parts such that the coefficients will become equal. See example Ex. 3.14

The preceding is fairly easy, but with Robin boundary conditions it becomes a lot more involved. For later reference, we will show this too, by means of an example.

Example 3.7 *In the case of the Robin boundary conditions*

$$\frac{du}{dx}(0) - \frac{\beta}{2}u(0) = 1 \text{ and } \frac{du}{dx}(L) + \frac{\alpha}{2}u(L) = -2$$

and $\mathcal{L} = -d^2/dx^2$ we have the bi-linear form (see (3.20b)) and its derivation,

$$a(v, u) = -\left(v, \frac{d^2u}{dx^2}\right) + v(L)\left(\frac{du}{dx}(L) + \frac{\alpha}{2}u(L)\right) - v(0)\left(\frac{du}{dx}(0) - \frac{\beta}{2}u(0)\right).$$

We want to know for which values of α and β one can show coercivity. After partial integration we obtain

$$a(v, u) = \left(\frac{dv}{dx}, \frac{du}{dx}\right) + \frac{\alpha}{2}v(L)u(L) + \frac{\beta}{2}v(0)u(0). \tag{3.26}$$

For coercivity we have to check whether there exists a positive $\hat{c}$ such that

$$a(u, u) = \frac{1}{2}\|\frac{du}{dx}\|^2 + \frac{1}{2}\left(\|\frac{du}{dx}\|^2 + \alpha u(L)^2 + \beta u(0)^2\right) \geq \hat{c}\|u\|_{H^1}^2,$$

where we have split the norm of the first derivative into equal parts, as in the previous example. We see from this expression that α and β should be at least non-negative in order to preclude that $a(u, u)$ can become negative. Even $\alpha = \beta = 0$ should be precluded since then the Robin conditions turn into Neumann boundary conditions and $a(u, u)$ does not change if we add an arbitrary constant to u, while $\|u\|_{H^1}^2 \equiv \|u\|^2 + \|\frac{du}{dx}\|^2$ does. Hence, the essential step is showing that for at least one of α and β being positive there exists a positive c such that

$$\alpha u(L)^2 + \beta u(0)^2 + \|\frac{du}{dx}\|^2 \geq c\|u\|^2.$$

Next, we substitute the Poincaré inequality saying that this is true if we can find a positive c such that

$$\alpha u(L)^2 + \beta u(0)^2 + ||u - u(y)||^2/L^2 \geq c||u||^2. \tag{3.27}$$

This in itself is true if we can find a positive c such that

$$\alpha u(L)^2 + \beta u(0)^2 + (||u||^2 - 2||u||||u(y)|| + ||u(y)||^2)/L^2 \geq c||u||^2$$

or

$$\alpha L^2 u(L)^2 + \beta L^2 u(0)^2 + ||u||^2 - 2||u||||u(y)|| + ||u(y)||^2 \geq cL^2||u||^2$$

or

$$(1 - cL^2)||u||^2 - 2||u||||u(y)|| + ||u(y)||^2 + \alpha L^2 u(L)^2 + \beta L^2 u(0)^2 \geq 0.$$

Note that $||u(y)|| = |u(y)|\sqrt{L}$. Suppose $\alpha > 0$. When we choose $y = L$, we split off a square leading to

$$\left(\sqrt{1 - cL^2}||u|| - \frac{\sqrt{L}}{\sqrt{1 - cL^2}}||u(L)||\right)^2 + \left(-\frac{L}{1 - cL^2} + L + \alpha L^2\right) u(L)^2 + \beta L^2 u(0)^2 \geq 0.$$

For $c < 1/L^2$, this is true if

$$c \leq \frac{\alpha}{L(1 + \alpha L)}.$$

Similarly we can choose $y = 0$ if $\beta > 0$. Since one of α and β is positive, we can choose the maximum of the two c's. So, the bi-linear form (3.26) is coercive.

As mentioned earlier, if both α and β are zero, we cannot find a positive c. In that case, we have to restrict our space to the space perpendicular to the constant function, which will make the solution unique again, that is, we have $||u - u(y)||^2 = ||u||^2 - 2u(y)(1, u) + ||u(y)||^2 = ||u||^2 + ||u(y)||^2 \geq ||u||^2$. Substituting this in the Poincaré inequality we get, like the Dirichlet condition, $||u|| \leq L||\frac{du}{dx}||$.

The preceding examples suggest that if we can bring a problem with a non-self adjoint operator in a weak form with $a(*, *)$ being coercive, then in general (the other conditions of the Lax–Milgram in general do not cause problems) the weak form is well-posed. Hence, in that case a weak solution of our non–self adjoint problem exists. This also means that by checking coercivity we can determine which boundary conditions make the problem well-posed. As we have seen in the examples, for second-order elliptic equations it is enough that on some part of the boundary a Dirichlet or, even weaker, a Robin condition is prescribed. This ensures that the problem is well-posed.

A crucial prerequisite of the Lax–Milgram theorem is that u and w must be in the same linear space. However, the space of all functions satisfying $u(0) = 1$ is not a linear space. The key to solving this problem is to re-write the problem such that the boundary condition becomes homogeneous. So, we write $u = u_e + \hat{u}$, where u_e is an arbitrarily chosen function such that $u_e(0) = 1$, for example $u_e \equiv 1$. If we have a PDE $\mathcal{L}u = f$ with natural boundary condition $\mathcal{B}u = g$ (3.20b) and essential boundary condition $u(0) = 1$, then this turns over in $\mathcal{L}\hat{u} = \hat{f}(\equiv f - \mathcal{L}u_e)$, $\mathcal{B}\hat{u} = \hat{g}(\equiv g - \mathcal{B}u_e)$, and $\hat{u}(0) = 0$. For this set we can write down the weak formulation to which we can apply the Lax–Milgram theorem.

One could think of more general weak forms associated to the PDE problem. For instance, find u such that

$$(\mathcal{L}u - f, w) + w(0)(u(0) - 1) + w(1)\left(\frac{du}{dx}(1) - 5\right) = 0 \qquad (3.28)$$

for all w, where the inner product is defined by $(f, g) = \int_0^1 fg\,dx$ and w is arbitrary. This is clearly related to $\mathcal{L}u = f$, $u(0) = 1$ and $\frac{du}{dx}(1) = 5$. However, to this form we cannot apply the Lax–Milgram theorem, but we can often rework the problem to one to which it does.

Example 3.8 *Suppose* $\mathcal{L}u = \frac{d}{dx}\left[(1 + x^2)\frac{du}{dx}\right]$. *Re-write (3.28) in a form that allows application of the Lax–Milgram theorem.*

The key problem is the bi-linear form: we have to re-write the equation such that the associated bi-linear form can be successful in the coercivity test. The bi-linear form of (3.28) is

$$a(u, w) = -(\mathcal{L}u, w) - w(0)u(0) - w(1)\frac{du}{dx}(1).$$

Let us first simply try to transform $(\mathcal{L}u, w)$ *to a symmetric form by partial integration; we obtain*

$$a(u, w) = \left(\frac{du}{dx}, (1 + x^2)\frac{dw}{dx}\right) - 2w(1)\frac{du}{dx}(1) + w(0)\frac{du}{dx}(0) - w(0)u(0) - w(1)\frac{du}{dx}(1).$$

We can combine the second and fifth terms in the right-hand side, yielding

$$a(u, w) = \left(\frac{du}{dx}, (1 + x^2)\frac{dw}{dx}\right) - 3w(1)\frac{du}{dx}(1) + w(0)\frac{du}{dx}(0) - w(0)u(0).$$

If we look to a(w, w), we see that the first term looks at least non-negative, but the remaining terms easily can become negative for appropriate choices of w. (Check this.) Note that we get rid of the third and fourth terms if we require that $w(0) = 0$. *So, if we want to apply the Lax–Milgram theorem, then u and w should be in the same linear space, hence also u(0) should be zero. This is currently not the case, but by setting* $u = 1 + \hat{u}$ *we can accomplish that.*

Note that the bi-linear form remains the same if we go to this homogeneous boundary condition. Now we are left with only the second term. We cannot set $w(1) = 0$ because then also $u(1) = 0$. Also note that this term originates from a combination of partial integration on the operator and the added term with the boundary condition at $x = 1$ in the original (3.28). Observe also that we can put an arbitrary non-zero factor in front of this term, perform the partial integration, and then pick the constant such that the term $w(1))\frac{du}{dx}(1)$ cancels out. So, having dealt with the condition at $x = 0$, we start off with

$$- (\mathcal{L}\hat{u} - f, w) + \alpha w(1) \left(\frac{d\hat{u}}{dx}(1) - 5 \right) = 0, \tag{3.29}$$

where $w(0) = \hat{u}(0) = 0$ and $\alpha \neq 0$, where the bi-linear form is

$$a(\hat{u}, w) = -(\mathcal{L}\hat{u}, w) + \alpha w(1) \frac{d\hat{u}}{dx}(1).$$

After partial integration we obtain

$$a(\hat{u}, w) = \left(\frac{d\hat{u}}{dx}, (1 + x^2) \frac{dw}{dx} \right) + (\alpha - 2)w(1) \frac{d\hat{u}}{dx}(1).$$

Hence, to get rid of the nasty term we have to choose $\alpha = 2$. So, (3.29) turns into

$$- (\mathcal{L}\hat{u} - f, w) + 2w(1) \left(\frac{d\hat{u}}{dx}(1) - 5 \right) = 0, \tag{3.30}$$

where $w(0) = \hat{u}(0) = 0$. The bi-linear and the linear forms are, respectively, given by

$$a(\hat{u}, w) = \left(\frac{d\hat{u}}{dx}, (1 + x^2) \frac{dw}{dx} \right), \tag{3.31a}$$

$$F(w) = -(f, w) + 10w(1). \tag{3.31b}$$

The bi-linear form is now a potential candidate for success in the coercivity test. In fact, it can be shown that the coercivity condition is satisfied.

In the preceding example we explained how to modify a weak form such that it fits the Lax–Milgram theorem. There is another way of dealing with the non-Dirichlet conditions leading to the same result, which often is found in other textbooks. We will explain that too in the same problem example.

Example 3.9 *Instead of (3.29), one sets off by posing*

$$- (\mathcal{L}\hat{u} - f, w) = 0 \tag{3.32}$$

and performs partial integration which leads to

$$\left(\frac{d\hat{u}}{dx}, (1 + x^2) \frac{dw}{dx} \right) - 2w(1) \frac{d\hat{u}}{dx}(1) + (f, w) = 0.$$

Now, the boundary condition $\frac{d\hat{u}}{dx}(1) = 5$ is substituted, leading to

$$\left(\frac{d\hat{u}}{dx}, (1+x^2)\frac{dw}{dx}\right) - 10w(1) + (f, w) = 0,$$

which has the bi-linear form (3.31a) and linear form (3.31b) as before. On the one hand this is more convenient, but on the other hand it obscures what the operator $\mathcal{A}$ is. This is not defined by the bi-linear form in (3.32) but by that in the last equation, so

$$(w, \mathcal{A}\hat{u}) = \left(\frac{d\hat{u}}{dx}, (1+x^2)\frac{dw}{dx}\right).$$

By a partial integration in reverse order it will split out in a part over the domain and one on the boundary. See example Ex. 3.15

Example 3.10 (Ginzburg–Landau equation) *Consider the steady-state equation, where we took the coefficient for the non-linear term zero,*

$$\gamma_1 A + \gamma_2 \frac{d^2 A}{dx^2} = 0$$

with boundary conditions $A(0) = 0$, $A(1) = 1$. Let us first get rid of the inhomogeneous boundary condition introducing $\tilde{A} = A - x$. This leads to the equation for $\tilde{A}$,

$$\gamma_1 \tilde{A} + \gamma_2 \frac{d^2 \tilde{A}}{dx^2} = -\gamma_1 x$$

with $\tilde{A}(0) = 0$; $\tilde{A}(1) = 0$. Herewith the associated bi-linear and linear forms are respectively

$$a(v, u) = -\left(v, \gamma_1 u + \gamma_2 \frac{d^2 u}{dx^2}\right), \quad F(v) = -\gamma_1(v, x).$$

The test on coercivity has to be performed on

$$a(u, u) = -\gamma_1 ||u||^2 + \gamma_2 ||\frac{du}{dx}||^2.$$

Assuming $\gamma_2 > 0$ and applying the Poincaré inequality in Corollary 3.1, it follows that

$$a(u, u) \geq (\gamma_2 C_P - \gamma_1)||u||^2.$$

As seen in Example 3.6, we should have kept a fraction of $||\frac{du}{dx}||^2$ to accommodate for the first derivative in $||u||_{H^1}$. The preceding shows that we can split off such a part if $(\gamma_2 C_P - \gamma_1) > 0$. We would like to use a sharp C_p in order to know when this is possible. And fortunately we know that value. It occurs for $u(x) = \sin(\pi x)$, so $C_P = \pi^2$. Now assume that $(\gamma_2 C_P - \gamma_1) > 0$; then we can find the c in the coercivity relation. To do this as sharp as possible, we first we split up the norm of the first derivative using a to-be-determined α in (0,1), that is,

$$a(u, u) = -\gamma_1 ||u||^2 + \gamma_2 ||\frac{du}{dx}||^2 = -\gamma_1 ||u||^2 + \gamma_2 \left((1-\alpha)||\frac{du}{dx}||^2 + \alpha||\frac{du}{dx}||^2\right),$$

and then apply the Poincaré inequality to the first one, leading to

$$a(u,u) \geq (\gamma_2(1-\alpha)C_P - \gamma_1)||u||^2 + \gamma_2\alpha||\frac{du}{dx}||^2$$

$$= \gamma_2\alpha||u||^2_{H^1},$$

where we chose α such that $\gamma_2\alpha = \gamma_2(1-\alpha)C_P - \gamma_1$. This leads to $\alpha = (\gamma_2 C_P - \gamma_1)/(\gamma_2(1 + C_P))$. So finally

$$a(u,u) \geq \frac{\gamma_2 C_P - \gamma_1}{1 + C_P}||u||^2_{H^1}.$$

Suppose we have a bi-linear form that is not coercive. What to do in such a case? Let us first discuss the finite-dimensional case. According to Theorem B.9 the coercivity condition of the Lax–Milgram theorem means that all eigenvalues are in the positive plane, and hence it is a sufficient condition for a matrix *not* to be singular. However, this is not a necessary condition. Eigenvalues may be anywhere in the plane, except in the origin, for a matrix to be non-singular. A general necessary condition is that the singular values of the matrix are all positive; the singular values are the eigenvalues of $\sqrt{A^T A}$. Hence, the matrix $A^T A$ should be positive definite, or $(\mathbf{x}, A^T A\mathbf{x}) > c^2(\mathbf{x}, \mathbf{x})$. This is equivalent to the condition that

$$\inf_{\mathbf{x}} \sup_{\mathbf{y}} \frac{(\mathbf{y}, A\mathbf{x})}{||\mathbf{y}||||\mathbf{x}||} > c. \tag{3.33}$$

From this we generalize to the infinite-dimensional form of this so-called *inf-sup condition, See example Ex. 3.16*

$$\inf_{u} \sup_{v} \frac{a(v,u)}{||v||||u||} > c. \tag{3.34}$$

So, if a bi-linear form is not coercive, then we can try to check whether the inf-sup condition is satisfied.

We did not discuss the Petrov–Galerkin approach here. This gives an extra freedom to study the well-posedness of PDEs. *See example Ex. 3.17*

Until now, we only looked to the one-dimensional case. In the multi-dimensional case, the concepts are the same, but it is slightly more technical to proof certain properties. For completeness we consider again the problem (3.14), where the operator $\mathcal{A}$ is defined in (3.12) and (3.12), and f by (3.13). Before considering the coercivity of the bi-linear form, we recall a generalization of the product rule which is useful in all cases where we have conservation laws in PDEs. For sufficient smooth arbitrary function ω and arbitrary vector field $\mathbf{v}$ we have

$$\nabla \cdot (\omega\mathbf{v}) = \omega\nabla \cdot \mathbf{v} + \mathbf{v} \cdot \nabla\omega. \tag{3.35}$$

Integrating the last line over a domain Ω with boundary Γ, using Gauss' theorem on the left part, we get an equally useful relation:

$$\int_{\Gamma} \omega \mathbf{v} \cdot \mathbf{n} d\Gamma = \int_{\Omega} \omega \nabla \cdot \mathbf{v} d\Omega + \int_{\Omega} \mathbf{v} \cdot \nabla \omega d\Omega. \tag{3.36}$$

This last relation can be used to replace the partial integrations needed to prove coercivity of the bilinear form. With $\mathbf{v} = A\nabla u$ and $\omega = v$ we obtain

$$
\begin{aligned}
a(v, u) &= -\int_{\Gamma} v(A\nabla u) \cdot \mathbf{n} d\Gamma + \int_{\Omega} \nabla v \cdot (A\nabla u) + qvu d\Omega + \int_{\Gamma_2} v((A\nabla u) \cdot \mathbf{n} + su) d\Gamma \\
&= \int_{\Omega} \nabla v \cdot (A\nabla u) + qvu d\Omega + \int_{\Gamma_2} vsu d\Gamma, \tag{3.37}
\end{aligned}
$$

where we have also used that $v = 0$ on $\Gamma_1 = \Gamma/\Gamma_2$. The last expression shows that for A symmetric positive definite, $s \geq 0$, and $q \geq 0$, the $a(u, u)$ will be non-negative. We see, in the case both s and q are zero, that a constant also would make $a(u, u) = 0$; but this constant must be zero if Γ_1, where $u = 0$ is imposed, is not empty. In that case, one can show with a multi-dimensional generalization of the Poincaré equation that the bi-linear form is coercive.

With this we have shown that (3.8a) will be a well-posed problem if f satisfies the boundedness condition in the Lax–Milgram theorem. Also the wave equation (3.8b) fits in the pattern. However, here A has -1 and 1 on its diagonal, and consequently it is indefinite. This means that we cannot show the well-posedness of the wave equation with the Lax–Milgram theorem.

3.6 Posedness of Common (Non)linear Equations

In this section, we show how the posedness theory related to the Lax–Milgram theorem can be used to find out which boundary conditions make some common PDEs well-posed. In fact, the theory is closely related to the *energy method* (Straughan, 2004). For non-linear equations the Lax–Milgram theorem is not applicable, but the energy method is. Also we can use the energy method to study for which part in parameter space the energy goes down monotonically as depicted in Fig. 2.1.

Additional Material

- For a recent paper on determining boundary conditions for fluid flow equations, also listing the relevant literature, see Nordström (2022).

3.6.1 *Linear Convection–Diffusion Equation*

In this section, we consider the linear convection–diffusion equation on $[0, 1]$

$$u_t = -\bar{u}u_x + \mu u_{xx} + f \tag{3.38}$$

which is a parabolic equation according to the terminology for second-order PDEs discussed in Section 3.3.1.

To determine the boundary conditions which make this problem well-posed we first consider the steady-state case. This brings us back to the realm of the Lax–Milgram theorem, i.e., we have to check the associated bi-linear form for coercivity. As before, it is extended by contributions of possible boundary conditions and given by

$$a(v, u) = -\left(v, -\bar{u}\frac{du}{dx} + \mu\frac{d^2u}{dx^2}\right) - v\left(\alpha u - \beta\frac{du}{dx}\right)\Big|_0^1 \qquad (3.39)$$

with general α and β. By standard integration by parts, we find that

$$a(u, u) = -\left(\left(\alpha - \frac{\bar{u}}{2}\right)u + (\mu - \beta)\frac{du}{dx}\right)u\Big|_0^1 + \mu\Big\|\frac{du}{dx}\Big\|^2. \qquad (3.40)$$

The first thing this expression shows is that, if we do not prescribe a Dirichlet condition at a boundary, then at that boundary we must have $\beta = \mu$ in order to preclude an undeterminate term. Additionally in that case, we need at that boundary that $\alpha - \bar{u}/2$ is semi-negative at $x = 1$ and semi-positive at $x = 0$. The latter also includes the Neumann condition $\alpha = 0$ which is only possible if $\bar{u}$ is negative at $x = 0$ or positive at $x = 1$. So, if $\bar{u}$ represents a velocity, then one can prescribe a Neumann boundary condition only at the outflow boundary.

Note that for $\alpha = \bar{u}$, $\beta = \mu$, we prescribe exactly the conserved quantity $-\bar{u}u + \mu\frac{du}{dx}$, called *flux*, of the steady case, which comes about when we write the steady case of (3.38) in divergence form:

$$0 = \frac{d}{dx}(-\bar{u}u + \mu\frac{du}{dx}) + f. \qquad (3.41)$$

When we apply this boundary condition to both sides, then the values of these fluxes must satisfy the integral of the last equation

$$(-\bar{u}u + \mu\frac{du}{dx})\Big|_0^1 = -\int_0^1 f(x)dx. \qquad (3.42)$$

And, if this relation is satisfied, the solution will not be unique since, if $\hat{u}$ is a solution of the equation, then one can add any solution $\tilde{u}$ which satisfies the homogeneous equations and boundary conditions. This boils down to $\tilde{u}$ being an arbitrary solution of $-\bar{u}u + \mu\frac{du}{dx} = 0$. The arbitrary solution of the last equation is $\tilde{u} = C\exp(\bar{u}x/\mu)$, where C is arbitrary. Note that for $\bar{u} = 0$ this turns $\tilde{u}$ in an arbitrary constant which is generally the solution we can add to that of a Poisson equation when we use Neumann boundary conditions. Note that imposing the flux at both boundaries does not lead to a coercive bi-linear form.

For the transient case, one can use the same conditions as in the steady state, but now we don't have the condition (3.42) because we integrate (3.38) over time, which allows that the integrals of u over the domain change with time. On the other hand, if this condition is violated, it is impossible to reach the steady state. Moreover, all the boundary conditions may now vary in time.

An initial condition on u and a Dirichlet, Neumann, or Robin condition at the left and the right boundaries will make the problem well-posed. However, in many cases the diffusion term is small with respect to the convection term. In such a case, one considers the extreme case where the diffusion term is absent in (3.38), resulting in the transport equation treated in Section 3.2.1. That equation needs a Dirichlet condition u at the inflow boundary, and that is just copied to the convection–diffusion equation. The transport equation does not have a condition at the other boundary, while the convection–diffusion has. According to (3.39) and (3.40), the Neumann boundary condition ($\alpha = 0$ and $\beta = \mu$) is most appropriate since then the condition will vanish if μ tends to zero. One usually takes $du/dx = 0$ since in practice one often does not know the value of this derivative. We get back to this matter when considering the discretization of this equation in Section 4.2.

3.6.2 Burgers' Equation

A non-linear variant of the convection–diffusion equation is Burgers' equation:

$$u_t = -uu_x + \mu u_{xx} + f. \tag{3.43}$$

Assume we have a Dirichlet condition $u(0) = u_0$ at $x = 0$. Hence, we can choose a function $\bar{u}$ that is independent of time such that $\bar{u}(0) = u_0$. Let us call the right-hand side of the Burgers' equation $r(u, f)$; then, on substituting $u = \bar{u} + \tilde{u}$, we obtain

$$\tilde{u}_t = -\bar{u}\tilde{u}_x - \tilde{u}\bar{u}_x + r(\tilde{u}, 0) + r(\bar{u}, f). \tag{3.44}$$

Note that the terms $\bar{u}\tilde{u}_x$ and $\tilde{u}\bar{u}$ enter due to the non-linearity of $r(u, f)$, so they would not be there if it were linear. At the right boundary we try the boundary condition $\gamma u^2(1) + \alpha u(1) - \mu \frac{du}{dx}(1) = g$, which we will write as $r_1(u, g) = 0$. After the substitution it becomes $2\gamma\bar{u}(1)\tilde{u}(1) + r_1(\tilde{u}, 0) + r_1(\bar{u}, g) = 0$. Now $a(v, \tilde{u})$ is given by

$$a(v, \tilde{u}) = -(v, -\bar{u}\tilde{u}_x - \tilde{u}\bar{u}_x + r(\tilde{u}, 0)) - v(1)\left(2\gamma\bar{u}(1)\tilde{u}(1) + r_1(\tilde{u}, 0)\right).$$

Note that $a(v, \tilde{u})$ is not bi-linear in this case, since it is not linear in $\tilde{u}$. Next, replacing r and r_1 by their respective definitions, we find after partial integration that for positivity we have to check

$$a(\tilde{u}, \tilde{u}) = \left(\left(\frac{1}{3} - \gamma\right)\tilde{u}(1) + \left(-2\gamma + \frac{1}{2}\right)\bar{u}(1) - \alpha\right)\tilde{u}^2(1) + (\bar{u}_x\tilde{u}, \tilde{u}) + \mu||\tilde{u}_x||^2.$$

At this point we can consider various choices for $\bar{u}$. Let us start with taking it constant. We see that we need that the whole coefficient in front of $\tilde{u}(1)$ should be positive, but that this coefficient depends on $\tilde{u}(1)$ if $\gamma \neq 1/3$. If we make this coefficient zero, we still have to choose the other freedom α such that the coefficient is positive, which appears to be always possible.

A common choice of boundary condition is to take a Neumann condition, that is, $\alpha = \gamma = 0$. In that case we have that this can be correct only if $\frac{1}{3}\bar{u}(1) + \frac{1}{2}\bar{u}(1) > 0$. Since we want to study the well-posedness for small $\tilde{u}$, we must have that $\bar{u}(1) > 0$ in that case. This means it should be an *outflow boundary*. It also reveals a difficulty of non-linear equations in numerical computations. During the process we need to check the sign of u, and one may even need to change the type of boundary condition if it changes sign, that is, if *back flow* occurs.

Now let us return to our time-dependent problem and study whether $\tilde{u}$, and consequently u will eventually be steady. It holds that

$$||\tilde{u}||\frac{d}{dt}||\tilde{u}|| = \frac{1}{2}\frac{d}{dt}||\tilde{u}||^2 = \frac{1}{2}\frac{d}{dt}(\tilde{u},\tilde{u})$$

$$= \left(\tilde{u}, \frac{d}{dt}\tilde{u}\right) = -a(\tilde{u},\tilde{u}) + (\tilde{u}, r(\bar{u},f)) + \tilde{u}(1)r_1(\bar{u},g).$$

For appropriate boundary conditions, $a(\tilde{u},\tilde{u}) \geq ||\tilde{u}_x||^2$ and therefore

$$||\tilde{u}||\frac{d}{dt}||\tilde{u}|| \leq -\mu||\tilde{u}_x||^2 + ||\tilde{u}||\,||r(\bar{u},f)|| + |\tilde{u}(1)|\,|r_1(\bar{u},g)|. \tag{3.45}$$

Using the Cauchy–Schwarz inequality, one can bound

$$|\tilde{u}(1)| = |\int_0^1 \tilde{u}_x dx| = |(\tilde{u}_x, 1)| \leq ||\tilde{u}_x||. \tag{3.46}$$

Combined with the Poincaré inequality in (3.45), we find

$$\frac{d}{dt}||\tilde{u}|| \leq \frac{||\tilde{u}_x||}{||\tilde{u}||}(-\mu||\tilde{u}_x|| + |r_1(\bar{u},g)|) + ||r(\bar{u},f)||$$

$$\leq \frac{||\tilde{u}_x||}{||\tilde{u}||}(-\mu\sqrt{c_p}||\tilde{u}|| + |r_1(\bar{u},g)|) + ||r(\bar{u},f)||, \tag{3.47}$$

where c_p is the appropriate Poincaré constant. Now if $\mu\sqrt{c_p}||\tilde{u}|| > |r_1(\bar{u},g)|)$, then the right-hand side will be less than $\sqrt{c_p}(-\mu\sqrt{c_p}||\tilde{u}|| + |r_1(\bar{u},g)|) + ||r(\bar{u},f)||$. This is negative if $||\tilde{u}|| \geq |r_1(\bar{u},g)|/(\mu\sqrt{c_p}) + ||r(\bar{u},f)||/c_p$. Hence, $||\tilde{u}||$ will decrease for the maximum of these two criteria, which is obviously the last one (it contains the first one), and thus eventually $\tilde{u}$ will enter a sphere with that radius, that is, eventually we have

$$||\tilde{u}|| \leq |r_1(\bar{u},g)|/(\mu\sqrt{c_p}) + ||r(\bar{u},f)||/c_p. \tag{3.48}$$

Hence, $\tilde{u}$ might be time dependent, but eventually $||\tilde{u}||$ is bounded. Note that for a homogeneous boundary condition we have $r_1(\bar{u},0) = 0$. Concluding, this gives us a sphere around $\bar{u}$ where u will eventually dwell. So, it may be a transient solution, but it will stay in the sphere.

This leads us to the choice of another special case. The steady solution of the Burgers' equation. The question is whether that solution is stable in the Dynamical Systems sense; see Section 2.3. So, we would like to know whether a perturbation of the steady solution will eventually return to the steady solution. Now we have to take into account the omitted term $(\bar{u}_x\tilde{u}, \tilde{u})$ in the bi-linear form, since in general the steady solution will not be constant. For the steady state $r(\bar{u}) = r_1(\bar{u}) = 0$. Assuming again that our boundary condition at $x = 1$ gives a non-negative contribution in $a(\tilde{u}, \tilde{u})$, we get now

$$||\tilde{u}||\frac{d}{dt}||\tilde{u}|| \le -(\bar{u}_x\tilde{u}, \tilde{u}) - \mu||\tilde{u}_x||^2 \le (-\min(\bar{u}_x) - c_p)||\tilde{u}||^2. \tag{3.49}$$

Consequently,

$$\frac{d}{dt}||\tilde{u}|| \le -(\min(\bar{u}_x) + c_p)||\tilde{u}||, \tag{3.50}$$

which shows that the steady state is stable if $\min(\bar{u}_x) > -c_p$. So, unlike the linear case, it is not true that we can prove stability of each steady state. *See example Ex. 3.3.d–g*

3.6.3 Energy Decay of Ginzburg–Landau Equation

In Section 3.5.3 we have already considered the linear problem and derived a condition for which the associated bi-linear form is coercive, and hence, for which the energy of the time-dependent problem will decay. Here we will analyse the full non-linear case (2.36), written as

$$\frac{\partial A}{\partial t} = \gamma_1 A + \gamma_2\frac{\partial^2 A}{\partial x^2} - \gamma_3 A|A|^2 \tag{3.51}$$

on [0,1] with $A(x,0) = f(x)$, $A(0,t) = 0$, and $A(1,t) = 1$. In the linear case, we created homogeneous boundary conditions by introducing $\tilde{A} = A - x$. Here, we just introduce a time-independent function B that satisfies the boundary conditions, that is, $B(0) = 0$ and $B(1) = 1$. So $\tilde{A} = A - B$. Later on, we can make various choices for this B, for example the steady-state solution of the equation. With this replacement, (3.51) turns into

$$\frac{\partial \tilde{A}}{\partial t} = \gamma_1\tilde{A} + \gamma_2\frac{\partial^2 \tilde{A}}{\partial x^2} - \gamma_3\left((\tilde{A}+B)|\tilde{A}+B|^2 - B|B|^2\right) + \gamma_1 B + \gamma_2\frac{\partial^2 B}{\partial x^2} - \gamma_3 B|B|^2$$

with

$$\tilde{A}(x,0) = f(x) - B(x); \ \tilde{A}(0,t) = 0; \ \tilde{A}(1,t) = 0.$$

Note that $(\tilde{A} + B)|\tilde{A} + B|^2 - B|B|^2$ is zero for $\tilde{A} = 0$. This leads to the following generalization of the bi-linear form,

$$a(v, u) = -\gamma_1(v, u) + \gamma_2 \left(\frac{\partial v}{\partial x}, \frac{\partial u}{\partial x}\right) + \gamma_3 \left(v, ((u + B)|u + B|^2 - B|B|^2)\right),$$

and the linear form becomes

$$F(v) = (v, r(B)), \text{ with } r(B) = \gamma_1 B + \gamma_2 \frac{\partial^2 B}{\partial x^2} - \gamma_3 B|B|^2.$$

Here the inner product is defined by $(u, v) = \int_0^1 \bar{u}v\,dx$, where the bar denotes the complex conjugate. Note that $r(B)$ is exactly the right-hand side of our equation evaluated at B. Hence, it is zero for the steady-state solution. We have to check for the positivity of

$$a(u, u) = -\gamma_1||u||^2 + \gamma_2||\frac{\partial u}{\partial x}||^2 + \gamma_3(u, [(u + B)|u + B|^2 - B|B|^2]).$$

Using

$$\begin{aligned}(u + B)|u + B|^2 - B|B|^2 &= (u + B)(|u|^2 + u\bar{B} + \bar{u}B + |B|^2) - B|B|^2 \\ &= u|u|^2 + u^2\bar{B} + |u|^2 B + u|B|^2 + B|u|^2 + u|B|^2 + \bar{u}B^2 \\ &= u|u|^2 + u^2\bar{B} + 2|u|^2 B + 2u|B|^2 + \bar{u}B^2\end{aligned}$$

in the inner product, we obtain

$$(u, (u + B)|u + B|^2 - B|B|^2) = ||u^2||^2 + 3(|u|^2, uB) + 2||uB||^2 + (u^2, B^2).$$

Suppose that $\text{Re}(\gamma_3)$ is positive; then we need a lower bound for the last expression. By the Cauchy–Schwarz inequality, the last term $(u^2, B^2) = (u\bar{B}, \bar{u}B) \geq -||\bar{u}B||^2 = -||uB||^2$ and also $(|u|^2, uB) \geq -||u^2||\,||uB||$. Next, we just look to the whole expression and find that a lower bound is

$$||u^2||^2 - 3||u^2||\,||uB|| + ||uB||^2 = \left(||u^2|| - \frac{3}{2}||uB||\right)^2 - \frac{5}{4}||uB||^2 \geq -\frac{5}{4}||B||^2||u||^2.$$

This leads to

$$a(u, u) \geq \left(\gamma_2 c_P - \gamma_1 - \text{Re}(\gamma_3)\frac{5}{4}||B||^2\right)||u||^2,$$

where c_P is the appropriate Poincaré constant, that is, $||\frac{\partial u}{\partial x}||^2 \geq c_P||u||^2$. So, if $\gamma_2 c_P - \gamma_1 - \text{Re}(\gamma_3)\frac{5}{4}||B||^2 > 0$, then $a(u, u)$ is positive and we expect a unique solution to exist.

Now consider the time-dependent case in order to investigate the stability of the steady solution in the Dynamical Systems sense, see Section 2.3. So, in particular we have $a(u, u) \geq c_p||u||^2$. This means that

$$2||\tilde{A}||\frac{d}{dt}||\tilde{A}|| = \frac{d}{dt}||\tilde{A}||^2 = \frac{d}{dt}(\tilde{A}, \tilde{A}) = (\tilde{A}, \frac{d}{dt}\tilde{A}) + \overline{(\tilde{A}, \frac{d}{dt}\tilde{A})}$$

$$= -a(\tilde{A}, \tilde{A}) + F(\tilde{A}) \overline{-a(\tilde{A}, \tilde{A}) + F(\tilde{A})}$$

$$\leq -2(\mathrm{Re}(\gamma_2)c_P - \mathrm{Re}(\gamma_1) - \mathrm{Re}(\gamma_3)\frac{5}{4}||B||^2)||\tilde{A}||^2 + 2||\tilde{A}|| \; ||r(B)||.$$

Finally, we get

$$\frac{d}{dt}||\tilde{A}|| \leq (\mathrm{Re}(\gamma_3)\frac{5}{4}||B||^2 + \mathrm{Re}(\gamma_1) - \mathrm{Re}(\gamma_2)c_P)||\tilde{A}|| + ||r(B)||.$$

So, $||\tilde{A}||$ will certainly decrease as long as the right-hand side in this inequality is negative. Thus, eventually

$$||\tilde{A}|| \leq ||r(B)||/(\mathrm{Re}(\gamma_2)c_P - \mathrm{Re}(\gamma_3)\frac{5}{4}||B||^2 - \mathrm{Re}(\gamma_1)).$$

In particular, this shows for the steady-state solution where $r(B) = 0$ that any perturbation of that will go to zero if $\mathrm{Re}(\gamma_2)c_P - \mathrm{Re}(\gamma_1) - \mathrm{Re}(\gamma_3)\frac{5}{4}||B||^2 > 0$. Hence, this shows that there are restrictions on the parameters here in order to have a stable solution. Increasing $\mathrm{Re}(\gamma_1)$ or $\mathrm{Re}(\gamma_3)$ may lead to unstable solutions.

To study the boundedness of the solution, we can choose B such that $||B||$ is arbitrarily small; for example, $B = x^n$ gives norm $||B|| = 1/(2n + 1)$. However, $||r(B)||$ will be dominated by the norm of the second derivative of B which is about $\mathrm{Re}(\gamma_2)n\sqrt{n/2}$. Of course, we would like to keep this as small as possible. The proper choice of n relies on the magnitude of $\mathrm{Re}(\gamma_2)c_P - \mathrm{Re}(\gamma_1)$. Once that is set, we can compute $||r(B)||$, which will give the sphere around B on or in which A will enter eventually, but it still may be transient then.

3.6.4 *Energy Decay and Boundary Conditions of Convection–Diffusion in an Incompressible Fluid*

In this section, we consider the well-posedness of the convection and diffusion of any material dissolved in an incompressible fluid, where it is assumed that the material itself has no influence on the fluid flow. Crucial will again be the positivity of the bi-linear form associated to the right-hand side and boundary conditions. This will show us which conditions will lead to a well-posed problem. In addition, we will also be able to say something about the magnitude of the solution and the stability of the steady state. Hence, as such it is a further application of the theory developed in Section 3.5.3. Moreover, the same part will re-occur when we consider the quest of the well-posedness of the Navier–Stokes equations.

Additional Material

- The *proofs* of the lemmas and theorems in this section and the following ones in this chapter can be found in Appendix A.

Consider the convection–diffusion equation

$$\frac{\partial \phi}{\partial t} = -\mathcal{C}\phi + \mu \Delta \phi + f, \tag{3.52}$$

where

$$\mathcal{C}\phi = \nabla \cdot (\phi \mathbf{u}). \tag{3.53}$$

Suppose we have a general Robin condition $\alpha\phi - \beta\frac{\partial\phi}{\partial n} = b$ on part of the boundary Γ_R and a *homogeneous* Dirichlet condition $\phi = 0$ elsewhere. Then, a bi-linear form associated to (3.52) is

$$a(\psi, \phi) = -(\psi, -\mathcal{C}\phi + \mu\Delta\phi)_\Omega - (\psi, \alpha\phi - \beta\frac{\partial\phi}{\partial n})_{\Gamma_R}.$$

Considering $a(\phi, \phi)$, it turns out that $(\phi, \mathcal{C}\phi)$ yields only a contribution on the boundary. In fact, the following lemma holds.

Lemma 3.1 *If $\nabla \cdot \mathbf{u} = 0$ and $\phi = 0$ on Γ/Γ_R, then for positivity one has to check*

$$a(\phi, \phi) = \mu||\nabla\phi||_\Omega^2 - \left(\phi, \left(-\frac{1}{2}\mathbf{u} \cdot \mathbf{n} + \alpha\right)\phi + (\mu - \beta)\frac{\partial\phi}{\partial n}\right)_{\Gamma_R}. \tag{3.54}$$

Corollary 3.2 *If Γ_R is non-void, then one finds well-posedness for $\beta = \mu$ and*

1. *$\alpha = 0$ at an outflow boundary ($\mathbf{u} \cdot \mathbf{n} \geq 0$), so a Neumann condition at outflow, or*
2. *a Robin condition $\alpha\phi - \mu\frac{\partial\phi}{\partial n} = b$ with $(\frac{1}{2}\mathbf{u} \cdot \mathbf{n} - \alpha) \geq 0$.*

Observe that the latter allows a Robin condition at inflow if α is sufficiently negative.

Lemma 3.2 *If $\nabla \cdot \mathbf{u} = 0$, then the change of energy of the system is given by*

$$\frac{d}{dt}(\phi, \phi)_\Omega = -2a(\phi, \phi) - 2(b, \phi)_{\Gamma_R} + 2(f, \phi)_\Omega. \tag{3.55}$$

Corollary 3.3

1. *In the absence of viscosity, that is, $\mu = 0$, and sources and sinks, $f \equiv 0$, energy only changes by incoming or outgoing fluxes over the boundary. In the case where the normal velocity at the boundaries and α are zero, the change is zero.*

2. *In the absence of sources and sinks, that is, $f \equiv 0$ and $b = 0$, the energy can only decrease when the boundary conditions of Corollary 3.1 are applied.*

Now still interesting questions remain. For instance, by what speed will the energy decay if the given velocity fields originate from a flow in a closed box, that is, normal velocities are zero at the walls, in the absence of sinks and sources in the transport equation? And can the energy grow unboundedly if f is not zero?

As will be clear, the first question is the same as asking whether the boundary conditions are such that the Laplace operator is coercive. The most difficult part of this question is the existence of a Poincaré inequality, which leads to the following theorem.

Theorem 3.4 *Suppose Γ_R is void, that is, $\phi = 0$ on the boundary everywhere; then there exists a $c_P > 0$ such that $(\nabla\phi, \nabla\phi)_\Omega \geq c_P(\phi, \phi)_\Omega$ and on a closed domain*

$$\frac{d}{dt}||\phi||_\Omega \leq -c_P\mu||\phi||_\Omega + ||f||_\Omega. \tag{3.56}$$

Corollary 3.4

1. *If $f \equiv 0$, then the energy decreases with speed $c_P\mu$.*
2. *Eventually, $||\phi||_\Omega \leq ||f||_\Omega/(c_P\mu)$.*
3. *The steady state of (3.52), with possibly in-homogeneous boundary conditions, is stable.*

The last statement follows simply from linearity of the equations. If $\bar{\phi}$ is a steady-state solution, then, due to linearity, $\tilde{\phi} = \phi - \bar{\phi}$ satisfies the homogeneous variant of the equation, which brings us back to case 1 of Corollary 3.4.

This shows that for the case of homogeneous Dirichlet boundary conditions, without any forcing present, the energy is decreasing monotonically to zero. And since this energy is a norm also ϕ will go to zero. Furthermore, in the case where we have in-homogeneous Dirichlet boundary conditions and some forcing we have that any perturbation of a steady state will vanish. In the next theorem, we consider more general boundary conditions as stated in Theorem 3.1 and Corollary 3.2.

Theorem 3.5 *Let Γ_R be non-void, $\beta = \mu$, $|(\phi, b)_{\Gamma_R}| \leq c_{b,0}||\phi||_\Omega + c_{b,1}||\nabla\phi||_\Omega$, and the conditions satisfy one of those in Corollary 3.2. Moreover, we have a Dirichlet condition $\phi = 0$ on Γ/Γ_R. Then*

$$\frac{d}{dt}||\phi||_\Omega \leq \sqrt{\frac{a(\phi,\phi)}{||\phi||_\Omega^2}}(-\sqrt{a(\phi,\phi)} + \frac{c_{b,1}}{\sqrt{\mu}}) + ||f||_\Omega + c_{b,0}$$

and eventually $||\phi||_\Omega \leq \max\left(\frac{c_{b,1}}{\sqrt{\mu c_c}}, \left(c_{b,1}\sqrt{\frac{c_c}{\mu}} + ||f||_\Omega + c_{b,0}\right)/c_c\right)$, where $c_c > 0$ is the constant in the bound $a(\phi,\phi)_\Omega \geq c_c||\phi||_\Omega^2$.

Corollary 3.5

1. *If there exists a bounded function $\bar{\phi}$ which satisfies the inhomogeneous Dirichlet boundary conditions, then the perturbation $\tilde{\phi} = \phi - \bar{\phi}$ is bounded and hence ϕ is bounded.*
2. *The steady state of (3.52) with possibly also an inhomogeneous Dirichlet boundary condition is stable.*

The last theorem shows the boundedness of the solution for quite general boundary conditions and hence its stability. It also shows that for homogeneous boundary conditions and zero forcing the solution will go to zero.

Remark The bound $|(\phi, b)_{\Gamma_R}| \leq c_{b,0}\|\phi\|_\Omega + c_{b,1}\|\nabla\phi\|_\Omega$ can be thought of as originating from Gauss' theorem applied to $(\phi, \mathbf{v} \cdot \mathbf{n})$, where $\mathbf{v} \cdot \mathbf{n} = b$ on Γ_R and arbitrary on Γ / Γ_R. Also, inside the domain Ω $\mathbf{v}$ is arbitrary. Recalling that $\phi = 0$ on Γ / Γ_R, this yields

$$|(\phi, b)_{\Gamma_R}| = |(\phi, \mathbf{v} \cdot \mathbf{n})_\Gamma| = |\int_\Omega \nabla \cdot \phi \mathbf{v} d\Omega|$$
$$= |(\phi, \nabla \cdot \mathbf{v})_\Omega + (\nabla\phi, \mathbf{v})_\Omega|$$
$$\leq \|\nabla \cdot \mathbf{v}\|_\Omega \|\phi\|_\Omega + \|\mathbf{v}\|_\Omega \|\nabla\phi\|_\Omega,$$

where in the third step we used (3.35). Next $\mathbf{v}$ should be chosen such that $\|\nabla \cdot \mathbf{v}\|_\Omega$ and $\|\mathbf{v}\|_\Omega$ are small. One could consider minimizing

$$\min_{\{\mathbf{v}|\mathbf{v}\cdot\mathbf{n}=b \text{ on } \Gamma_R\}} \|\nabla \cdot \mathbf{v}\|_\Omega^2 + \|\mathbf{v}\|_\Omega^2.$$

In general, such a minimum will exist, say $\hat{\mathbf{v}}$, and then one could pick $c_{b,0} = \|\nabla \cdot \hat{\mathbf{v}}\|_\Omega$ and $c_{b,1} = \|\hat{\mathbf{v}}\|_\Omega$. It is clear from the minimization that the coefficients will tend to zero if b tends to zero, which corresponds with the term being absent if $b = 0$.

Due to the linearity of the convection–diffusion equation, the theorems in this section show us that the initial value problem is stable and hence well-posed.

3.6.5 *Energy Decay and Boundary Conditions Incompressible Navier–Stokes Equations*

The non-linearity and the fact that we have to deal with a system of PDEs make the Navier–Stokes equations slightly more complicated to analyze than the convection–diffusion equation (3.52). In terms of $\mathcal{C}$ defined in (3.53), the equations can be written as

$$\frac{\partial}{\partial t}\mathbf{u} = (I \otimes (-\mathcal{C} + \mu\Delta))\mathbf{u} - \nabla p + \mathbf{f},$$
$$0 = \mathcal{C}1. \tag{3.57}$$

Here $\otimes$ is just denoting the Kronecker notation, by which we can indicate that $(-\mathcal{C} + \mu\Delta)$ is applied to each component of the velocity $\mathbf{v}$ separately. The 1 in the continuity equation can be thought of as the constant density we have in this incompressible case.

We keep the boundary condition general of the form $B(\mathbf{u}, p) = \mathbf{b}$, with property $B(0, 0) = 0$. Furthermore, we introduce the projection $(I - \mathbf{n}\mathbf{n}^T)$ which projects a vector to the space orthogonal to $\mathbf{n}$.

This means that the weak form of the equation is given by $a(\mathbf{u}, \mathbf{v}) = F(\mathbf{v})$, where

$$a(\mathbf{u}, \mathbf{v}) = -((I \otimes (-\mathcal{C} + \mu\Delta))\mathbf{u} - \nabla p, \mathbf{v})_\Omega - (B(\mathbf{u}, p), \mathbf{v})_\Gamma \tag{3.58a}$$
$$F(\mathbf{v}) = (\mathbf{f}, \mathbf{v})_\Omega - (\mathbf{b}, \mathbf{v})_\Gamma. \tag{3.58b}$$

The expression (3.58a) turns into the form we want to study for positiveness if we take $\mathbf{v} = \mathbf{u}$ and can be simplified to the one in the next lemma.

Lemma 3.3 *With the general boundary conditions as just defined, the expression*

$$a(\mathbf{u}, \mathbf{u}) = (|\mathbf{u}|^2/2 + p, \mathbf{u} \cdot \mathbf{n})_\Gamma + \mu ||(I \otimes \nabla)\mathbf{u}||_\Omega^2 - \mu(\frac{\partial}{\partial n}\mathbf{u}, \mathbf{u})_\Gamma$$
$$-(B(\mathbf{u}, p), \mathbf{u})_\Gamma \tag{3.59}$$

has to be checked on positivity.

Theorem 3.6 *The following boundary conditions make (3.59) non-negative:*

$$\mathbf{u} = 0 \text{ (no-slip wall)} \tag{3.60a}$$

$$\mathbf{u} \cdot \mathbf{n} = 0, \ \frac{\partial}{\partial n}(I - \mathbf{n}\mathbf{n}^T)\mathbf{u} = 0 \text{ (slip wall)} \tag{3.60b}$$

$$(p - \mu\nabla\mathbf{u}) \cdot \mathbf{n} = \mathbf{b} \text{ (pressure boundary condition)}. \tag{3.60c}$$

Note that if (3.60b) and/or (3.60c) are prescribed on the entire boundary, then the level of the velocity tangential to the wall is not fixed for the former and the level of the whole velocity vector is not fixed for the latter. So, on part of the boundary one needs other conditions that fix the remaining freedoms.

Other boundary conditions are possible. For instance, we can have

$$\mathbf{u} \cdot \mathbf{n} = 0, (I - \mathbf{n}\mathbf{n}^T)\mathbf{u} = \mathbf{c} \text{ (sliding wall)}, \tag{3.61}$$

where $\mathbf{n} \cdot \mathbf{c} = 0$; for details see, for example, Nordström (2022). We like to know whether this boundary condition will result in conditions stating that a steady-state solution is only conditionally stable.

The analysis gets more complicated because the unknowns are also in the operator $\mathcal{C}$. Let us just re-name $\mathcal{C}$ to $\mathcal{C}(\mathbf{u})$ and note that it is also linear in this argument so $\mathcal{C}(\mathbf{u} + \mathbf{v}) = \mathcal{C}(\mathbf{u}) + \mathcal{C}(\mathbf{v})$.

Let $(\bar{\mathbf{u}}, \bar{p})$ be a function independent of t of the equations satisfying the boundary conditions and the divergence equation, that is, $\mathcal{C}(\bar{\mathbf{u}})1 = 0$. Replacing in the Navier–Stokes equation $\mathbf{u} = \bar{\mathbf{u}} + \tilde{\mathbf{u}}$ and $p = \bar{p} + \tilde{p}$, we end up with the equation

$$
\begin{aligned}
\frac{\partial}{\partial t}\tilde{\mathbf{u}} = &-(I \otimes \mathcal{C}(\tilde{\mathbf{u}}))\tilde{\mathbf{u}} - (I \otimes \mathcal{C}(\tilde{\mathbf{u}}))\tilde{\mathbf{u}} - (I \otimes \mathcal{C}(\tilde{\mathbf{u}}))\bar{\mathbf{u}} + \mu(I \otimes \Delta)\tilde{\mathbf{u}} \\
&+ \nabla\tilde{p} + \mathbf{g}, \\
0 = &\ \mathcal{C}(\tilde{\mathbf{u}})1,
\end{aligned}
\tag{3.62}
$$

where

$$
\mathbf{g} = \mathbf{f} + [I \otimes (-\mathcal{C}(\bar{\mathbf{u}}) + \mu\Delta)]\bar{\mathbf{u}} + \nabla\bar{p}.
$$

This equation shows some extra linear terms entering $a(\cdot, \cdot)$. Moreover, $\mathbf{g}$ is the residual occurring after substituting $(\bar{\mathbf{u}}, \bar{p})$ into the right-hand side of the momentum equations of (3.57).

We only consider the closed wall case for analyzing the energy and a bound on $||\tilde{\mathbf{u}}||_{\Omega}$. Other cases can be treated in a similar way; see Straughan (2004) for further results.

For the following, we introduce $U = \max_{\Omega} |\bar{\mathbf{u}}|$, that is, the maximum velocity over the domain. Let us define a Reynolds number based on the constant c_P holding for the Poincaré inequality defined in the following theorem: $\mathrm{Re}_{c_P} \equiv \frac{U}{\mu\sqrt{c_P}}$.

Theorem 3.7 *On a domain with closed walls, let on Γ_{B_t} the normal derivative of the tangential velocities be zero (slip wall) and on its complement the tangential velocity be prescribed (sliding wall), making $\tilde{\mathbf{u}} = 0$. Furthermore, the following Poincaré inequality holds: $||(I \otimes \nabla)\mathbf{u}||^2_{\Omega} \geq c_P||\mathbf{u}||^2_{\Omega}$. Then the energy of a perturbation $\tilde{\mathbf{u}}$ is bounded by*

$$
\frac{d}{dt}||\tilde{\mathbf{u}}||_{\Omega} \leq -\mu||(I \otimes \nabla)\tilde{\mathbf{u}}||_{\Omega}\left(\frac{||(I \otimes \nabla)\tilde{\mathbf{u}}||_{\Omega}}{||\tilde{\mathbf{u}}||_{\Omega}} - \frac{U}{\mu}\right) + ||\mathbf{g}||_{\Omega}.
\tag{3.63}
$$

This energy can be bounded by

$$
\frac{d}{dt}||\tilde{\mathbf{u}}||_{\Omega} \leq -\mu||(I \otimes \nabla)\tilde{\mathbf{u}}||_{\Omega}\left(\sqrt{c_P} - \frac{U}{\mu}\right) + ||\mathbf{g}||_{\Omega}.
\tag{3.64}
$$

If $\mathrm{Re}_{c_P} < 1$, then this can be bounded by

$$
\frac{d}{dt}||\tilde{\mathbf{u}}||_{\Omega} \leq -\mu c_P||\tilde{\mathbf{u}}||_{\Omega}(1 - \mathrm{Re}_{c_P}) + ||\mathbf{g}||_{\Omega}.
\tag{3.65}
$$

Corollary 3.6

1. *If all boundary conditions are homogeneous, then we can take $\bar{\mathbf{u}} = 0$, so $U = 0$. Hence, in that case we have that eventually the solution is bounded by $\|\mathbf{u}\|_\Omega \le \min_s \|\mathbf{f} - \nabla s\|_\Omega / (\mu \sqrt{c_P})$ and hence will go to zero for a conservative forcing field.*
2. *If $Re_{cp} < 1$, then $\tilde{\mathbf{u}}$ is bounded and eventually*

$$\|\tilde{\mathbf{u}}\|_\Omega \le \|\mathbf{g}\|_\Omega / (\mu c_P (1 - Re_{cp})). \tag{3.66}$$

3. *If $\bar{\mathbf{u}}$ is the steady Stokes solution, that is, the solution satisfying the steady equations without the non-linear terms, then $\mathbf{g} = -(I \otimes \mathcal{C}(\bar{\mathbf{u}}))\bar{\mathbf{u}}$ and eventually the perturbation can be bounded by (3.66), but it does not necessarily become stationary.*
4. *If $\bar{\mathbf{u}}$ is a steady solution of the equation, then $\mathbf{g} = 0$ and from (3.65) we have that it is provably stable for any perturbation if $Re_{cp} < 1$.*

The first result (3.63) in the theorem is also interesting for what it means for the magnitude derivatives can have eventually. It indicates that as long as $\frac{\|(I \otimes \nabla)\tilde{\mathbf{u}}\|_\Omega}{\|\tilde{\mathbf{u}}\|_\Omega} > \frac{U}{\mu}$), we have decrease. Eventually, decrease will possibly stop if $\frac{\|(I \otimes \nabla)\tilde{\mathbf{u}}\|_\Omega}{\sqrt{c_P}\|\tilde{\mathbf{u}}\|_\Omega} < Re_{cp}$. This limits the magnitude of the derivatives of the perturbation with respect to that of the magnitude of the perturbation itself. So, if the perturbation goes to zero, then the derivative will too.

The fourth result is relevant for the lid-driven cavity problem described in Section 1.2. It also shows that the solution might become unstable when the speed of the lid becomes too high.

An example of a conservative forcing field is the gravity field. So, stirring coffee in a closed cup and leaving it for a while results in a fluid at rest.

3.6.6 Differentially Heated Cavity

In this case the lateral walls are heated; see Section 1.5. The problem is governed by the Boussinesq equations, which are a combination of the incompressible Navier–Stokes equation (3.57) and (3.52) extended with a feedback of the temperature on the Navier–Stokes part:

$$\begin{aligned}
\frac{\partial}{\partial t}\mathbf{u} &= (I \otimes (-\mathcal{C} + \mu\Delta))\mathbf{u} - \nabla p + \mathbf{f} + \alpha g T \mathbf{e}_z, \\
0 &= \mathcal{C}1, \\
\frac{\partial T}{\partial t} &= -\mathcal{C}T + \mu\Delta T + f_T,
\end{aligned} \tag{3.67}$$

where $\alpha > 0$. The energy we consider here is $\|\mathbf{u}\|_\Omega^2 + \gamma^2 \|T\|_\Omega^2$. This means that we have

$$a([\mathbf{u};T],[\mathbf{u};T]) = a_{NS}(\mathbf{u},\mathbf{u}) + \gamma^2 a_{CD}(T,T) + \alpha g(\mathbf{u}, T\mathbf{e}_z),$$

where a_{NS} is given by (3.59) and a_{CD} by (3.55).

The boundary conditions that can be used here are the same as those for the incompressible Navier–Stokes equations and for the convection–diffusion equation.

Let us study now in more detail the stability of a steady solution the problem for closed no-slip walls.

Theorem 3.8 *Let* $\nu \equiv \min(\mu, \kappa)$ *and* $Re_{cP} \equiv U/(\nu\sqrt{c_P}) < 1$; *then*

$$\frac{d}{dt}||[\tilde{\mathbf{u}}; \gamma\tilde{T}]||_\Omega \leq -\nu c_P ||[\tilde{\mathbf{u}}; \gamma\tilde{T}]||_\Omega (1 - Re_{cP})$$

$$+ \frac{\alpha g}{|\gamma|}||[\tilde{\mathbf{u}}, \gamma\tilde{T}]||_\Omega + \max(||g_T||_\Omega, ||\mathbf{g}||_\Omega)$$

$$\leq \left(-\nu c_P (1 - Re_{cP}) + \frac{\alpha g}{|\gamma|}\right) ||[\tilde{\mathbf{u}}, \gamma\tilde{T}]||_\Omega + \max(||g_T||_\Omega, ||\mathbf{g}||_\Omega).$$

This theorem shows that the solution will eventually be bounded if

$$-\nu c_P (1 - Re_{cP}) + \frac{\alpha g}{|\gamma|} < 0.$$

It also shows that under this condition the steady state is stable. Note that in this case a conduction only, without a flow, is not a steady state. This makes $g_T = 0$, but since $\nabla \times \bar{T}\mathbf{e}_z = [\bar{T}_y; -\bar{T}_x; 0]$ will not be zero for a linearly increasing T, there exists no $\bar{p}$ such that $\nabla\bar{p} = \alpha g\bar{T}\mathbf{e}_z$; hence $\bar{\mathbf{u}}$ must be non-zero to make $\mathbf{g}$ zero.

There is still a parameter γ which should be fixed. It is a parameter by which one can weigh velocity and temperature differently in the norm $||[\tilde{\mathbf{u}}; \gamma\tilde{T}]||$, which may be necessary if the average values of velocities and temperature, respectively, differ greatly.

3.6.7 Rayleigh–Bénard Problem

A special case is the *Rayleigh–Bénard* convection problem; see Section 1.4, where T_H and T_C are the temperatures of the lower and upper horizontal plates of the container, respectively, and d is the vertical height of the container. Usually, the temperature of the dynamical system is expressed as a sum of the linear profile of the conduction state and the temperature fluctuation $\theta(x, t)$, such as $T(x, t) = T_H - \frac{T_H - T_C}{d}z + \theta(x, t)$. Then the Boussinesq equations turn into

$$\frac{\partial}{\partial t}\mathbf{u} = (I \otimes (-\mathcal{C} + \mu\Delta))\mathbf{u} - \nabla p + \mathbf{f} + \alpha g\theta\mathbf{e}_z,$$
$$0 = \mathcal{C}1, \tag{3.68}$$
$$\frac{\partial\theta}{\partial t} = -\mathcal{C}\theta + \kappa\Delta\theta + \frac{T_H - T_C}{d}(\mathbf{e}_z \cdot \mathbf{u}),$$

where $\mathbf{f} = \alpha g\left(T_H - \frac{T_H - T_C}{d}z\right)\mathbf{e}_z$. Since the rotation of $\mathbf{f}$ is zero, we can incorporate $\mathbf{f}$ in p, that is, we introduce $\hat{p} = p - \alpha g(T_H z - \frac{T_H - T_C}{2d}z^2)$. The subtracted part is called the hydrostatic pressure, hence, $\hat{p}$ is the deviation of that. The energy we consider here is $|\mathbf{u}|^2 + \gamma^2 T^2$, where $\gamma^2 > 0$. This means that we have

$$
a([\mathbf{u}; \theta], [\mathbf{u}; \theta]) = a_{NS}(\mathbf{u}, \mathbf{u}) + \gamma^2 a_{CD}(\theta, \theta) - \alpha g(\mathbf{u}, \theta \mathbf{e}_z) - \gamma^2 \frac{T_H - T_C}{d} (\theta, \mathbf{e}_z \cdot \mathbf{u})
$$

$$
= a_{NS}(\mathbf{u}, \mathbf{u}) + \gamma^2 a_{CD}(\theta, \theta) - \left(\alpha g + \gamma^2 \frac{T_H - T_C}{d} \right) (\theta, \mathbf{e}_z \cdot \mathbf{u}),
$$

where a_{NS} is given by (3.59) and a_{CD} by (3.55).

The structure of this expression is the same as for the differentially heated cavity. This leads to bounded solutions if

$$
-\nu c_P(1 - Re_{c_P}) + \frac{\alpha g + \gamma^2 \frac{T_H - T_C}{d}}{|\gamma|} < 0.
$$

Moreover, steady states are stable in this regime, especially the zero solution which is the conduction solution.

3.7 Summary

- For linear equations well-posedness leads to classes of PDEs, where each PDE in a certain class is equivalent to one normal form PDE. If we know the boundary and initial conditions which make this normal form PDE well-posed, we know how to get these conditions for all PDEs in the class.
- For first- and second-order PDEs, it is important to distinguish the classes of elliptic, parabolic, and hyperbolic PDEs. Note that every second-order (system of) PDE(s) can be re-written into a system of first-order PDEs.
- Projections allow one to approximate a PDE on a finite-dimensional space as well as approximating a finite-dimensional problem by a problem of lower dimension.
- A linear PDE and its linear boundary conditions can formally be written in the form $\mathcal{A}u = \text{f}$, where u belongs to a certain sub-space; for example, if we have $\mathcal{L}u = f$ on Ω and $\mathcal{B}u = g$ on Γ_C and $u = 0$ on Γ_D with $\Gamma_D \oplus \Gamma_C = \Gamma$, then $\mathcal{A}$ is implicitly defined on a function space, called $\mathcal{V}$, of functions u, v all being zero on Γ_D and $(v, \mathcal{A}u)_\Omega = (v, \mathcal{L}u)_\Omega + \alpha(v, \mathcal{B}u)_{\Gamma_C}$, where α should be chosen such that we can compute the adjoint of $\mathcal{A}$ indicated by $\mathcal{A}^*$ and defined by $(\mathcal{A}^*v, u)_\Omega = (v, \mathcal{A}u)_\Omega$. Consistent with the definition of $\mathcal{A}$, f is implicitly defined by $(v, \text{f})_\Omega = (v, f)_\Omega + \alpha(v, g)_{\Gamma_C}$. The weak form of the problem can now be written as:
  Find $u \in \mathcal{V}$ such that $(v, \mathcal{A}u - \text{f})_\Omega = 0$ for all $v \in \mathcal{V}$.

- If $\mathcal{A}$ is self-adjoint and coercive (a slightly stronger statement than positive definiteness) on the space $\mathcal{V}$, then one can also find the weak form from a minimization. For this, one defines the functional $J(u) = \frac{1}{2}(u, \mathcal{A}u)_\Omega - (u, \mathrm{f})_\Omega$ and one should find $\min_{u \in \mathcal{V}} J(u)$.
- By partial integration of the bi-linear form one can distribute the derivatives over both functions, that is, $(v, \mathcal{A}u)_\Omega = (\mathcal{A}_1 u, \mathcal{A}_2 v)_\Omega + (\cdot, \cdot)_{\Gamma_C}$. By doing this, one observes that the actual order of differentiation for the solution space can be taken lower, allowing one to find a solution for a wider class of forcings f and g. The function space is the Sobolev space H^k defined by functions which are square integrable up to the kth derivative.
- The weak form is well-posed if it satisfies the Lax–Milgram theorem. The most important part is the coercivity condition requiring that $(u, \mathcal{A}u)_\Omega \geq C||u||^2_{H^k}$, for some $C > 0$ where k is the integer equal or immediately bigger than half the order of $\mathcal{A}$.
- In many cases, it is difficult to show that $(v, \mathcal{A}u)_\Omega \geq c||u||^2_{H^0} = c||u||^2_\Omega$. If $\mathcal{A}$ is a second-order operator, one can in general show that $(u, \mathcal{A}u)_\Omega \geq c_1||\nabla u||^2_\Omega$. To bound the latter from below, one needs the Poincaré inequality saying that $||\nabla u||^2_\Omega \geq c_P||u||^2_\Omega$ if on a non-empty part of the boundary we have that $u = 0$. Here, $c_P > 0$ depends on the domain size and the length of the boundary where $u = 0$.
- For linear equations the superposition property holds, and therefore stability of a stationary zero solution means stability of any stationary solution. For non-linear equations this property does not hold, and a zero solution may be stable while a non-zero solution is stable only when it satisfies certain conditions.
- Energy decay in time-dependent equations is related to coercivity. Even in non-linear cases one can look to coercivity, and this leads to boundary conditions that make the time-dependent PDE well-posed.

3.8 Exercises

Exercise 3.1 *Consider the problems*

$$\begin{bmatrix} 1 & -1 \\ -1 & 1 \end{bmatrix} \begin{bmatrix} x_1 \\ x_2 \end{bmatrix} = \begin{bmatrix} 1 \\ -1 \end{bmatrix}, \quad \begin{bmatrix} 1 & -1 \\ -1 & 1 \end{bmatrix} \begin{bmatrix} x_1 \\ x_2 \end{bmatrix} = \begin{bmatrix} 1 \\ 1 \end{bmatrix},$$

$$\begin{bmatrix} 1 & -1 \\ -1 & 0 \end{bmatrix} \begin{bmatrix} x_1 \\ x_2 \end{bmatrix} = \begin{bmatrix} 1 \\ 1 \end{bmatrix}, \quad \begin{bmatrix} 1+\varepsilon & -1 \\ -1 & 1+\varepsilon \end{bmatrix} \begin{bmatrix} x_1 \\ x_2 \end{bmatrix} = \begin{bmatrix} 1 \\ 1 \end{bmatrix}.$$

Which of these problems are not well-posed? What is the flaw and is it curable?

Exercise 3.2 *Consider $u_t = -au_x$ for $x, t > 0$.*

a. *Let $a = 1$, $u(x, 0) = 0$, and $u(0, t) = 1$. Make a sketch of the solution for $t = 0, 1, 2$ and explain by that the properties of the solution.*

b. *Consider a river with a length of 100 km and $a = 2$ m/s. Here, u denotes a pollutant dissolved in the water. Starting at $t = 0$, we have a discharge at $x = 20$ km, leading to a concentration of the following form.*

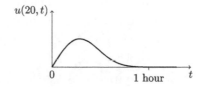

Make a picture of what you will observe at $x = 80$ km starting at $t = 0$?

Exercise 3.3 *The linearized shallow-water equations are given by*

$$u_t = -\bar{u}u_x - g\zeta_x, \tag{3.69}$$

$$\zeta_t = -\bar{u}\zeta_x - Hu_x, \tag{3.70}$$

where H is the depth, g is the gravity constant, u the velocity, ζ the wave height, and $\bar{u}$ the average flow speed. The shallow-water equations hold for waves with wavelengths much longer than the depth (hence, the word shallow), such as tidal waves in the North Sea.

a. *Transform the linearized shallow-water equations to diagonal form and determine which boundary conditions have to be applied.*

b. *Which boundary conditions have to be imposed in order to let any initial condition vanish from the domain without any reflection?*

c. *Let $\bar{u} = 1$ m/s, $H = 10$ m, $g = 10$ m/s^2, and the length of the domain be 1 km. Using the boundary conditions of the previous part, when will the solution (u, ζ) become identically equal to zero for any initial condition?*

d. *Define the inner product $(u, v) = \int_0^L uv\,dx$ and consider the norm $\sqrt{\|u\|^2 + \alpha\|\zeta\|^2}$, with $\alpha > 0$. For which value of α does the time derivative of this norm depend only on values at the boundary?*

e. *Let the new variables that occur in the diagonal form of part a be denoted by R_1 and R_2. Check that the norm $\sqrt{\|R_1\|^2 + \beta\|R_2\|^2}$, with $\beta > 0$, has only contributions on the boundary for all β. Express R_1 and R_2 in u and ζ and show that the norm is equivalent to the norm in the previous part.*

f. *Show that the previous step allows one to show that for the sub-critical case one may prescribe ζ at an inflow boundary; for example $\bar{u} > 0$ at the left boundary with $\bar{u} < \sqrt{gH}$.*

g. *Do the boundary conditions derived in parts a, b, and c cause the time derivative of the expression in part e to be non-positive?*

h. *Give a second-order scalar variant of the linearized shallow-water equations with $\bar{u} = 0$ and observe that it is also hyperbolic according to the definition for second-order scalar PDEs.*

Exercise 3.4 *Consider the PDEs: (i) $u_{xy} = f$, (ii) $u_{xx} + 2u_{yy} + 2u_{xy} = f$, and (iii) $a(x, y)u_{xx} + 2b(x, y)u_{xy} + c(x, y)u_{yy} = f(x, y, u, u_x, u_y)$.*

a. *For these PDEs, determine the matrix J as defined in Section 3.3.1 and deduce from that the type.*

b. *Suppose we have a square domain. Where should the boundary conditions be imposed for the respective equations?*

c. *Next, suppose that these boundary conditions are of Dirichlet type. Which of the preceding have a self-adjoint operator as defined in Section 3.5.1?*

Exercise 3.5 *Show that discretizing the inner product $(u, v) \equiv \int_0^1 u(x)v(x)dx$ defined on a function space using the composite midpoint rule leads to a factor times the standard inner product for vectors.*

Exercise 3.6

a. *Show that the entries of the symmetric matrix (3.10) are given by $(V(x), V(x))_{ij} = \int_0^1 x^{i+j-2}dx$. Next, using (3.10), compute the projection of $\exp(x)$ on the space V.*

b. *Set up the system which will give the projection of the function $e^{\sin(\pi x)}$ on $[0, 1]$ on the space $\mathrm{span}\{1, \sin(\pi x), \cos(\pi x)\}$, using the same inner product as before. To actually solve it you can use a package for symbolic computation, for example Mathematica or Maple.*

Exercise 3.7 *Consider $-u_{xx} = (\sin(2\pi(x - 1/4)))^2$ on $[0, 1]$ with periodic boundary conditions where f is periodic. Determine the system that follows from the Galerkin projection of this problem on the space defined by $V(x) = [\sin(2\pi x), \cos(2\pi x)]$.*

Exercise 3.8 *Let $\delta(x)$ be the Dirac δ-function defined by*

$$\delta(x) = \begin{cases} `\infty` & at \ x = 0 \\ 0 & otherwise \end{cases}$$

such that $\int_{-\varepsilon}^{\varepsilon} \delta(x)dx = 1$ for $\varepsilon > 0$ and small.

a. *Is this a function?*

b. *Let $f(x) = \delta\left(x - \frac{1}{2}\right)$. Evaluate $\int_0^1 vf dx$ in (3.19). What should be the conclusion about the difference between (3.19) and (3.20)?*

c. *In 1D, the Dirac delta function can be seen as the limit of ϵ to zero of the block function:
   $h_\epsilon(x) = 1/\epsilon$ for $|x| \le \epsilon/2$ and zero elsewhere. What is the corresponding function in
   2D?*
d. *Consider Section 3.5.1. How does one pick a Dirac delta function in the 2D case to get
   the identity $(u, v)_\Gamma = (u, v\delta_\Gamma)_\Omega$.*
e. *Use the previous part to show that one can write $\mathcal{A} = \mathcal{L} + \delta_{\Gamma_2}\mathcal{B}$.*

Exercise 3.9 *Define the inner product $\int_0^1 uvdx$.*

a. *Show by partial integration that the adjoint of the difference operator d/dx without
   boundary conditions is the operator $-d/dx$ on the domain and the boundary operator
   is minus and plus identity at 0 and 1, respectively.*
b. *Derive the adjoint operator of $(d/dx)^2$. Next, distribute the boundary conditions such
   that we can define $\mathcal{A}$ and $\mathcal{A}^*$ in the prescribed form.*
c. *Show, by changing the order of integration, that the integral operator as defined in
   Section 3.5.1 has the adjoint $\int_x^1 u(t)dt$.*
d. *Which of the operators in parts a, b, and c are skew or self-adjoint? Show that the
   integral operator is a skew-adjoint operator if the average of the functions it acts on is
   zero, that is, $\int_0^1 u(x)dx = 0$.*
e. *Show that (3.12) defines a self-adjoint operator. Hint: use (3.36).*

Exercise 3.10 *Consider the functions $J_1(\mathbf{x}) = \|\mathbf{x} - \mathbf{z}\|_A$, $J_2(\mathbf{x}) = \|\mathbf{b} - A\mathbf{x}\|^2$, and $J_3(\mathbf{x}) = \frac{1}{2}(\mathbf{x}, A\mathbf{x}) - (\mathbf{b}, \mathbf{x})$ with $A \in R^{n \times n}$ symmetric positive definite and $\mathbf{x} \in R^n$. Moreover, let $\mathcal{V}$ be
an m-dimensional sub-space of R^n.*

a. *Determine the minimum of these functions for $\mathbf{x} \in \mathcal{V}$ by using (i) the directional derivative, (ii) by requiring $\nabla \hat{J}(\mathbf{x}) = 0$. Show that this leads to a solution $\mathbf{x}_\mathcal{V} \in \mathcal{V}$ satisfying
   $(\mathbf{z} - \mathbf{x}_\mathcal{V}, A\mathbf{y}) = 0\ \forall \mathbf{y} \in \mathcal{V}$ for J_1 and J_3 and to $(A\mathbf{x}_\mathcal{V} - \mathbf{b}), A\mathbf{y}) = 0\ \forall \mathbf{y} \in \mathcal{V}$ for J_2.*
b. *Show that for J_1, if we know $A\mathbf{z}$, we can compute the A-orthogonal projection of $\mathbf{z}$ on $\mathcal{V}$
   without knowing $\mathbf{z}$.*
c. *For J_1 and J_3, suppose $\mathbf{w}$ is A-orthogonal to $\mathcal{V}$ and $\mathbf{x}_\mathcal{V}$ is known. Which scalar equation
   needs to be solved to find the A-orthogonal projection of $\mathbf{x}$ on $\mathcal{V} \cup span\{\mathbf{w}\}$?*
d. *To which equations does computing the stationary point lead if A is non-symmetric? In
   which of the cases is this still a minimum?*

Exercise 3.11 *Let $J(v) = \frac{1}{2}a(v, v) - F(v)$. Show that for symmetric $a(u, w)$ it holds for all
u and w: $J[u + \varepsilon w] = J[u] + \varepsilon(a(w, u) - F(w)) + \frac{1}{2}\varepsilon^2 a(w, w)$. What does this expression
say if u is a stationary point of $J(u)$?*

Exercise 3.12

a. *Show that for a symmetric positive definite matrix A it holds that $(\mathbf{x}, A\mathbf{x}) \ge c\ (\mathbf{x}, \mathbf{x})$,
   where c is the smallest eigenvalue of A.*
b. *Show that an arbitrary real matrix cannot be singular if $(\mathbf{x}, A\mathbf{x}) \ge c\ (\mathbf{x}, \mathbf{x})$ for some
   positive c. What does this mean if we want to solve the problem $A\mathbf{x} = \mathbf{b}$?*

c. *Show that if a matrix is positive definite, then for any Galerkin approximation it is non-singular.*

d. *Given the indefinite 2×2 diagonal matrix with 1 and -1 on the diagonal. Determine the one-dimensional sub-space for which the Galerkin approximation of this matrix, that is, the matrix reduced to this sub-space, is singular. Is that also possible if we take a Petrov–Galerkin approach with the test space A times the search space?*

Exercise 3.13 Stability weak form *Show that if we have the problems $a(w, u) = F(w)$ and $a(w, \tilde{u}) = F(w) + \delta(w)$, where the bi-linear form satisfies the coercivity condition, that $\|u - \tilde{u}\|_V \leq K_\delta/c$. So if K_δ tends to zero, then $\tilde{u}$ tends to u, hence we have stability. (Here $\delta(w)$ is also a linear functional.)*

Exercise 3.14 *Make a splitting of $\|du/dx\|^2$ such that the coefficients as indicated in Example 3.6 are equal.*

Exercise 3.15

a. *Consider the ODE*

$$-\frac{d^2u}{dx^2} + (1 + x)u = x^2 \text{ on } \Omega = [0, 1].$$

Set up the associated weak forms that can be subjected to the criteria of the Lax–Milgram theorem for the following boundary conditions:

1	$u(0) = 0$	$u(1) = 0$
2	$u(0) = 1$	$u(1) = 5$
3	$\frac{du}{dx}(0) = 3$	$u(1) = 0$
4	$\frac{du}{dx}(0) = 3$	$u(1) = 5$
5	$\frac{du}{dx}(0) + 9u(0) = 3$	$u(1) = 0$
6	$\frac{du}{dx}(0) - 9u(0) = 3$	$u(1) = 0$

b. *Do the same for the problem*

$$-\frac{d}{dx}\left(\exp(x)\frac{du}{dx}\right) = \sin(x)$$

on $(-1, 0)$, with boundary conditions $u(-1) = 0$ and $du/dx(0) = 5$.

c. *Give the ODEs and boundary conditions of the weak forms defined in the following table.*

	$a(v, u)$	$F(v)$	$\mathcal{V}$
1	$5\left(\frac{dv}{dx}, \frac{du}{dx}\right) + 3v(0)u(0)$	$(v, \cos(x)) - 5v(0)$	$\{v \in H^1[0, 1] \| v(1) = 0\}$
2	$5\left(\frac{dv}{dx}, \frac{du}{dx}\right) - 3v(0)u(0)$	$(v, \cos(x)) - 5v(0)$	$\{v \in H^1[0, 1] \| v(1) = 0\}$
3	$\left((1 + \cos(x))\frac{dv}{dx}, \frac{du}{dx}\right)$	$(v, \exp(x))$	$\{v \in H^1[0, 1] \| v(0) = v(1) = 0\}$
4	$\left(\frac{dv}{dx}, u + \frac{du}{dx}\right)$	$(v, \exp(x))$	$\{v \in H^1[0, 1] \| v(0) = v(1) = 0\}$

d. Check the positive definiteness of all weak forms, either given in part c or derived in parts a and b.

e. Which of the problems in parts a, b, and c can be written as a minimization problem? Make the associated minimization form explicit in integrals.

Exercise 3.16 Show that (3.33) is equivalent to $(\mathbf{x}, A^T A\mathbf{x}) > c^2(\mathbf{x}, \mathbf{x})$.

Exercise 3.17

a. Which boundary conditions would you suggest for the following differential equations such that the Lax–Milgram theorem could possibly be satisfied (you do not have to prove this):

$$(i) \ L_1 u = -\frac{d}{dx}e^x \frac{du}{dx} = f,$$

$$(ii) \ L_2 u = e^{-x}L_1 u = -\frac{d^2 u}{dx^2} - \frac{du}{dx} = e^{-x}f.$$

Observe that, in fact, the same equation is solved here. However, there is a difference in outcome. Explain in general which freedom is added here to set up a weak form. Also explain that this is, in fact, a Petrov–Galerkin approach.

b. Consider the linear convection–diffusion equation (3.38) and observe that for the steady state it is of type (ii). Determine the exponential by which you can write it in the form of type (i). Show also that the coercivity constant is pushed to zero if $\bar{u}/\mu$ tends to infinity and explain what this means for the conditioning.

4

Discretization of PDEs

Infiniteness is an abstraction;
we can only solve finite problems.

As made clear in Fig. 3.1 of the previous chapter, we have to make a finite representation/approximation of the system of PDEs because the computer is a finite apparatus. There are, however, a great many ways to do so. The process of making the PDEs, defined on a continuous domain, finite is called discretization because the problem will be approximated on a discrete sub-set of the domain. In the discretization process, one can distinguish space and time discretization. It is common to do the space discretization first and thereafter the time integration. We will start with the space discretizations based on the finite-difference, finite-volume, and finite-element methods, respectively. Next, in the case of transient (= time-dependent) problems we will consider the time discretization.

4.1 Finite-Difference and Finite-Volume Methods for Elliptic Equations

In this section, we will give an introduction to the finite-difference and finite-volume methods. For finite differences, Taylor series are the basic tool for the derivation.

4.1.1 Finite-Difference Approximations in One Dimension

Suppose we want to solve a 1D-PDE numerically on an interval $\Omega = [0, 1]$. We define the grid points x_i of an equidistant grid by

$$x_i = ih, \quad i = 0, 1, \ldots, N; \quad h = 1/N.$$

Furthermore, define

$$u_i = u(x_i), \quad i = 0, 1, \ldots, N.$$

Now assume that $u \in C^4[0, 1]$. Using the Taylor series of u, we can now derive a finite-difference approximation for the derivatives of u in the grid points x_i using

$$u_{i+1} = u(x_i + h) = u_i + h \left.\frac{du}{dx}\right|_{x_i} + \frac{h^2}{2!} \left.\frac{d^2u}{dx^2}\right|_{x_i} + \frac{h^3}{3!} \left.\frac{d^3u}{dx^3}\right|_{x_i} + \frac{h^4}{4!} \left.\frac{d^4u}{dx^4}\right|_{x_i+\xi_1 h} \qquad (4.1)$$

and

$$u_{i-1} = u(x_i - h) = u_i - h \left.\frac{du}{dx}\right|_{x_i} + \frac{h^2}{2!} \left.\frac{d^2u}{dx^2}\right|_{x_i} - \frac{h^3}{3!} \left.\frac{d^3u}{dx^3}\right|_{x_i} + \frac{h^4}{4!} \left.\frac{d^4u}{dx^4}\right|_{x_i-\xi_2 h}, \qquad (4.2)$$

where $\xi_1, \xi_2 \in [0, 1]$. Add both equations, re-arrange, and use the continuity of the fourth derivative to obtain

$$\left.\frac{d^2u}{dx^2}\right|_{x_i} = \frac{u_{i+1} - 2u_i + u_{i-1}}{h^2} - \frac{h^2}{12} \left.\frac{d^4u}{dx^4}\right|_{x_i+\xi h}, \quad \xi \in [-1, 1]. \qquad (4.3)$$

In the same way we can subtract (4.1) and (4.2) to obtain

$$\left.\frac{du}{dx}\right|_{x_i} = \frac{u_{i+1} - u_{i-1}}{2h} - \frac{h^2}{3!} \left.\frac{d^3u}{dx^3}\right|_{x_i+\tau h}, \quad \tau \in [-1, 1]. \qquad (4.4)$$

The first term in each of the right-hand sides can now be used as an approximation to the derivative in the left-hand side. The error is then the second term in the right-hand side called *local truncation error*. The truncation errors are of $O(h^2)$, which means that the truncation error, in an absolute sense, is less than Ch^2 for sufficiently small h, C being a positive constant. We call the first term in the right-hand side of each equation the second-order accurate central difference for the derivative of the left-hand side of the equation. The adjective 'central' refers to the symmetry of the discretizations with respect to the grid point x_i. By the order of accuracy we mean the power p of h occurring in the local truncation error; hence, here both discretizations are second-order accurate.

In the same way we can derive central differences for the third- and fourth-order derivatives:

$$\left.\frac{d^3u}{dx^3}\right|_{x_i} = \frac{u_{i+2} - 2u_{i+1} + 2u_{i-1} - u_{i-2}}{2h^3} + O(h^2),$$

$$\left.\frac{d^4u}{dx^4}\right|_{x_i} = \frac{u_{i+2} - 4u_{i+1} + 6u_i - 4u_{i-1} + u_{i-2}}{h^4} + O(h^2).$$

Non-central discretizations can also easily be obtained. For example the forward approximation of the first derivative of u can be obtained directly from the expansion (4.1), yielding

$$\left.\frac{du}{dx}\right|_{x_i} = \frac{u_{i+1} - u_i}{h} + O(h). \qquad (4.5)$$

Similarly, a backward difference can be obtained from (4.2):

$$\left.\frac{du}{dx}\right|_{x_i} = \frac{u_i - u_{i-1}}{h} + O(h). \tag{4.6}$$

Suppose we want to increase the order of accuracy to four instead of two. We can achieve this by bringing the Taylor series for $u_{i+2} = u(x_i+2h)$ and $u_{i-2} = u(x_i-2h)$ into play. We need to make a linear combination of four Taylor series, such that we only get second derivatives, while undesired derivates are cancelled out. In this case, the linear combination is given by

$$-16u_{i+1} - 16u_{i-1} + u_{i+2} + u_{i-2}.$$

After some re-writing, we obtain the following central difference formula for the second derivative of u:

$$\left.\frac{d^2u}{dx^2}\right|_{x_i} = \frac{-u_{i+2} + 16u_{i+1} - 30u_i + 16u_{i-1} - u_{i-2}}{12h^2} + O(h^4). \tag{4.7}$$

This is one way of getting higher-order variants of a discretization. Other routes are also possible.

Example 4.1 (Solving a 1D elliptic equation) *The numerical solution of the equation of the boundary value problem $-\frac{d^2u}{dx^2} = f(x)$, $u(0) = u(1) = 0$, proceeds as follows. We require that in every point x_i the differential equation is satisfied where we replace the derivatives in those points by the difference approximations. This yields a system of difference equations for the values in the grid points. The solution of the difference equation will be denoted by U_i which approximates u_i. It is called grid function, since it is defined only at the grid points. Likewise, we restrict f to the grid points $f_i = f(x_i)$. This yields the linear system*

$$h^{-2} \begin{bmatrix} 2 & -1 & & & \\ -1 & 2 & -1 & & \\ & \ddots & \ddots & \ddots & \\ & & -1 & 2 & -1 \\ & & & -1 & 2 \end{bmatrix} \begin{bmatrix} U_1 \\ U_2 \\ . \\ U_{N-2} \\ U_{N-1} \end{bmatrix} = \begin{bmatrix} f_1 \\ f_2 \\ . \\ f_{N-2} \\ f_{N-1} \end{bmatrix}.$$

After this system is solved, we have found the desired approximations of u_i at the grid points x_i, $i = 1, 2, \ldots, N-1$. See example Ex. 4.1a

The approximation in the example can be made more accurate by increasing the number of grid points, thereby increasing the size of the linear system, resulting in a higher computation time and larger memory consumption. This shows that in the numerical solution of PDEs, we always have to find a trade-off between the accuracy and the amount of computer time and memory we want to spend. The

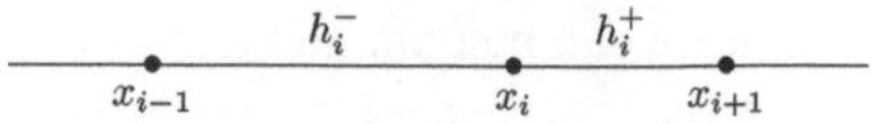

Figure 4.1 Illustration of a non-uniform one-dimensional grid.

challenge is, of course, to get the accuracy as high as possible by using accurate difference schemes for the lowest possible amount of computer time and/or memory usage. This example also shows that we run into problems if fourth-order accurate discretizations need to be used at the boundary. Usually, we have to accept lower-order accuracy near the boundaries. The posedness of the set of difference equations often dictates what we can do there.

Finally, we consider difference approximations for the first and second derivative on non-uniform grids. Consider three subsequent grid points,

$$x_{i-1}, x_i, x_{i+1}, \quad \text{with } x_i - x_{i-1} = h_i^- \text{ and } x_{i+1} - x_i = h_i^+,$$

as indicated in Fig. 4.1. Note that $h_i^+ = h_{i+1}^-$.

Analogous to (4.1) and (4.2), we find now the Taylor expansions

$$u(x_i + h_i^+) = u_i + h_i^+ \frac{du}{dx}\Big|_{x_i} + \frac{(h_i^+)^2}{2!} \frac{d^2u}{dx^2}\Big|_{x_i} + \frac{(h_i^+)^3}{3!} \frac{d^3u}{dx^3}\Big|_{x_i} + \frac{(h_i^+)^4}{4!} \frac{d^4u}{dx^4}\Big|_{x_i+\theta_1 h_i^+},$$
(4.8a)

$$u(x_i - h_i^-) = u_i - h_i^- \frac{du}{dx}\Big|_{x_i} + \frac{(h_i^-)^2}{2!} \frac{d^2u}{dx^2}\Big|_{x_i} - \frac{(h_i^-)^3}{3!} \frac{d^3u}{dx^3}\Big|_{x_i} + \frac{(h_i^-)^4}{4!} \frac{d^4u}{dx^4}\Big|_{x_i-\theta_2 h_i^-}.$$
(4.8b)

After multiplication by, respectively, h_i^- and h_i^+, we find after some algebraic manipulation

$$\frac{d^2u}{dx^2}\Big|_{x_i} = \frac{2u_{i+1}}{h_i^+(h_i^- + h_i^+)} - \frac{2u_i}{h_i^- h_i^+} + \frac{2u_{i-1}}{h_i^-(h_i^- + h_i^+)} + \frac{h_i^- - h_i^+}{3} \frac{d^3u}{dx^3}\Big|_{x_i} + \text{h.o.t.,} \quad (4.9)$$

and in a similar way

$$\frac{du}{dx}\Big|_{x_i} = \frac{h_i^- u_{i+1}}{h_i^+(h_i^- + h_i^+)} + \frac{(h_i^+ - h_i^-)u_i}{h_i^- h_i^+} - \frac{h_i^+ u_{i-1}}{h_i^-(h_i^- + h_i^+)} - \frac{h_i^- h_i^+}{6} \frac{d^3u}{dx^3}\Big|_{x_i} + \text{h.o.t,} \quad (4.10)$$

where h.o.t. stands for higher-order terms, in this case some combination of $(h_i^-)^2$, $h_i^- h_i^+$, and $(h_i^+)^2$. When the maximum mesh size is indicated by h (so $h_i \leq h$), then we see that (4.10) yields an $O(h^2)$ approximation to the first derivative, but that (4.9) only gives an $O(h)$ approximation to the second derivative. However, often the non-uniform grid will occur as a transformation of a uniform grid. So $x_i = g(\xi_i)$, where $\xi_{i+1} - \xi_i = h$. If that is the case and $g \in C^2$, then

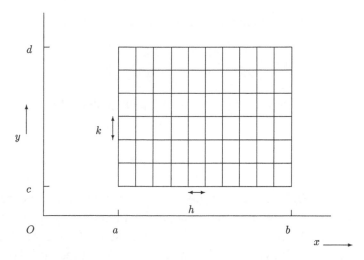

Figure 4.2 Uniform grid on rectangular domain.

$$h_i^+ - h_i^- = x_{i+1} - 2x_i + x_{i-1} = h^2 \frac{g(\xi_{i+1}) - 2g(\xi_i) + g(\xi_{i-1})}{h^2}$$

$$= h^2 \frac{d^2 g}{d\xi^2}(\xi_i + \tau h), \quad \tau \in [-1, 1],$$

and hence the approximation (4.9) will be second-order accurate. *See example Ex. 4.1b–c*

4.1.2 Finite-Difference Approximations in Two Dimensions

Now that we know how to apply the finite difference method in one dimension, we can easily do the same in two dimensions. Assuming $\Omega = \{(x, y) \in \mathbb{R}^2;$ $a < x < b, c < y < d\}$, we define our grid points to be (see also Fig. 4.2)

$$(x_i, y_i) = (a + ih, c + jk), \; i = 0, 1, \ldots, m; \, j = 0, 1, \ldots, n; \; h = \frac{b - a}{m}, \; k = \frac{d - c}{n}.$$

We use the results of the one-dimensional part to obtain the discretizations

$$\left.\frac{\partial^2 u}{\partial x^2}\right|_{i,j} = \frac{u_{i+1,j} - 2u_{i,j} + u_{i-1,j}}{h^2} + O(h^2), \tag{4.11}$$

and

$$\left.\frac{\partial^2 u}{\partial y^2}\right|_{i,j} = \frac{u_{i,j+1} - 2u_{i,j} + u_{i,j-1}}{k^2} + O(k^2), \tag{4.12}$$

where $u_{i,j} = u(x_i, y_j)$. New are the mixed derivatives, for example u_{xy}. This one can be discretized in two steps. We have

$$\left.\frac{\partial^2 u}{\partial x \partial y}\right|_{i,j} = \frac{\partial}{\partial x}\left[\frac{\partial u}{\partial y}\right]_{i,j}$$

which is, using central differences for a first derivative, equal to

$$\left.\frac{\partial^2 u}{\partial x \partial y}\right|_{i,j} = \frac{1}{2h}\left[\left.\frac{\partial u}{\partial y}\right|_{i+1,j} - \left.\frac{\partial u}{\partial y}\right|_{i-1,j}\right] + O(h^2).$$

Using also central discretizations for the remaining derivatives gives the full discretization:

$$\left.\frac{\partial^2 u}{\partial x \partial y}\right|_{i,j} = \frac{1}{4hk}\left[u_{i+1,j+1} - u_{i+1,j-1} - u_{i-1,j+1} + u_{i-1,j-1}\right] + O(h^2) + O(k^2).$$

If we replace in a PDE, say $\mathcal{L}u = f$, the partial derivatives by difference approximations, then we obtain also a difference approximation for $\mathcal{L}$ denoted by $\mathcal{L}_h$. Very often, $\mathcal{L}_h$ is schematically written as a *difference molecule* or *stencil*. Here, the occurring weights are shown according to the geometry of the grid. For example, for the Laplace operator Δ on a uniform grid with $h = k$, we find with (4.11) and (4.12) a difference approximation Δ_h with error $O(h^2)$, hence $\Delta = \Delta_h + O(h^2)$. The difference molecule is then shown as

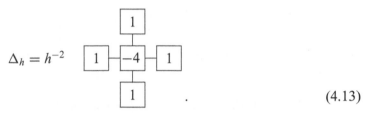

$$(4.13)$$

We call this the five-point stencil of the Laplace operator.

For the discretization on a non-uniform grid, we will consider a linear elliptic PDE $\mathcal{L}u = f$, where

$$\mathcal{L}u \equiv -\mu u_{xx} - \nu u_{yy} + \bar{u}u_x + \bar{v}u_y + qu. \qquad (4.14)$$

With μ and ν positive and $q \geq 0$. In the remaining part, we will assume $\mu, \nu, \bar{u}, \bar{v}$ to be constant to ease the notation. We use the discretization mentioned in the 1D non-uniform grid part. If we take a five-point approach (see Fig. 4.3), we find

$$C_P U_{i,j} - C_W U_{i-1,j} - C_S U_{i,j-1} - C_E U_{i+1,j} - C_N U_{i,j+1} = f_P.$$

This leads to the following equations for the coefficients:

$$C_W = \frac{2\mu + h_i^+ \bar{u}}{h_i^-(h_i^- + h_i^+)}, \quad C_S = \frac{2\nu + k_j^+ \bar{v}}{k_j^-(k_j^- + k_j^+)}, \quad C_E = \frac{2\mu - h_i^- \bar{u}}{h_i^+(h_i^- + h_i^+)},$$

$$(4.15)$$

$$C_N = \frac{2\nu - k_j^- \bar{v}}{k_j^+(k_j^- + k_j^+)}, \quad C_P = \frac{2\mu + (h_i^+ - h_i^-)\bar{u}}{h_i^- h_i^+} + \frac{2\nu + (k_j^+ - k_j^-)\bar{v}}{k_j^- k_j^+} + q.$$

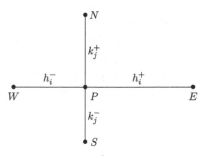

Figure 4.3 Non-uniform grid in 2D.

Thus, $C_P = C_W + C_S + C_E + C_N + q$, and with restriction $q \geq 0$ we have that $C_P \geq C_W + C_S + C_E + C_N$. If we take the mesh sizes small enough, all coefficients will be positive. This holds if

$$\max_i(h_i^-) \leq \frac{2\mu}{|\bar{u}|} \quad \text{and} \quad \max_j(k_j^-) \leq \frac{2v}{|\bar{v}|}. \tag{4.16}$$

Recall that $h_i^+ = h_{i+1}^-$ and similar for k_j^+. The matrix-vector form of this discretization has the following shape,

$$
\begin{bmatrix}
 & & & & & \\
 & & & & & \\
 & & & & & \\
 & & & & & \\
 & & & & & \\
 & & & & & \\
 & & & & & \\
 & & & & & \\
\end{bmatrix}
\begin{bmatrix}
U_{11} \\ \vdots \\ U_{m1} \\ U_{12} \\ \vdots \\ U_{m2} \\ \vdots \\ \vdots \\ U_{1n} \\ \vdots \\ U_{mn}
\end{bmatrix}
=
\begin{bmatrix}
f_{11} \\ \vdots \\ f_{m1} \\ f_{12} \\ \vdots \\ f_{m2} \\ \vdots \\ \vdots \\ \vdots \\ f_{1n} \\ \vdots \\ f_{mn}
\end{bmatrix},
$$

$$\tag{4.17}$$

where the unknowns are ordered line by line in Fig. 4.2 (called lexicographical ordering). On a typical row, the coefficients on the five diagonals of the matrix are C_S, C_W, C_P, C_E, and C_N, respectively. Near boundaries the coefficients may differ, but in general the structure is as given here. We call the matrix a penta-diagonal

matrix. The distance of the outer diagonals with respect to the main diagonal is
m, that is, the number of unknowns on a line, and the two diagonals closest to the
main diagonal are on distance 1 from the main diagonal. If the restrictions on the
mesh size are satisfied and there are Dirichlet boundary conditions on part of the
boundary, the matrix is usually *weakly diagonally dominant*, meaning for a matrix
of order mn, as we have here, that

$$|a_{ii}| \geq \sum_{j=1}^{mn} |a_{ij}| \text{ for } i = 1, \cdots, mn$$

with inequality for at least one value of i. Moreover, the matrix should be irreducible, meaning that every unknown depends on every other unknown, either directly
or via intermediate unknowns (see Section B.1 for more details). The diagonal-
dominance property is favourable for iterative methods to solve the set of coupled
equations (4.17).

If one or both of the coefficients, $\bar{u}$ and $\bar{v}$, in front of the first derivatives in $\mathcal{L}$
is big, then the condition on the mesh sizes (4.16) can lead to a very large system
(4.17). In order to prevent this, one could use an *upwind discretization* for the first
derivative. The direction of the wind is given by $(\bar{u}, \bar{v})$, where it is crucial that the
signs in front of the second-order derivatives are negative in (4.14). Here upwind
means that we pick all the information in the direction of the source of the wind,
that is, opposite the wind direction. If we do so, the diagonal entry increases and a
weakly diagonally dominant matrix is obtained for each mesh size. In more detail,
we take for the term $\bar{u}u_x$

$$\bar{u}\frac{U_{i,j} - U_{i-1,j}}{h_i^-} \text{ if } \bar{u} > 0$$

and

$$\bar{u}\frac{U_{i+1,j} - U_{i,j}}{h_i^+} \text{ if } \bar{u} < 0,$$

similarly for $\bar{v}u_y$. The price of this approach is that we only have a first-order accurate discretization and on top of that we have introduced *artificial diffusion*. This can
be seen using the Taylor expansion (4.8b)

$$\frac{U_{i,j} - U_{i-1,j}}{h_i^-} = u_x|_{i,j} + h_i^- u_{xx}|_{i,j} + O((h_i^-)^2).$$

So, the term $\bar{u}u_x$ introduces an extra diffusion term with coefficient $\bar{u}h_i^-$ which may
be rather big with respect to the real diffusion coefficient μ.

In order to diminish the artificial diffusion one can also be more subtle by
adding just as much artificial diffusion as needed to keep the off-diagonal coefficients positive. In this case, we just increase μ by μ_a so much that the condition
$\max_i(h_i^-) \leq \frac{2(\mu+\mu_a)}{|\bar{u}|}$ is satisfied. So, $\mu_a = \max(|\bar{u}| \max_i(h_i^-)/2 - \mu, 0)$. It is also

possible to add a discretization of a higher-order dissipative term, for example $-h^2 u_{xxxx}$, to the convection term in order to prevent the problems of losing diagonal dominance (Veldman, 2010).

4.1.3 Discretization near the Boundary

First, consider a square area Ω with boundary Γ and a uniform grid. If we consider the five-point formula

$$C_P U_{i,j} - C_W U_{i-1,j} - C_S U_{i,j-1} - C_E U_{i+1,j} - C_N U_{i,j+1} = f_{ij}, \qquad (4.18)$$

we see that when $i = 1, m - 1$ or $j = 1, n - 1$ in Fig. 4.2, at least one of the terms picks up a value of U at the boundary. For instance, if we have a Dirichlet condition $u(a, y) = r(y)$ on the y-axis, we can replace the term $C_W U_{0,j}$ by $C_W r(y_j)$. We then move this known term to the right and obtain

$$C_P U_{1,j} - C_S U_{1,j-1} - C_E U_{2,j} - C_N U_{1,j+1} = f_{ij} + C_W r(y_j).$$

If we have homogeneous Neumann boundary conditions, we have to use fictive grid points (x_{-1}, y_j), $j = 1, \ldots, n - 1$. With the help of these points we also use the five-point formula at the boundary. Next, one eliminates the fictive grid points in the formula using the fact that $\frac{U_{1,j} - U_{-1,j}}{2h} = 0$. For a point P on the boundary we then find that

$$C_P U_{0,j} - C_S U_{0,j-1} - (C_E + C_W) U_{1,j} - C_N U_{0,j+1} = f_j.$$

If we have Robin boundary conditions, we can do the same thing. Note that in both cases the diagonal dominance (see Definition B.8) of the matrix remains intact. We will see later that this property ensures a unique solution, at least if at one point a Dirichlet condition or, even weaker, a Robin condition is applied. At corner points, for example $(i, j) = (1, 1)$, of the domain there is a boundary condition from each boundary, both of which will affect the discretization at that point. For a non-rectangular domain, the boundary may intersect the grid arbitrarily. In that case we also have to use fictive points combined with interpolation and extrapolation. In the following example we will also consider a positioning of the boundary in the middle between two grid points. *See example Ex. 4.1d–f*

Example 4.2 (Linear convection–diffusion equation) *The preceding way of dealing with boundary conditions is in general fine, but in case of a fluid flow we have to be careful in case the convection is much more important than the diffusion. Consider again Equation (3.38), that is,*

$$-\bar{u} u_x + \mu u_{xx} = f(x).$$

On [0,1] then a central discretization of this equation is

$$-\bar{u} \frac{u_{i+1} - u_{i-1}}{2h} + \mu \frac{u_{i+1} - 2u_i + u_{i-1}}{h^2} = f_i.$$

If we multiply by h^2/μ and define the mesh-Péclet number by $P = \bar{u}h/\mu$, we get

$$-P(u_{i+1} - u_{i-1})/2 + u_{i+1} - 2u_i + u_{i-1} = \frac{h^2}{\mu}f_i.$$

In general, one could say that as long as we have diagonal dominance in the equation we can impose the Dirichlet, Neumann, and Robin boundary conditions as explained in the previous section. This is the case, if $|P| < 2$.

Note that on the continuous equation, see Section 3.6.1, one can impose Dirichlet, Neumann, and Robin conditions. However, if $\bar{u}/\mu$ tends to infinity, this is queer since the convection-only case needs only one boundary condition at one of the end points of the domain. Let us re-write the discretization into a form which suits better the case that P is big:

$$-(u_{i+1} - u_{i-1})/2 + (u_{i+1} - 2u_i + u_{i-1})/P = hf_i. \tag{4.19}$$

If $P < 0$, then the coefficient of u_{i+1}, namely $(-1/2 + 1/P)$, is bigger than that of u_{i-1}, namely $(1/2+1/P)$), hence, the equation needs a Dirichlet condition at the right boundary. Now say that at the left $(x = 0)$ boundary we want to impose the boundary condition $u(0) = g$. So, $u_0 = g$, but this conflicts with the idea that the convective part does not need any condition on the left, especially when P tends to infinity. We can avoid that by taking a one-sided difference for $i = 1$. So we replace $(u_2 - u_0)/(2h)$ by $(u_2 - u_1)/h$. This leaves us with u_0 only in the diffusion term, where we employ the Dirichlet boundary condition. So the final discretization at the left boundary is

$$-(u_2 - u_1) + (u_2 - 2u_1 + g)/P = hf_1. \tag{4.20}$$

After this we could also ask the question to which extrapolation this discretization corresponds. This means that we just have to subtract it from Equation (4.19) at $i = 1$, after which we obtain

$$-(u_2 - 2u_1 + u_0)/2 + u_0/P = g/P. \tag{4.21}$$

We see that this is a second-order accurate representation of the Dirichlet boundary condition $u(0) = g$.

The equation can be re-written into extrapolation form; u_0 is obtained from

$$\left(1 - \frac{2}{P}\right)u_0 = -\frac{2}{P}g + 2u_1 - u_2,$$

which is well defined for all negative P. Note that, though we explicitly advocate such a treatment for large negative P, it also will work well for low values of P. In that case, the diffusion will dominate and the extrapolation tends to our original boundary condition. Note that in an actual implementation one can choose between implementing the boundary condition (4.21) explicitly (hence having also u_0 as an unknown) or eliminating u_0 and use (4.20).

Another example is the condition $u_x(0) = g$. The analogue of (4.21) is here the boundary condition $-(u_2 - 2u_1 + u_0)/2 + (u_1 - u_0)/P = gh/P$, where the boundary is now *in the middle between the grid points x_0 and x_1. Subtracting it from Equation (4.19) for $i = 1$ such that u_0 cancels out gives now $-2(u_2 - u_1) + (u_2 - u_1)/P = hf_1 - gh/P$ or $(-2 + 1/P)(u_2 - u_1) = hf_1 - gh/P$. The last equation shows that for negative P the coefficient $-2 + 1/P$ can never become zero and the equation shows clear weak diagonal dominance.*

4.1.4 Maximum Principles and Monotony

Consider the Laplace equation $\Delta u = 0$. Then from Gauss' theorem we have

$$\int_\Gamma (\nabla u, \mathbf{n}) d\Gamma = 0,$$

where n is the outward normal. This says that, if $(\nabla u, \mathbf{n}) \neq 0$ on the whole boundary, it must be on parts positive and on other parts negative. This holds for the boundary of any volume in the definition domain of the Laplace equation. Hence, if we take this volume very small around a point, then we see that this point cannot be an extremum since in some directions u is increasing and in others it is decreasing. As this holds in any point in the internal of the domain, this means that we cannot have an extremum inside the domain, hence it must be on the boundary.

Consider now the Poisson equation $-\Delta u = f$ with $f \leq 0$. Then from Gauss' theorem we have

$$\int_\Gamma (\nabla u, \mathbf{n}) d\Gamma \geq 0. \tag{4.22}$$

If we now contract the volume to a point we see that this point can be a minimum since u may be increasing in all directions away from it. But, it cannot be a maximum because then $(\nabla u, \mathbf{n})$ must be negative in any direction which conflicts with (4.22). So, in this case we cannot have a maximum in the internal domain. Similarly, if $f \geq 0$ we cannot have a minimum in the internal domain.

Next, consider the case $-\Delta u + qu = f$, where $q \geq 0$ and $f \leq 0$; then we have that

$$\int_\Gamma (\nabla u, \mathbf{n}) d\Gamma \geq \int_\Omega qu d\Omega.$$

Now again we contract the volume and see, using the preceding reasoning, that if u is positive, no maximum is possible. Similarly, for positive f no negative minimum is possible.

Now we consider a general discrete case:

$$C_P U_{i,j} - C_W U_{i-1,j} - C_S U_{i,j-1} - C_E U_{i+1,j} - C_N U_{i,j+1} = f_P,$$

where $C_P \geq C_W + C_S + C_E + C_N$ and all the coefficients are non-negative. (At a Neumann boundary some of them may be zero.) This can be re-written as

$$(C_P - C_W - C_S - C_E - C_N)U_{i,j}$$
$$+ C_W(U_{i,j} - U_{i-1,j}) + C_S(U_{i,j} - U_{i,j-1}) + C_E(U_{i,j} - U_{i+1,j}) + C_N(U_{i,j} - U_{i,j+1})$$
$$= f_P,$$

which is a discrete analogue of Gauss' theorem. Now consider again three cases.

If $f_P = 0$ for all internal points of the computational domain and C_P is equal to the sum of the other coefficients, then the first coefficient in the equation cancels and we have that $U_{i,j}$ cannot be an extremum for any internal point, since if one of the differences is positive, then another must be negative.

If $f_P \leq 0$ for all internal points of the computational domain and C_P is equal to the sum of the other coefficients, then again the first coefficient in the equation cancels. Reasoning the same as in the continuous case, we cannot have a maximum in the internal domain. Similarly, we cannot have a minimum if $f_P \geq 0$.

If $f_P \leq 0$ for all internal points of the computational domain and C_P is bigger than the sum of the other coefficients, then we have a positive coefficient in front of $U_{i,j}$. And reasoning the same as in the continuous case now leads to the observation that no positive maximum is possible; similarly, no negative minimum is possible if $f_P \geq 0$.

The consequence of this is that if on a Dirichlet boundary we prescribe a positive value and furthermore if the right-hand side is positive, then the solution must be positive, since if it would become negative, it must have a negative minimum in the internal or at the Neumann boundary, which conflicts with the preceding assertions.

The maximum principles are related to the concept of non-negative matrices as given in Definition B.13.

Example 4.3 (Monotony for 1D linear convection–diffusion equation) *One can also apply this theory to (4.19). One can re-write it as*

$$\left(-\frac{1}{2} + \frac{1}{P}\right)(u_i - u_{i+1}) + \left(\frac{1}{2} + \frac{1}{P}\right)(u_i - u_{i-1}) = -hf_i, \quad (4.23)$$

which shows that we cannot have an extremum in the inside of the domain if $f_i \equiv 0$ and $|P| < 2$. If $f_i > 0$ at some point i, then for $|P| < 2$, $u_i - u_{i+1}$ and $u_i - u_{i-1}$ may both be negative, so there can be a minimum at that point, but there cannot be a maximum. Consider also the discretization (4.20) near the boundary, which can be re-written as

$$\left(1 + \frac{1}{P}\right)(u_1 - u_2) + (u_1 - g)/P = -hf_1. \quad (4.24)$$

If $f_1 > 0$, then u_1 cannot be a maximum. Now suppose that additionally $u_{N+1} = s$ is given and that (4.23) holds for $i = N$. Then we can state that, if $|P| < 2$ and $f_i \geq 0$ for *i = 1, ⋯, N and g, s ≤ 0, then $u_i ≤ 0$ for i = 1, ⋯, N. The proof is by contradiction, since, if $u_i > 0$ for some i, then due to the negative boundary conditions it must have a maximum somewhere, which it hasn't.*

4.1.5 The Finite-Volume Discretization

The finite-volume approach can only be used in special cases; for example, when $\mathcal{L}$ assumes the form $\mathcal{L}u = \nabla \cdot (\mathcal{M}u)$, where $\mathcal{M}$ is an arbitrary operator that transforms a scalar function into a vector function with two or three entries, depending on the dimension, for example $\mathcal{M}u = \nabla u$. This divergence form of $\mathcal{L}$ is common in fluid dynamics because its determining equations are in fact conservation laws, that is, conservation of mass, momentum, and so on. The aim of the finite-volume discretization is to retain the conservation property of the original PDE in the discretization.

If we consider the problem $u_t = \mathcal{L}u - f$, then in the continuous case we have that from $\int_\Omega (u_t - \nabla \cdot (\mathcal{M}u) + f)d\Omega = 0$ it follows by applying Gauss' theorem that

$$\frac{d}{dt}\int_\Omega u\, d\Omega = \int_\Gamma (\mathcal{M}u, \mathbf{n})d\Gamma - \int_\Omega f d\Omega,$$

where $\mathcal{M}u$ is the *flux* and $(\mathcal{M}u, \mathbf{n})$ the flux normal to the boundary Γ; here $(\mathcal{M}u, \mathbf{n})$ is just the inner product of the flux vector $\mathcal{M}u$ and the normal vector $\mathbf{n}$. Hence, we see that the quantity $\int_\Omega u d\Omega$ will not change if both the integral of the flux over the boundary and the integral of f over the domain are zero, or if the sum of both integrals is zero. These are the properties we would like to maintain for a discretization. *See example Ex. 4.2*

In the general discretization process for the steady case of the preceding example equation, the domain is covered with n disjunct *control volumes* (also called finite volumes); see Fig. 4.4. The starting point is the integral over control volume Ω_i where Gauss' theorem has been applied:

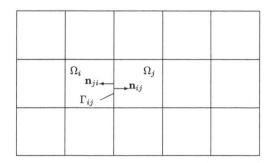

Figure 4.4 Illustration of naming in finite-volume discretization.

$$\int_{\Gamma_i} (\widehat{\mathcal{M}u}, \mathbf{n}) d\Gamma = \int_{\Omega_i} f d\Omega.$$

If we use for the interface between Ω_i and Ω_j the notation Γ_{ij}, then one can write $\Gamma_i = \sum_{j \in \mathcal{N}(i)} \Gamma_{ij}$, where $\mathcal{N}(i)$ is the set of indices of control volumes neighbouring Ω_i. Note that $\Gamma_{ji} = \Gamma_{ij}$. Likewise, we indicate the outward-pointing normal on the boundary Γ_{ij} of Ω_i by $\mathbf{n}_{ij}$. In order to have the desired conservation property, on the interface Γ_{ij} it should hold that the normal fluxes for both control volumes are equal up to a sign, that is, $(\widehat{\mathcal{M}u}, \mathbf{n}_{ij}) = -(\widehat{\mathcal{M}u}, \mathbf{n}_{ji})$.

Now let us consider the conservation property we discussed for the continuous case earlier. Note that

$$\int_{\Omega} (\nabla \cdot (\widehat{\mathcal{M}u}) d\Omega = \sum_{i=1}^{n} \int_{\Omega_i} (\nabla \cdot (\widehat{\mathcal{M}u}) d\Omega = \sum_{i=1}^{n} \sum_{j \in \mathcal{N}(i)} \int_{\Gamma_{ij}} (\widehat{\mathcal{M}u}, \mathbf{n}) d\Gamma$$

$$= \sum_{k \in K} \int_{\Gamma_k} (\widehat{\mathcal{M}u}, \mathbf{n}) d\Gamma = \int_{\Gamma} (\widehat{\mathcal{M}u}, \mathbf{n}) d\Gamma,$$

where K contains only the indices of the boundary parts Γ_k that are on the outer boundary of the domain. Hence, we have

$$\frac{d}{dt} \int_{\Omega} u d\Omega = \int_{\Gamma} (\widehat{\mathcal{M}u}, \mathbf{n}) d\Gamma - \int_{\Omega} f d\Omega.$$

So, without more details about the discretization we already have the same property as for the continuous case. Due to such a conservation property, the finite-volume method is the most popular approach in computational fluid dynamics.

Additional Material

- More details on the finite-volume method and its applications in computational fluid dynamics can be found in Peyret and Taylor (1983) and Veldman (2010).

In the remainder of this section, we will become more specific about the discretization for the steady-state equation $\mathcal{L}u = f$, with $\mathcal{L}$ the operator defined by

$$\mathcal{L}u = -\operatorname{div}(A\nabla u).$$

Now it is trivial that on any part of the domain $\Omega_1 \subset \Omega$ we have

$$\int_{\Omega_1} \mathcal{L}u \, d\Omega_1 = \int_{\Omega_1} f \, d\Omega_1.$$

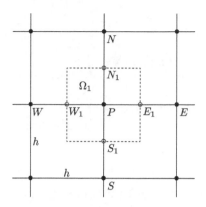

Figure 4.5 A control volume.

Using Gauss' theorem on the left-hand side, we obtain

$$-\int_{\Gamma_1} (A\nabla u, \mathbf{n})\, d\Gamma_1 = \int_{\Omega_1} f\, d\Omega_1, \qquad (4.25)$$

where Γ_1 is the boundary of Ω_1.

Assume that Ω is covered by a uniform grid with a mesh size h in both the x- and y-direction. In Fig. 4.5 a part of the grid around the point $P = (x_i, y_j)$ is drawn. Let Ω_1 be the dashed square from which the sides are in the middle of the lines connecting P and its neighbours N, E, S, and W. So (4.25) turns into

$$
-\int_{x_i-\frac{1}{2}h,y=y_j-\frac{1}{2}h}^{x_i+\frac{1}{2}h} (A\nabla u, -\mathbf{e}_2)\, dx - \int_{y_j-\frac{1}{2}h,x=x_i+\frac{1}{2}h}^{y_j+\frac{1}{2}h} (A\nabla u, \mathbf{e}_1)\, dy +
$$

$$
-\int_{x_i-\frac{1}{2}h,y=y_j+\frac{1}{2}h}^{x_i+\frac{1}{2}h} (A\nabla u, \mathbf{e}_2)\, dx - \int_{y_j-\frac{1}{2}h,x=x_i-\frac{1}{2}h}^{y_j+\frac{1}{2}h} (A\nabla u, -\mathbf{e}_1)\, dy = \qquad (4.26)
$$

$$
\int_{x_i-\frac{1}{2}h}^{x_i+\frac{1}{2}h} \int_{y_j-\frac{1}{2}h}^{y_j+\frac{1}{2}h} f\, dx\, dy,
$$

where $\mathbf{e}_1 = [1, 0]^T$ and $\mathbf{e}_2 = [0, 1]^T$. We will now consider the specific case where $A = \text{diag}([a, c])$, so we try to solve the PDE

$$-\frac{\partial}{\partial x}\left[a\frac{\partial u}{\partial x}\right] - \frac{\partial}{\partial y}\left[c\frac{\partial u}{\partial y}\right] = f. \qquad (4.27)$$

With this, (4.26) becomes

$$
+ \int_{x_i-\frac{1}{2}h,y=y_j-\frac{1}{2}h}^{x_i+\frac{1}{2}h} c(x,y)u_y \, dx - \int_{y_j-\frac{1}{2}h,x=x_i+\frac{1}{2}h}^{y_j+\frac{1}{2}h} a(x,y)u_x \, dy +
$$

$$
- \int_{x_i-\frac{1}{2}h,y=y_j+\frac{1}{2}h}^{x_i+\frac{1}{2}h} c(x,y)u_y \, dx + \int_{y_j-\frac{1}{2}h,x=x_i-\frac{1}{2}h}^{y_j+\frac{1}{2}h} a(x,y)u_x \, dy = \qquad (4.28)
$$

$$
\int_{x_i-\frac{1}{2}h}^{x_i+\frac{1}{2}h} \int_{y_j-\frac{1}{2}h}^{y_j+\frac{1}{2}h} f dx dy.
$$

For the integrals occurring in (4.28) we apply the midpoint rule. This yields the following partial discretization:

$$
\left[-a\frac{\partial u}{\partial x}\Big|_{E_1} + a\frac{\partial u}{\partial x}\Big|_{W_1} \right] h + \left[-c\frac{\partial u}{\partial y}\Big|_{N_1} + c\frac{\partial u}{\partial y}\Big|_{S_1} \right] h = f_P h^2.
$$

Note that no matter how we discretize the remaining derivatives, if we discretize such that $au_x|_{E_1}$ in the current cell is equal to $au_x|_{W_1}$ at the cell at its right, and similar on all other boundaries of the control volume, then if we sum all the equations, all the internal fluxes cancel and only the flux over the outer boundary of the domain remains.

To continue the discretization we use the central discretization

$$
\frac{\partial u}{\partial x}\Big|_{E_1} = (U_{i+1,j} - U_{i,j})/h \qquad (4.29)
$$

which yields the discretization

$$
C_P U_{i,j} - C_W U_{i-1,j} - C_S U_{i,j-1} - C_E U_{i+1,j} - C_N U_{i,j+1} = f_P h^2 \qquad (4.30)
$$

with

$$
C_W = a(x_i - \tfrac{1}{2}h, y_j), \quad C_S = c(x_i, y_j - \tfrac{1}{2}h),
$$
$$
C_E = a(x_i + \tfrac{1}{2}h, y_j) \quad C_N = c(x_i, y_j + \tfrac{1}{2}h),
$$
$$
C_P = C_W + C_S + C_E + C_N.
$$

From the fact that C_N is equal to C_S of the volume on top of this one (similar for the horizontal direction), it follows that the matrix will be symmetric. This fact can be exploited in the solution process. Note that by dividing (4.30) by h^2 we get a formula that also could have been derived using finite differences (see Section 4.1.2). *See example Ex. 4.3*

Example 4.4 (Burgers' equation) *Non-linearity is not very difficult to handle. Let us again consider the Burgers equation (3.43):*

$$
-u\frac{du}{dx} + \mu\frac{d^2u}{dx^2} = f(x),
$$

where we impose $u(1) = -1$ and $\frac{du}{dx}(0) = g$. Moreover, we assume that f and g are such that $u < 0$ on [0,1]. The divergence form of this equation is

$$-\frac{d}{dx}\left(\frac{1}{2}u^2 - \mu\frac{du}{dx}\right) = f(x) \qquad (4.31)$$

and a discretization can be written as

$$F_{i+\frac{1}{2}} - F_{i-\frac{1}{2}} = hf_i,$$

where $F_{i-\frac{1}{2}} = u^2_{i-\frac{1}{2}} - \mu(u_i - u_{i-1})/h$ in which $u_{i-\frac{1}{2}} = (u_i + u_{i-1})/2$. This leads to

$$-\left(\frac{u_i + u_{i+1}}{2}\right)^2 - \left(\frac{u_i + u_{i-1}}{2}\right)^2 + \frac{\mu}{h}(u_{i+1} - 2u_i + u_{i-1}) = hf_i. \qquad (4.32)$$

Analogous to (4.20), for $i = 1$, we replace $(\frac{u_1+u_2}{2})^2 - (\frac{u_1+u_0}{2})^2$ by a one-sided variant $2\left((\frac{u_1+u_2}{2})^2 - (\frac{u_1}{2})^2\right)$, leading to the discretization

$$-2\left(\left(\frac{u_1 + u_2}{2}\right)^2 - u_1^2\right) + \frac{\mu}{h}(u_2 - 2u_1 + u_0) = hf_1. \qquad (4.33)$$

By this step the Neumann condition $u_x(0) = g$ only affects the diffusive part. Assuming that the condition is imposed in the middle between x_0 and x_1, it becomes $(u_1 - u_0)/h = g$. We could eliminate u_0 from (4.33), leading to the discretization at $i = 1$:

$$-2\left(\frac{u_1 + u_2}{2}\right)^2 - u_1^2\right) + \frac{\mu}{h}(u_2 - u_1) = hf_1 + \mu g. \qquad (4.34)$$

 Out of curiosity, one would like to know which boundary condition should be applied to (4.32) for $i = 1$ in order to obtain (4.34). By subtracting the latter from the former, we find that we actually employed the following boundary condition:

$$-\left(\frac{u_1 + u_2}{2}\right)^2 + 2u_1^2 - \left(\frac{u_0 + u_1}{2}\right)^2 + \frac{\mu}{h}(u_1 - u_0) = \mu g.$$

Observing that the first part is a discretization of $-(h/2)^2 du^2/dx^2$, this is a second-order non-linear approximation to the Neumann boundary condition. We note that in practice often one has no clue about the derivative of u. So, one simply takes $g = 0$. For the Dirichlet condition at the right boundary we can just employ the standard approaches. So, either $u = -1$ is prescribed at a grid point or halfway between two points.

 There is a lot more to say about the discretization of the Burgers equation; we refer to Hirsch (1994) for more details.

Example 4.5 *[Ginzburg–Landau equation] For the finite-volume discretization of the Ginzburg–Landau equation on [0,1] we first have to write it in divergence form:*

$$\frac{\partial A}{\partial t} = \gamma_1 A + \frac{\partial}{\partial x}\left(\gamma_2\frac{\partial A}{\partial x}\right) - \gamma_3 A|A|^2.$$

This shows that the quantity A is conserved only if $\gamma_1 = \gamma_3 = 0$. They add to the quantity A if γ_1 is positive and γ_3 is negative and extract if they are negative and positive, respectively. On an equidistant grid the discretization just proceeds as for the Burgers equation and becomes

$$h\frac{dA_i}{dt} = h\gamma_1 A_i + \frac{\gamma_2}{h}(A_{i+1} - 2A_i + A_{i-1}) - h\gamma_3 A_i |A_i|^2 \text{ for } i = 1 \cdots m,$$

where $h = 1/(m+1)$ and $x_i = ih$. The boundary conditions $A(0,t) = 0$ and $A(1,t) = 1$ are discretized by $A_0(t) = 0$ and $A_{m+1}(t) = 1$, respectively. The initial condition $A(x,0) = f(x)$ becomes $A_i(0) = f(ih)$ for $i = 1 \cdots m$.

For this equation and its boundary conditions it may also be attractive to discretize on a grid which is refined towards $x = 1$ for the following reason. Observe that $A(x,t) = 0$ is a steady state of the equation in two cases: (1) if $\gamma_2 = 0$ and (2) if the boundary condition $A(1,t) = 0$ instead of $A(1,t) = 1$. If γ_2 is small, relative to one of γ_1 and γ_3, then for the biggest part of the interval [0,1] the solution will be close to 0, but in the vicinity of $x = 1$ it will rapidly grow to its boundary value 1. In fluid dynamics applications, such a rapid growth is called a boundary layer, the thickness of which depends on the ratios of the coefficients γ_1, γ_2, and γ_3. For the numerical approximation of the equation, such a boundary layer means that for a large part of the domain only a few grid points will be needed to capture the solution, while close to $x = 1$ it will need a lot to capture the solution's sudden growth. So, for such a problem, it is attractive to use a non-equidistant grid. Therefore, let us also derive the finite-volume discretization for that.

Let x_i follow from a transformation $x = g(\xi)$, where $\xi_i = ih$ with $h = 1/(m+1)$, as discussed in Section 4.1.1. We define $h_{i+\frac{1}{2}} = x_{i+1} - x_i$ and we take the interfaces halfway between the grid points, which we define by $x_{i+\frac{1}{2}} = (x_{i+1} + x_i)/2$. Hence, the length of a volume is in this case $x_{i+\frac{1}{2}} - x_{i-\frac{1}{2}} = (x_{i+1} + x_i)/2 - (x_i + x_{i-1})/2 = (x_{i+1} - x_{i-1})/2 = (x_{i+1} - x_i + x_i - x_{i-1})/2 = (h_{i+\frac{1}{2}} + h_{i-\frac{1}{2}})/2$ which we will define as h_i. So we integrate the equation and obtain

$$\int_{x_{i-\frac{1}{2}}}^{x_{i+\frac{1}{2}}} \frac{\partial A}{\partial t}dx = \int_{x_{i-\frac{1}{2}}}^{x_{i+\frac{1}{2}}} \gamma_1 A dx + \gamma_2 \frac{\partial A}{\partial x}\Big|_{x_{i-\frac{1}{2}}}^{x_{i+\frac{1}{2}}} - \int_{x_{i-\frac{1}{2}}}^{x_{i+\frac{1}{2}}} \gamma_3 A|A|^2 dx.$$

Now we make the approximations as before and obtain

$$h_i\frac{dA_i}{dt} = h_i\gamma_1 A_i + \gamma_2[(A_{i+1} - A_i)/h_{i+\frac{1}{2}} - (A_i - A_{i-1})/h_{i-\frac{1}{2}}] - h_i\gamma_3 A_i|A_i|^2 \text{ for } i = 1 \cdots m,$$

where $A_0(t) = 0$ and $A_{m+1}(t) = 1$ and $A_i(0) = f(x_i)$ for $i = 1 \cdots m$. In matrix-vector form this assumes the form

$$M\frac{d}{dt}\mathbf{A} = (\gamma_1 M + \gamma_2 D)\mathbf{A} - M\gamma_3|\mathbf{A}|^2\mathbf{A} + \mathbf{g}, \tag{4.35}$$

where $(|\mathbf{A}|^2\mathbf{A})_i = |A_i|^2 A_i$. The vector $\mathbf{g}$ is zero except for the last entry which is $g_m = \gamma_2/h_{i+\frac{1}{2}}$. Moreover, $(D\mathbf{A})_i = (A_{i+1} - A_i)/h_{i+\frac{1}{2}} - (A_i - A_{i-1})/h_{i-\frac{1}{2}}$ and M is a diagonal matrix with entries $M_{ii} = h_i$ for $i = 1 \cdots m$.

4.1.6 Staggered Grids

For a number of systems of PDEs it appears to be advantageous to not define the various unknowns occurring in it at the same grid points. So, if in (4.27) a and c would depend on another variable, say v, then it would be nice if that variable would be defined at the midpoints of the control volume faces (see Fig. 4.5), so at N_1, E_1, S_1, and W_1. Of course, there is another equation for v that needs to be discretized. Staggered grids are often applied in computational fluid dynamics; see Peyret and Taylor (1983) and Veldman (2010).

The idea of staggered grids can be nicely introduced by discretizing the Cauchy–Riemann equations. The Cauchy–Riemann equations are given by

$$u_x + v_y = 0, \tag{4.36a}$$

$$u_y - v_x = 0, \tag{4.36b}$$

representing conservation of mass and irrotationality of a flow, respectively. We remark that the equations also admit the existence of a velocity potential ϕ which defines velocities by $[u, \ v]^T = \nabla\phi$. Substituting this in the equations shows that ϕ should satisfy the Laplace equation.

The equations (4.36) can be written in divergence form:

$$\nabla \cdot [u; v] = 0, \tag{4.37a}$$

$$\nabla \cdot [-v; u] = 0. \tag{4.37b}$$

Integrating over a volume as depicted in Fig. 4.6 and using Gauss' theorem, we get

$$\int_{y=y_l}^{y_u} u(x_r, y) - u(x_l, y)dy + \int_{x-x_l}^{x_r} v(x, y_u) - v(x, y_l)dx = 0, \tag{4.38a}$$

$$-\int_{y=y_l}^{y_u} v(x_r, y) - v(x_l, y)dy + \int_{x=x_l}^{x_r} u(x, y_u) - u(x, y_l)dx = 0. \tag{4.38b}$$

Suppose we use the same control volume for both equations (see Fig. 4.6). If again, we employ the midpoint rule for the integrals, then for the first equation we need u in the middle of the vertical faces and v in the middle of the horizontal faces of the control volume. For the second equation this is just the other way around. Note that for this choice the unknowns occurring in the first equation will not occur in the second equation, and vice versa. If we fill a plane with these control volumes, this will generalize to two sets of completely decoupled equations. This is, of course, not in line with the physics. Note, however, that if we use two different control volumes as depicted in Fig. 4.6, this problem will not occur. With this we also obtain that nearest neighbours on the grid are connected. The discretization is

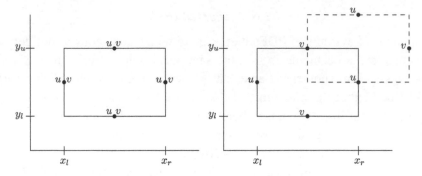

Figure 4.6 Left, one control volume for both equations, u, v co-located; right, each equation with its own control volume, u, v staggered.

$$0 = k_j \left(u_{i+\frac{1}{2},j} - u_{i-\frac{1}{2},j} \right) + h_i \left(v_{i,j+\frac{1}{2}} - v_{i,j-\frac{1}{2}} \right) \tag{4.39a}$$

$$0 = -k_{j+\frac{1}{2}} \left(v_{i+1,j+\frac{1}{2}} - v_{i,j+\frac{1}{2}} \right) + h_{i+\frac{1}{2}} \left(u_{i+\frac{1}{2},j+1} - u_{i+\frac{1}{2},j} \right), \tag{4.39b}$$

where $h_i = x_{i+\frac{1}{2}} - x_{i-\frac{1}{2}}$, $k_j = y_{j+\frac{1}{2}} - y_{j-\frac{1}{2}}$, $h_{i+\frac{1}{2}} = (h_{i+1} + h_i)/2$ and $k_{j+\frac{1}{2}} = (k_{j+1} + k_j)/2$. Choosing it like this gives that $v_{i,j+\frac{1}{2}}$ with respect to the x-direction is precisely located in the middle, so $x_i = (x_{i+\frac{1}{2}} + x_{i-\frac{1}{2}})/2$ and $y_j = (y_{j+\frac{1}{2}} + y_{j-\frac{1}{2}})/2$.

4.2 Conservation Properties

In this section we will describe the energy-conserving discretization of some common equations. It is a follow-up of the discussion in Section 3.6.

4.2.1 Difference Operators

Expanding discretizations of flow equations results in long expressions. This can be kept at bay by introducing difference operators. They share properties with differential operators, but there are, of course, subtle differences. The most important is that one should also take into account an averaging operator, which also has, like the difference operator, a product rule.

Let the averaging operator, indicated by μ, be implicitly defined by $\mu u_i \equiv (u_{i-\frac{1}{2}} + u_{i+\frac{1}{2}})/2$ and the differencing operator δ by $\delta u_i \equiv u_{i+\frac{1}{2}} - u_{i-\frac{1}{2}}$, where we have assumed that $u_{i+\frac{1}{2}}$ exists for an appropriate set of indices i. These definitions seem to indicate that the variable is defined halfway between two gridpoints. This can occur on a staggered grid. We will get exactly at grid points again if we apply the difference operator twice: $\delta^2 u_i = u_{i-1} - 2u_i + u_{i+1}$, and similar for the averaging operator. In fact, $\delta^j \mu^k u_i$ will pick up unknowns at the grid points if $j + k$ is even and halfway between the grid points if it is odd, for example

$$\frac{du}{dx}(x_i) \approx \frac{1}{h}\mu\delta u_i = \frac{1}{h}\delta\mu u_i = (u_{i+1} - u_{i-1})/2h.$$

Also note that the operators δ and μ commute.

Example 4.6 (Cauchy–Riemann equations) *The discretization (4.39b) of the Cauchy–Riemann equations can also be cast into operators δ and μ by indicating in the subscript in which direction we take the difference (and average). This gives*

$$0 = k_j\delta_x u_{i,j} + h_i\delta_y v_{i,j},$$

$$0 = -k_{j+\frac{1}{2}}\delta_x v_{i+\frac{1}{2},j+\frac{1}{2}} + h_{i+\frac{1}{2}}\delta_y u_{i+\frac{1}{2},j+\frac{1}{2}}.$$

There are some rules and equivalences when applying these operators, which are given by

$$\delta(uv) = \mu u\delta v + \delta u\mu v, \tag{4.40a}$$

$$\mu(uv) = \mu u\mu v + \frac{1}{4}\delta u\delta v, \tag{4.40b}$$

$$\mu^2 u = u + \frac{1}{4}\delta^2 u, \tag{4.40c}$$

$$\delta(u\mu v) = \delta u\mu^2 v + \mu u\delta\mu v = v\delta u + \mu(u\delta v). \tag{4.40d}$$

The subscripts are missing because they are the same everywhere. So, one can add subscript i or $i + \frac{1}{2}$ to all the variables in this expression. With these relations we can now perform the equivalent of partial integration with differences.

Lemma 4.1 *A partial integration rule in terms of diferences is given by*

$$\sum_{i=1}^{N} v_i\delta u_i = v_N u_{N+\frac{1}{2}} - v_1 u_{\frac{1}{2}} - \sum_{i=1}^{N-1}\delta v_{i+\frac{1}{2}} u_{i+\frac{1}{2}}. \tag{4.41}$$

Proof Using (4.40d), one can write

$$\sum_{i=1}^{N} v_i\delta u_i = \sum_{i=1}^{N}\delta(\mu v_i u_i) - \mu(\delta v_i u_i)$$

$$= \mu v_{N+\frac{1}{2}}\delta u_{N+\frac{1}{2}} - \mu v_{\frac{1}{2}}\delta u_{\frac{1}{2}} - \sum_{i=1}^{N}\mu(\delta v_i u_i).$$

The last sum can be simplified using

$$\sum_{i=1}^{N}\mu u_i = \frac{1}{2}\sum_{i=1}^{N}(u_{i-\frac{1}{2}} + u_{i+\frac{1}{2}}) = \frac{1}{2}(u_{\frac{1}{2}} + u_{M+\frac{1}{2}}) + \sum_{i=1}^{N-1}u_{i+\frac{1}{2}}.$$

So,

$$\sum_{i=1}^{N} v_i \delta u_i = -\left(\frac{1}{2}\delta v_{N+\frac{1}{2}} - \mu v_{N+\frac{1}{2}}\right) u_{N+\frac{1}{2}} - \left(\frac{1}{2}\delta v_{\frac{1}{2}} + \mu v_{\frac{1}{2}}\right) u_{\frac{1}{2}} - \sum_{i=1}^{N-1} \delta v_{i+\frac{1}{2}} u_{i+\frac{1}{2}}$$

$$= v_N u_{N+\frac{1}{2}} - v_1 u_{\frac{1}{2}} - \sum_{i=1}^{N-1} \delta v_{i+\frac{1}{2}} u_{i+\frac{1}{2}},$$

which shows the assertion.   ∎

When considering energy conservation properties for difference schemes, the derivative $(u\phi)_x$ plays an important role, cf. (3.53). If we consider the 1D case, then a finite-volume discretization of this derivative leads to

$$C_i \phi_i \equiv \delta(u_i \mu \phi_i). \tag{4.42}$$

To derive the adjoint of C_i using a discrete inner product, the following theorem is relevant.

Theorem 4.1

$$\delta\left(u_i\left(\mu\psi_i\mu\phi_i - \frac{1}{4}\delta\psi_i\delta\phi_i\right)\right) = \psi_i\delta(u_i\mu\phi_i) + \mu(u_i\delta\psi_i)\phi_i. \tag{4.43}$$

Proof Using (4.40), we have

$$\delta(u_i\mu\psi_i\mu\phi_i) = \mu^2\psi_i\delta(u_i\mu\phi_i) + \mu(u_i\mu\phi_i)\delta\mu\psi_i$$

$$= \left(\psi_i + \frac{1}{4}\delta^2\psi_i\right)\delta(u_i\mu\phi_i) + \left(\mu u_i\mu^2\phi_i + \frac{1}{4}\delta u_i\delta\mu\phi_i\right)\delta\mu\psi_i$$

$$= \left(\psi_i + \frac{1}{4}\delta^2\psi_i\right)\delta(u_i\mu\phi_i) + \left(\mu u_i\phi_i + \frac{1}{4}(\mu u_i\delta^2\phi_i + \delta u_i\delta\mu\phi_i)\right)\delta\mu\psi_i$$

$$= \left(\psi_i + \frac{1}{4}\delta^2\psi_i\right)\delta(u_i\mu\phi_i) + \left(\mu u_i\phi_i + \frac{1}{4}\delta(u_i\delta\phi_i)\right)\delta\mu\psi_i.$$

Moreover, by an application of (4.40a), we have

$$\delta(u_i\delta\psi_i\delta\phi_i) = \mu(u_i\delta\phi_i)\delta^2\psi_i + \delta(u_i\delta\phi_i)\mu\delta\psi_i \tag{4.44}$$

and, using (4.40d),

$$\mu(u_i\delta\phi_i) = \delta(u_i\mu\phi_i) - \delta u_i\phi_i. \tag{4.45}$$

Consequently,

$$\delta(u_i(\mu\psi_i\mu\phi_i - \frac{1}{4}\delta\psi_i\delta\phi_i)) = \psi_i\delta(u_i\mu\phi_i) + \phi_i\mu u_i\delta\mu\psi_i + \frac{1}{4}\phi_i\delta u_i\delta^2\psi_i$$

$$= \psi_i\delta(u_i\mu\phi_i) + \mu(u_i\delta\psi_i)\phi_i. \tag{4.46}$$

∎

Note that (4.43) also holds for non-equidistant grids, since h is not occurring in δ and μ. For later use it is also relevant to compute the symmetric and skew-symmetric parts of $\mathcal{C}_i$. Therefore, we need the adjoint of this operator. To find it, we need the relevant inner product, which is given by $(\boldsymbol{\phi}, \boldsymbol{\psi}) \equiv \sum_{i=1}^{N} h_i \phi_i \psi_i$, where h_i is the length of the control volume around grid point x_i, and boundary conditions. Here we consider closed walls at both sides given by the conditions $u_{\frac{1}{2}} = u_{N+\frac{1}{2}} = 0$. The approximation of $(u\phi)_x$ is $(\mathcal{C}_i\phi_i)/h_i$. To find the adjoint operator, we have to consider $\sum_{i=1}^{N} h_i (\mathcal{C}_i\phi_i)/h_i \psi_i = \sum_{i=1}^{N} (\mathcal{C}_i\phi_i)\psi_i$. Using (4.43), this leads to $-\sum_{i=1}^{N} \phi_i \mu(u_i \delta \psi_i)$, hence

$$\mathcal{C}_i^* \phi_i = -\mu(u_i \delta \phi_i). \tag{4.47}$$

Now we can construct the symmetric and skew-symmetric parts of the operator on a domain with closed walls. The symmetric part follows from

$$\frac{1}{2}(\mathcal{C}_i + \mathcal{C}_i^*)\phi_i = [\delta(u_i\mu\phi_i) - \mu(u_i\delta\phi_i)]/2$$

$$= \left[\delta u_i \mu^2 \phi_i - \frac{1}{4}\delta u_i \delta^2 \phi_i\right]/2$$

$$= \frac{1}{2}\delta u_i \phi_i. \tag{4.48}$$

This will be zero anywhere in the domain if u_i is constant. Later on, $\delta u_i = 0$ will generalize to the requirement that the discrete divergence should be zero. The skew-symmetric part of the operator $\mathcal{C}$ is

$$\frac{1}{2}(\mathcal{C}_i - \mathcal{C}_i^*)\phi_i = \mathcal{C}_i\phi_i + \delta u_i \phi_i = \frac{1}{2}(u_{i+\frac{1}{2}}\phi_{i+1} - u_{i-\frac{1}{2}}\phi_{i-1}). \tag{4.49}$$

Observe that there is no term with ϕ_i in this expression, which is a necessity for any operator to be skew symmetric. Moreover, if we omit in $\mathcal{C}_i\phi_i$ the term with ϕ_i, we get the skew-symmetric operator.

4.2.2 2D Convection–Diffusion Equation

We will now consider the discrete case of the convection–diffusion equation discussed in Section 3.6.4 for energy conservation. Let

$$\phi_t = -\nabla \cdot (\phi \mathbf{u}) + \nu \Delta \phi,$$

with $\nabla \cdot \mathbf{u} = 0$. The finite-volume discretization on a grid with spacing $h_{i+\frac{1}{2}}$ and $k_{j+\frac{1}{2}}$ between grid points x_i and x_{i+1} and between grid points y_j and y_{j+1}, respectively, and $h_i \equiv (h_{i+\frac{1}{2}} + h_{i-\frac{1}{2}})/2$, $k_j \equiv (k_{j+\frac{1}{2}} + k_{j-\frac{1}{2}})/2$, leads to

$$h_i k_j \frac{d}{dt}\phi_{i,j} = k_j \delta_x u_{i,j} \mu_x \phi_{i,j} + h_i \delta_y v_{i,j} \mu_y \phi_{i,j} + \nu \left(k_j \delta_x \frac{\delta_x \phi_{i,j}}{h_i} + h_i \delta_y \frac{\delta_y \phi_{i,j}}{k_j}\right), \tag{4.50}$$

where the subscripts on δ and μ indicate the direction in which they are applied. Using (4.48), we have for a closed domain that the symmetric part of the first two terms in the right-hand side is

$$(k_j\delta_x u_{i,j} + h_i\delta_y v_{i,j})\phi_{i,j} \tag{4.51}$$

which will be zero if

$$k_j\delta_x u_{i,j} + h_i\delta_y v_{i,j} = 0. \tag{4.52}$$

This is precisely the discretization of the continuity equation for incompressible flows, and we will consider it to be zero in what follows.

For this discretization, we study the conservation of the discrete energy $\|\boldsymbol{\phi}\|^2 = (\boldsymbol{\phi}, \boldsymbol{\phi})$, where the inner product is defined by

$$(\boldsymbol{\phi}, \boldsymbol{\psi}) \equiv \sum_{i=1}^{M}\sum_{j=1}^{N} h_i k_j \phi_{i,j}\psi_{i,j}. \tag{4.53}$$

Like in the continuous case, the energy dissipates in the domain by diffusion and can only be fed or further extracted at the boundary.

Theorem 4.2 *The change of energy in a closed domain is governed by the equation*

$$\|\boldsymbol{\phi}\|\frac{d}{dt}\|\boldsymbol{\phi}\| = -\nu\sum_{j=1}^{N} k_j\left(\phi_{1,j}\frac{\delta_x\phi_{\frac{1}{2},j}}{h_{\frac{1}{2}}} - \phi_{M,j}\frac{\delta_x\phi_{M+\frac{1}{2},j}}{h_{M+\frac{1}{2}}}\right)$$
$$-\nu\sum_{i=1}^{M} h_i\left(\phi_{i,1}\frac{\delta_y\phi_{i,\frac{1}{2}}}{k_{\frac{1}{2}}} - \phi_{i,N}\frac{\delta_y\phi_{i,N+\frac{1}{2}}}{k_{N+\frac{1}{2}}}\right)$$
$$-\nu\sum_{i=1}^{M-1}\sum_{j=1}^{N} k_j\frac{\left(\delta_x\phi_{i+\frac{1}{2},j}\right)^2}{h_{i+\frac{1}{2}}} - \nu\sum_{i=1}^{M}\sum_{j=1}^{N-1} h_i\frac{\left(\delta_y\phi_{i,j+\frac{1}{2}}\right)^2}{k_{j+\frac{1}{2}}}. \tag{4.54}$$

Proof First, set $\nu = 0$. Then, for the associated norm we have, using (4.43),

$$\frac{d}{dt}\|\boldsymbol{\phi}\|^2 = 2\left(\boldsymbol{\phi}, \frac{d}{dt}\boldsymbol{\phi}\right)$$
$$= 2\sum_{i=1}^{M}\sum_{j=1}^{N} \phi_{i,j}(k_j\mathcal{C}_{xi,j} + h_i\mathcal{C}_{yi,j})\phi_{i,j}$$
$$= \frac{1}{2}\sum_{i=1}^{M}\sum_{j=1}^{N} \phi_{i,j}\left(k_j(\mathcal{C}_{xi,j} + \mathcal{C}^*_{x\,i,j}) + h_i(\mathcal{C}_{yi,j} + \mathcal{C}^*_{y\,i,j})\right)\phi_{i,j}$$
$$= \sum_{i=1}^{M}\sum_{j=1}^{N} \phi_{i,j}^2(k_j\delta_x u_{i,j} + h_i\delta_y v_{i,j}) = 0$$

where in the third line the skew-symmetric part will cancel out. In the next line, we use (4.48) and finally (4.52).

For $\nu > 0$, we have to add, apart from a factor 2ν,

$$\sum_{i=1}^{M}\sum_{j=1}^{N} \phi_{i,j}\left(k_j\delta_x\frac{\delta_x\phi_{i,j}}{h_i} + h_i\delta_y\frac{\delta_y\phi_{i,j}}{k_j}\right).$$

Now we apply Lemma 4.1 to both terms. For the first term this gives

$$\sum_{i=1}^{M}\sum_{j=1}^{N} \phi_{i,j}k_j\delta_x\frac{\delta_x\phi_{i,j}}{h_i} = -\sum_{j=1}^{N}k_j\left(\phi_1\frac{\delta_x\phi_{\frac{1}{2},j}}{h_{\frac{1}{2}}} - \phi_M\frac{\delta_x\phi_{M+\frac{1}{2},j}}{h_{M+\frac{1}{2}}}\right) - \sum_{i=1}^{M-1}\sum_{j=1}^{N}k_j\frac{\left(\delta_x\phi_{i+\frac{1}{2},j}\right)^2}{h_{i+\frac{1}{2}}},$$

and a similar result holds for the second term. ∎

From the theorem we deduce that without the diffusion term, that is, $\nu = 0$, and zero velocities at the walls we also have conservation of the discrete energy. We have shown it here for a rectangular domain, but in fact it holds for arbitrary-shaped domains.

Now consider the case $\nu > 0$. It is clear that the sums over the domain are non-positive. So, we have to consider the terms at the boundary, which are all of similar shape. At the left boundary, that is, $i = \frac{1}{2}$ (omitting the subscripts j and x), we have to show that the following part is non-positive ($h_{\frac{1}{2}}$ can be pulled out of the expression):

$$-\phi_1\delta\phi_{\frac{1}{2}} = -\phi_1^2 + \phi_1\phi_0. \tag{4.55}$$

This is non-positive if $\phi_0 = \alpha\phi_1$ with $\alpha \leq 1$. We can relate this to Robin conditions of the form $(\gamma\mu + \delta)\phi_{\frac{1}{2}} = 0$, where $\gamma = 2(1 - \alpha)/(1 + \alpha)$. Here γ goes from 0 to $-\infty$ if α goes from 1 to -1. The former corresponds with a Neumann boundary condition and the latter with a Dirichlet condition at the control volume interface.

4.2.3 Discretization of the Navier–Stokes Equations

As a starting point for the discretization, we will use the incompressible Navier–Stokes equations in conservation form in 2D:

$$\frac{\partial u}{\partial t} + \frac{\partial uu}{\partial x} + \frac{\partial vu}{\partial y} = -\frac{1}{\rho}\frac{\partial p}{\partial x} + \nu\nabla^2 u + f_x, \tag{4.56a}$$

$$\frac{\partial v}{\partial t} + \frac{\partial uv}{\partial x} + \frac{\partial v}{\partial y} = -\frac{1}{\rho}\frac{\partial p}{\partial y} + \nu\nabla^2 v + f_y, \tag{4.56b}$$

$$\frac{\partial u}{\partial x} + \frac{\partial v}{\partial y} = 0. \tag{4.56c}$$

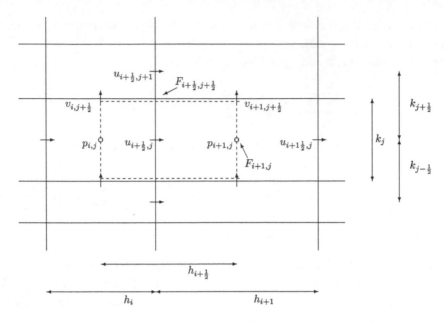

Figure 4.7 Two-dimensional stretched grid and control volume for $u_{i+\frac{1}{2},j}$.

The generalization to 3D is straightforward. The equations are discretized using a finite-volume approach on the staggered grid shown in Fig. 4.7 and will be explained by considering the x-momentum equation and the continuity equation. One defines $h_{i+\frac{1}{2}} = \frac{1}{2}(h_i+h_{i+1})$ and $k_{j+\frac{1}{2}} = \frac{1}{2}(k_j+k_{j+1})$. The x-momentum equation is first written in divergence form,

$$\frac{\partial u}{\partial t} = \nabla \cdot \left(\mathcal{F}(u, v) + \begin{bmatrix} p \\ 0 \end{bmatrix} \right), \tag{4.57}$$

where $\mathcal{F}(u, v)$ is a short for the convective and diffusive fluxes of the x-momentum equation which are given by

$$\mathcal{F}(u, v) = \begin{bmatrix} F(u, v) \\ G(u, v) \end{bmatrix} = \begin{bmatrix} uu - vu_x \\ uv - vu_y \end{bmatrix}.$$

The semi-discrete x-momentum equation is

$$h_{i+\frac{1}{2}}k_j\frac{du_{i+\frac{1}{2},j}}{dt} + k_j(\hat{F}_{i+1,j} - \hat{F}_{i,j}) + h_{i+\frac{1}{2}}\left(\hat{G}_{i+\frac{1}{2},j+\frac{1}{2}} - \hat{G}_{i+\frac{1}{2},j-\frac{1}{2}} \right)$$
$$= -k_j(p_{i+1,j} - p_{i,j}), \tag{4.58}$$

where the discrete fluxes are given by

$$\hat{F}_{i,j} = u_{i,j}\frac{u_{i+\frac{1}{2},j} + u_{i-\frac{1}{2},j}}{2} - v\frac{u_{i+\frac{1}{2},j} - u_{i-\frac{1}{2},j}}{h_i}$$

and

$$\hat{G}_{i+\frac{1}{2},j+\frac{1}{2}} = v_{i+\frac{1}{2},j+\frac{1}{2}}\frac{u_{i+\frac{1}{2},j+1}+u_{i+\frac{1}{2},j}}{2} - v\frac{u_{i+\frac{1}{2},j+1}-u_{i+\frac{1}{2},j}}{k_{j+\frac{1}{2}}}.$$

According to the staggering shown in Fig. 4.7, the velocities $u_{i,j}$ and $v_{i+\frac{1}{2},j+\frac{1}{2}}$ do not exist. They can be computed by taking an average of their surrounding values:

$$u_{i,j} \equiv \frac{1}{2}\left(u_{i-\frac{1}{2},j}+u_{i+\frac{1}{2},j}\right) \quad , \quad v_{i+\frac{1}{2},j+\frac{1}{2}} \equiv \frac{1}{2}\left(h_i v_{i,j+\frac{1}{2}}+h_{i+1}v_{i+1,j+\frac{1}{2}}\right)/h_{i+\frac{1}{2}}. \quad (4.59)$$

For the continuity equation, we take as a control volume the standard grid cell, yielding for the cell around $p_{i,j}$

$$k_j\left(u_{i+\frac{1}{2},j}-u_{i-\frac{1}{2},j}\right) + h_i\left(v_{i,j+\frac{1}{2}}-v_{i,j-\frac{1}{2}}\right) = 0.$$

In order to study the energy properties of the preceding scheme, we first write the equations in terms of δ and μ. Equation (4.58) can be written as

$$h_{i+\frac{1}{2}}k_j\frac{du_{i+\frac{1}{2},j}}{dt} = -k_j\delta_x\hat{F}_{i+\frac{1}{2},j} - h_{i+\frac{1}{2}}\delta_y\hat{G}_{i+\frac{1}{2},j} - k_j\delta_x p_{i+\frac{1}{2},j} \quad (4.60)$$

with

$$\hat{F}_{i,j} = u_{i,j}\mu_x u_{i,j} - \frac{v}{h_i}\delta_x u_{i,j},$$

$$\hat{G}_{i+\frac{1}{2},j+\frac{1}{2}} = v_{i+\frac{1}{2},j+\frac{1}{2}}\mu_y u_{i+\frac{1}{2},j+\frac{1}{2}} - \frac{v}{k_{j+\frac{1}{2}}}\delta_y u_{i+\frac{1}{2},j+\frac{1}{2}}.$$

We will consider the case of a closed rectangular domain, where $u_{\frac{1}{2},j} = u_{M+\frac{1}{2},j} = 0$ for $j = 1\cdots N$ and $v_{i,\frac{1}{2}} = v_{i,N+\frac{1}{2}} = 0$ for $j = 1\cdots M$. The relevant energy here is the discrete kinetic energy defined by

$$E_k \equiv \sum_{i=1}^{M-1}\sum_{j=1}^{N}h_{i+\frac{1}{2}}k_j u_{i+\frac{1}{2},j}^2 + \sum_{i=1}^{M}\sum_{j=1}^{N-1}k_{j+\frac{1}{2}}h_i v_{i,j+\frac{1}{2}}^2, \quad (4.61)$$

that is, the sum over the energy in all the control volumes inside the domain.

Theorem 4.3 *For a rectangular domain with no flow through the boundaries, the time derivative of the energy is given by*

$$\frac{dE_k}{dt} = -v\sum_{j=1}^{N}k_j\left(\frac{u_{M-\frac{1}{2},j}^2}{h_M} + \frac{u_{\frac{3}{2},j}^2}{h_1} + \sum_{i=2}^{M-1}\frac{(\delta_x u_{i,j})^2}{h_i}\right)$$

$$-v\sum_{i=1}^{M-1}h_{i+\frac{1}{2}}\left(-u_{i+\frac{1}{2},N}\frac{\delta_y u_{i+\frac{1}{2},N+\frac{1}{2}}}{k_{N+\frac{1}{2}}} + u_{i+\frac{1}{2},1}\frac{\delta_y u_{i+\frac{1}{2},\frac{1}{2}}}{k_{\frac{1}{2}}} + \sum_{j=1}^{N-1}\frac{(\delta_y u_{i+\frac{1}{2},j+\frac{1}{2}})^2}{k_{j+\frac{1}{2}}}\right)$$

$$- \nu \sum_{j=1}^{N-1} k_{j+\frac{1}{2}} \left(-v_{M,j+\frac{1}{2}} \frac{\delta_x v_{M+\frac{1}{2},j+\frac{1}{2}}}{h_{M+\frac{1}{2}}} + v_{1,j+\frac{1}{2}} \frac{\delta_x v_{\frac{1}{2},j+\frac{1}{2}}}{h_{\frac{1}{2}}} + \sum_{i=1}^{M-1} \frac{(\delta_x v_{i+\frac{1}{2},j+\frac{1}{2}})^2}{h_{i+\frac{1}{2}}} \right)$$

$$- \nu \sum_{i=1}^{M} h_i \left(\frac{v_{i,N-\frac{1}{2}}^2}{k_N} + \frac{v_{i,\frac{3}{2}}^2}{k_1} + \sum_{j=2}^{N-1} \frac{(\delta_y v_{i,j})^2}{k_j} \right).$$

Proof First, we consider the convective part of (4.60):

$$k_j \delta_x(u_{i+\frac{1}{2},j} \mu_x u_{i+\frac{1}{2},j}) + h_{i+\frac{1}{2}} \delta_y(v_{i+\frac{1}{2},j} \mu_y u_{i+\frac{1}{2},j}).$$

In the symmetric part of this operator, we have, similar to (4.51), the part multiplying u_i:

$$k_j \delta_x u_{i+\frac{1}{2},j} + h_{i+\frac{1}{2}} \delta_y v_{i+\frac{1}{2},j}. \tag{4.62}$$

Now we use the definitions (4.59), written here as

$$u_{i,j} \equiv \mu_x u_{i,j} \text{ and } v_{i+\frac{1}{2},j+\frac{1}{2}} \equiv (\mu_x(h_{i+\frac{1}{2}} v_{i+\frac{1}{2},j+\frac{1}{2}}))/h_{i+\frac{1}{2}},$$

and find that (4.62) is equal to

$$\mu_x(k_j \delta_x u_{i+\frac{1}{2},j} + h_{i+\frac{1}{2}} \delta_y v_{i+\frac{1}{2},j}),$$

which is precisely the x-average of the discrete continuity equation, hence zero.

For the diffusion terms there are two essential parts to be considered. The first originates from the differences in x-direction in the u-momentum equation. For readability we omit the subscript j:

$$\sum_{i=1}^{M-1} u_{i+\frac{1}{2}} \delta_x(\frac{\delta_x u_{i+\frac{1}{2}}}{h_{i+\frac{1}{2}}}) = -\sum_{i=1}^{M-2} \frac{(\delta_x u_{i+1})^2}{h_{i+1}} + u_{M-\frac{1}{2}} \frac{\delta_x u_M}{h_M} - u_{\frac{3}{2}} \frac{\delta_x u_1}{h_1}$$

$$= -\sum_{i=2}^{M-1} \frac{(\delta_x u_i)^2}{h_i} - \frac{u_{M-\frac{1}{2}}^2}{h_M} - \frac{u_{\frac{3}{2}}^2}{h_1},$$

where we have applied Lemma 4.1 in the first step and applied the Dirichlet conditions in the second step. The second essential part originates from the differences in y-direction in the u-momentum equation. For readability we omit the subscript i:

$$\sum_{j=1}^{N-1} u_j \delta_y(\frac{\delta_y u_j}{k_j}) = -\sum_{j=1}^{N-1} \frac{(\delta_y u_{j+\frac{1}{2}})^2}{k_{j+\frac{1}{2}}} + u_N \frac{\delta_y u_{N+\frac{1}{2}}}{k_{N+\frac{1}{2}}} - u_1 \frac{\delta_y u_{\frac{1}{2}}}{k_{\frac{1}{2}}}.$$

A new term here is that related to the pressure. This term will vanish, as can be shown from

$$\sum_{i=1}^{M-1}\sum_{j=1}^{N} k_j u_{i+\frac{1}{2},j}\delta_x p_{i+\frac{1}{2},j} + \sum_{i=1}^{M}\sum_{j=1}^{N-1} h_i v_{i,j+\frac{1}{2}}\delta_y p_{i,j+\frac{1}{2}}$$

$$= \sum_{i=1}^{M}\sum_{j=1}^{N} k_j \mu_x(u_{i,j}\delta_x p_{i,j}) + h_i \mu_y(v_{i,j}\delta_y p_{i,j})$$

$$= \sum_{i=1}^{M}\sum_{j=1}^{N} k_j \delta_x(u_{i,j}\mu_x p_{i,j}) + h_i \delta_y(v_{i,j}\mu_y p_{i,j})$$

$$- p_{i,j}(k_j \delta_x u_{i,j} + h_i \delta_y v_{i,j}) = 0,$$

where in the first step we used that normal velocities are zero on the boundary. In the second step we used (4.40d), and finally we used again that the normal velocities are zero on the boundary and in the second term that the discrete continuity equation is zero. ∎

Observe that on the second and third lines in the expression in Theorem 4.3 we find on the boundaries the discrete equivalent of the tangential velocity times the normal derivative of that velocity, which also occurred in the continuous case. The terms are actually similar to (4.55), where it is shown that a Neumann and a Dirichlet condition gives a negative contribution and also which Robin condition makes it negative.

These relations make it possible to show similar properties for the energy, as we did in Section 3.6.5.

We note that the preceding discretization can have a decoupling of velocities at odd and even grid points at high Reynolds numbers, resulting in so-called wiggles in the solution; but by refining the grid at those locations we can make the coupling stronger and get rid of the wiggles. See also Example 4.3 on linear convection diffusion, discussing the same issue in terms of monotonicity of the solution.

When we collect all velocities $u_{i+\frac{1}{2},j}$ and $v_{i,j+\frac{1}{2}}$ on the grid in the vectors in $\mathbf{u}$ and $\mathbf{v}$, respectively, and define $\mathbf{x} \equiv [\mathbf{u}; \mathbf{v}]$, then the discretized Navier–Stokes equations can also be written as the system

$$M\frac{d}{dt}\mathbf{x} = N(\mathbf{x}, \mathbf{x}) + L\mathbf{x} + G\mathbf{p} + \mathbf{f},$$
$$0 = D\mathbf{x}, \tag{4.63}$$

where $N(\mathbf{x}, \mathbf{x})$, L, G, and D represent the discretization of the convection terms, the discrete Laplace operator, the discrete gradient operator, and discrete divergence operator, respectively. Moreover, $\mathbf{f}$ represents the forcing.

4.3 Convergence Theory

4.3.1 Stability and Consistency

In the previous section we have seen how a PDE $\mathcal{L}u = f$ defined on a domain Ω, with boundary conditions on $\partial\Omega$, can be approximated by a system of difference equations for the internal points $L_h U = f_h$ and the boundary points $B_h U = g_h$. For the following analysis, we eliminate the unknowns determined by the boundary conditions from the equations such that we end up with a linear system: $A_h U = F_h$.

We will write the restriction of the exact solution u of our PDE problem to the grid as the vector $\mathbf{u}$. In general, the exact solution $\mathbf{u}$ will not satisfy the discrete equation and the residual

$$\tau_h \equiv A_h\mathbf{u} - \mathbf{F}_h = A_h(\mathbf{u} - \mathbf{U}) \tag{4.64}$$

is called *local discretization error*. It indicates the local approximation errors, since each of its components depends only on values of u in a few surrounding grid points. This error could be estimated using Taylor expansions expressed in the mesh size h and the partial derivatives of u.

Definition 4.1 (Consistency) *When for $h \to 0$ also $\tau_h \to 0$, the discretization is consistent. If $\tau_h = O(h^p)$, then the discretization is consistent of order p.*

Note that the vectors occurring here become longer when h tends to zero. As a consequence of the local discretization error, the solution of the discretization $\mathbf{U}$ also will differ from the exact solution $\mathbf{u}$. The difference

$$\mathbf{e}_h = \mathbf{u} - \mathbf{U} \tag{4.65}$$

is called the *global discretization error*, since the error is a result of local discretization errors all over the domain.

Definition 4.2 (Convergence) *When for $h \to 0$ the global discretization error $\mathbf{u} - \mathbf{U}$ tends to zero, then the numererical solution is convergent.*

To show that the global discretization error is bounded by the local discretization error we need the operator A_h to be stable. So in general, if we consider $A_h\mathbf{U} = \mathbf{F}_h$ and the same problem with a slight perturbation in the right-hand side $A_h\tilde{\mathbf{U}} = \mathbf{F}_h + \boldsymbol{\delta}_h$, then the difference $\boldsymbol{\epsilon}_h = \tilde{\mathbf{U}} - \mathbf{U}$, which is a solution of $A_h\boldsymbol{\epsilon}_h = \boldsymbol{\delta}_h$, should be bounded in $\boldsymbol{\delta}_h$. It follows from (4.64) that the restriction of u to the grid is a solution of the perturbed problem $A_h\mathbf{u} = \mathbf{F}_h + \tau_h$; hence, from stability we will find convergence. Stability follows from coercivity or here from the Poincaré inequality (see also Theorem 3.2). Suppose $\mathbf{y} = (y_0, y_1, \cdots y_{M+1})^T \in \mathbb{R}^{M+2}$ and similar for $\mathbf{z}$;

then $(\mathbf{y}, \mathbf{z})_h \equiv \sum_{i=1}^{M} y_i z_i h_i$, where h_i is the size of the control volume. Then stability follows if

$$(\mathbf{y}, A_h \mathbf{y})_h \geq c ||\mathbf{y}||_h^2 \qquad (4.66)$$

for arbitrary vectors $\mathbf{y}$ (see (3.24)). In fact, c can be chosen the minimum eigenvalue of the symmetric part of the matrix A_h. Applying the Cauchy–Schwarz inequality to (4.66) gives $c||\mathbf{y}||_h^2 \leq ||\mathbf{y}||_h ||A_h \mathbf{y}||_h$ which after division by $||\mathbf{y}||_h$ gives $c||\mathbf{y}||_h \leq ||A_h \mathbf{y}||_h$. So, in the special case of $\mathbf{y} = \boldsymbol{\epsilon}_h$ we find $||\boldsymbol{\epsilon}_h||_h \leq \frac{1}{c}||\boldsymbol{\delta}_h||_h$. In our stability study, this yields

$$||\mathbf{e}_h||_h \leq \frac{1}{c}||\boldsymbol{\tau}_h||_h. \qquad (4.67)$$

It appears quite general that convergence can be proven from stability and consistency. The theorem making this claim precise is named the Lax and Richtmyer *equivalence theorem*. So in general 'Consistency and Stability' = 'Convergence'.

Additional Material

- For more details on the theory of convergence, stability, and consistency, see the famous book of Richtmyer and Morton (1967).

We have seen that for the continous case, stability follows from the Poincaré inequality (see Section 3.5.3). Since, in contrast to the finite element discretization, the finite difference and finite volume discretization do not inherit the Poincaré inequality from the continuous case, we need a discrete variant of it, and we will show it for a non-equidistant 1D grid. First, we define $\delta v_{i-\frac{1}{2}} = v_i - v_{i-1}$, $h_{i-\frac{1}{2}} = x_i - x_{i-1}$ and h_i is the size of the contol volume, for example $h_i = (h_{i-\frac{1}{2}} + h_{i+\frac{1}{2}})/2$. Define

$$L = \max(\sum_{i=1}^{M+1} h_{i-\frac{1}{2}}, \sum_{i=1}^{M} h_i),$$

$$||\mathbf{v}||_h \equiv \sqrt{\sum_{i=1}^{M} v_i^2 h_i}$$

$$||\frac{\delta \mathbf{v}}{\delta x}||_{\hat{h}} \equiv \sqrt{\sum_{l=2}^{M} (\frac{\delta v_{l-\frac{1}{2}}}{h_{l-\frac{1}{2}}})^2 h_{l-\frac{1}{2}}},$$

and, moreover, the vector $\mathbf{1}$ each element of which is one. Then we have the following theorem regarding the discrete Poincaré inequality.

Theorem 4.4 (Discrete Poincaré inequality) *We have that*

$$||\mathbf{v} - v_i\mathbf{1}||_h \le L||\frac{\delta \mathbf{v}}{\delta x}||_{\hat{h}} \text{ for } i = 1, \cdots M. \tag{4.68}$$

When $i = 0$ is added, we have

$$||\mathbf{v} - v_i\mathbf{1}||_h \le L\sqrt{\frac{(\delta v_{\frac{1}{2}})^2}{h_{\frac{1}{2}}} + ||\frac{\delta \mathbf{v}}{\delta x}||_{\hat{h}}^2} \text{ for } i = 0, \cdots M. \tag{4.69}$$

Instead, when $i = M + 1$ is added, we have

$$||\mathbf{v} - v_i\mathbf{1}||_h \le L\sqrt{\frac{(\delta v_{M+\frac{1}{2}})^2}{h_{M+\frac{1}{2}}} + ||\frac{\delta \mathbf{v}}{\delta x}||_{\hat{h}}^2} \text{ for } i = 1, \cdots M + 1. \tag{4.70}$$

When both $i = 0$ and $i = M + 1$ are added, we have

$$||\mathbf{v} - v_i\mathbf{1}||_h \le L\sqrt{\frac{(\delta v_{\frac{1}{2}})^2}{h_{\frac{1}{2}}} + \frac{(\delta v_{M+\frac{1}{2}})^2}{h_{M+\frac{1}{2}}} + ||\frac{\delta \mathbf{v}}{\delta x}||_{\hat{h}}^2} \text{ for } i = 0, \cdots M + 1 \tag{4.71}$$

Proof Let $j \ge i$; then

$$v_j - v_i = \sum_{l=i+1}^{j} \delta v_{l-\frac{1}{2}}.$$

Now, for $1 \le i, j \le M$, we have

$$|v_j - v_i| \le \sum_{l=i+1}^{j} |\delta v_{l-\frac{1}{2}}| \le \sum_{l=2}^{M} |\delta v_{l-\frac{1}{2}}| = \sum_{l=2}^{M} \sqrt{h_{l-\frac{1}{2}}} \frac{|\delta v_{l-\frac{1}{2}}|}{h_{l-\frac{1}{2}}} \sqrt{h_{l-\frac{1}{2}}}$$

$$\le \sqrt{\sum_{l=2}^{M} h_{l-\frac{1}{2}}} \sqrt{\sum_{l=2}^{M} (\frac{\delta v_{l-\frac{1}{2}}}{h_{l-\frac{1}{2}}})^2 h_{l-\frac{1}{2}}} \le \sqrt{L}||\frac{\delta \mathbf{v}}{\delta x}||_{\hat{h}}, \tag{4.72}$$

where the third inequality follows from the Cauchy–Schwarz inequality. If $j < i$, we can just interchange the roles of i and j and we find the same inequality. Our first result follows now from

$$||\mathbf{v} - v_i\mathbf{1}||_h^2 = \sum_{j=1}^{M} (v_j - v_i)^2 h_j \le \sum_{j=1}^{M} L||\frac{\delta \mathbf{v}}{\delta x}||_{\hat{h}}^2 h_j \le L||\frac{\delta \mathbf{v}}{\delta x}||_{\hat{h}}^2 \sum_{j=1}^{M} h_j \le L^2||\frac{\delta \mathbf{v}}{\delta x}||_{\hat{h}}^2.$$

If now $i = 0$, then (4.72) becomes

$$|v_j - v_i| \le \sqrt{L}\sqrt{\frac{(\delta v_{\frac{1}{2}})^2}{h_{\frac{1}{2}}} + ||\frac{\delta \mathbf{v}}{\delta x}||_{\hat{h}}^2},$$

which will lead in a similar way as the previous case to the second result. Similarly, if $i = M + 1$, then (4.72) becomes

$$|v_j - v_i| \leq \sqrt{L}\sqrt{\frac{(\delta v_{M+\frac{1}{2}})^2}{h_{M+\frac{1}{2}}} + ||\frac{\delta \mathbf{v}}{\delta x}||_{\hat{h}}^2},$$

which will lead to the third result. ∎

Theorem 4.5 *If at the left boundary we have* $\gamma v_0 + (1 - \gamma)v_1 = 0$ *with* $0 < \gamma \leq 1$, *then*

$$||\mathbf{v}||_h \leq L\sqrt{\frac{(\delta v_{\frac{1}{2}})^2}{h_{\frac{1}{2}}} + ||\frac{\delta \mathbf{v}}{\delta x}||_{\hat{h}}^2}. \tag{4.73}$$

Proof We have

$$||\mathbf{v}||_h = ||\mathbf{v} - (\gamma v_0 + (1 - \gamma)v_1)\mathbf{1}||_h = ||\gamma(\mathbf{v} - v_0\mathbf{1}) + (1 - \gamma)(\mathbf{v} - v_1\mathbf{1})||_h$$

$$\leq \gamma||\mathbf{v} - v_0\mathbf{1})||_h + (1 - \gamma)||\mathbf{v} - v_1\mathbf{1})||_h \leq L\sqrt{\frac{(\delta v_{\frac{1}{2}})^2}{h_{\frac{1}{2}}} + ||\frac{\delta \mathbf{v}}{\delta x}||_{\hat{h}}^2},$$

where the first inequality holds because of the triangle inequality. ∎

This theorem does contain the standard cases where the Dirichlet boundary condition is imposed at the grid point, $v_0 = 0$, and the one where it is imposed in the middle between two grid points, $v_0 + v_1 = 0$.

If Dirichlet conditions are being described at all boundaries at the grid points, then there will be no approximations at those points, and the preceding can be straightforwardly applied to prove convergence. However, if we need an interpolation to fix a Dirichlet condition between grid points, or if we have Neumann or Robin conditions, we get an approximation error which needs a modification of the preceding approach. We will discuss this issue in the following example.

Example 4.7 *In this example we will show second-order convergence for a simple equation. Consider on the interval [0,1] the differential equation*

$$\frac{du}{dx} - \frac{d}{dx}(\exp(x)\frac{du}{dx}) = 0 \tag{4.74}$$

with boundary conditions $u(0) = 1$ *and* $\frac{du}{dx}(1) = 2$. *Suppose we make a finite volume discretization of this equation on an equidistant grid with mesh size h and where each boundary condition is applied at the interface of a control volume. For the grid, we assume that* $x_j = (j - \frac{1}{2})h$ *with* $x_{n+1} = 1 + h/2$, *so* $h = 1/n$. *This yields the discretization*

$$F_{j+\frac{1}{2}} - F_{j-\frac{1}{2}} = 0 \text{ for } j = 1, \ldots, n \tag{4.75}$$

with

$$F_{j+\frac{1}{2}} = \frac{U_{j+1} + U_j}{2} - \exp(x_{j+\frac{1}{2}})\frac{U_{j+1} - U_j}{h} \text{ for } j = 0, \ldots, n,$$

where $x_{j+\frac{1}{2}} = jh$. The discretizations of the boundary conditions are $U_0 + U_1 = 2$ and $U_{n+1} - U_n = 2h$. We substitute the boundary conditions leading to $F_{\frac{1}{2}} = -2U_1/h$ and $F_{n+\frac{1}{2}} = U_n$. In these two expressions we did not incorporate the constant parts, since they will not occur in the bi-linear form.

The bi-linear form attached to the discrete problem in the previous part is given by

$$a(\mathbf{V}, \mathbf{U}) \equiv \sum_{j=1}^{n} V_j(F_{j+\frac{1}{2}} - F_{j-\frac{1}{2}})$$

$$= -V_1 F_{\frac{1}{2}} + V_n F_{n+\frac{1}{2}} - \sum_{j=2}^{n} (V_j - V_{j-1})F_{j-\frac{1}{2}}$$

$$= 2V_1 U_1/h + V_n U_n$$
$$- \sum_{j=2}^{n} (V_j - V_{j-1})[(U_j + U_{j-1})/2 + (U_j - U_{j-1})\exp(x_{j-\frac{1}{2}})/h],$$

where the elements of the vector $\mathbf{U}$ are $U_1, U_2, \ldots, U_n$, similar for $\mathbf{V}$.

Now we need to show that $a(\mathbf{U}, \mathbf{U})$ is non-negative:

$$a(\mathbf{U}, \mathbf{U}) = 2U_1^2/h + U_n^2 - \sum_{j=2}^{n} (U_j - U_{j-1})(U_j + U_{j-1})/2 + (U_j - U_{j-1})^2 \exp\left(x_{j-\frac{1}{2}}\right)/h$$

$$= 2U_1^2/h + U_n^2 - \frac{1}{2}\sum_{j=2}^{n} U_j^2 - U_{j-1}^2 + \sum_{j=2}^{n}(U_j - U_{j-1})^2 \exp\left(x_{j-\frac{1}{2}}\right)/h$$

$$= 2U_1^2/h + U_n^2 + \frac{1}{2}U_1^2 - \frac{1}{2}U_n^2 + \sum_{j=2}^{n}(U_j - U_{j-1})^2 \exp\left(x_{j-\frac{1}{2}}\right)/h$$

$$= \left(\frac{1}{2} + \frac{2}{h}\right)U_1^2 + \frac{1}{2}U_n^2 + \sum_{j=2}^{n}(U_j - U_{j-1})^2 \exp\left(x_{j-\frac{1}{2}}\right)/h \geq 0.$$

Even more can be shown, using that $U_1 = -U_0$:

$$a(\mathbf{U}, \mathbf{U}) \geq \frac{2}{h}U_1^2 + \sum_{j=2}^{n}(U_j - U_{j-1})^2 \exp\left(x_{j-\frac{1}{2}}\right)/h$$

$$= \frac{1}{2h}(U_1 - U_0)^2 + \sum_{j=2}^{n}(U_j - U_{j-1})^2 \exp\left(x_{j-\frac{1}{2}}\right)/h$$

$$\geq \frac{1}{2}\sum_{j=1}^{n}(U_j - U_{j-1})^2/h \geq \frac{1}{2}\|\mathbf{U}\|_h^2,$$

where in the last step we use Theorem 4.5, with $\gamma = \frac{1}{2}$ and $L = 1$. This relation holds independent of n (recall $h = 1/n$), hence we have a stable discretization. Next, we have to determine the vector τ_h. Due to the central discretizations used, we easily deduce that $(\tau_h)_j = O(h^2)$ for $j = 2, \cdots n - 1$. However, the exact solution will not precisely fit the boundary conditions. So $(u(x_0) + u(x_1))/2 - 1 = (u(x_0) + u(x_1))/2 - u(x_{\frac{1}{2}}) = h^2/4 d^2u/dx^2 = O(h^2)$. After substitution in (4.75) for $j = 1$, we encounter that we have to divide this approximation error by h^2 so we will get $(\tau_h)_1 = O(1)$. Similarly, we get at the other boundary $(\tau_h)_n = O(h)$. This is not enough to show that the global error is $O(h^2)$. Therefore, instead of considering convergence to the exact solution we consider convergence to a perturbation of that, that is, $\hat{u}(x) \equiv u(x) + h^2 w(x, h)$ where w is chosen such that $\hat{u}(x)$ satisfies both discrete boundary conditions exactly. So, $(\hat{u}(x_0) + \hat{u}(x_1))/2 = 1$ and $\hat{u}(x_{n+1}) - \hat{u}(x_n) = 2h$. Consequently, the elimination from the boundary conditions will not give a contribution to the truncation error. However, the truncation error of (4.75) will change to

$$(\tau_h(\hat{u}))_j = (\tau_h(u))_j + h^2 \left(\frac{dw}{dx} - \frac{d}{dx} \left(\exp(x) \frac{dw}{dx} \right) \right)_{x=x_j} + O(h^4)$$

for $j = 1, \ldots, n$. Consequently, due to stability, $\mathbf{u}$ converges to $\hat{u}$ with second-order accuracy, and $\hat{u}$ will by definition converge to u with second-order accuracy. Hence, $\mathbf{u}$ converges with second-order accuracy when h tends to zero. Note that we should be able to find a twice-differentiable function $w(x, h)$ such that $w(x_0) + w(x_1) = -(u(x_0) + u(x_1) - 2)/h^2 = -(u(x_0) + u(x_1) - 2u(x_{\frac{1}{2}}))/h^2 = -d^2u/dx^2(0) + O(h^2)$ and $(w(x_{n+1}) - w(x_n))/h = (2 - (u(x_{n+1}) - u(x_n))/h)/h^2 = \frac{1}{12} d^3u/dx^3(1) + O(h^2)$. In this case, we can choose a linear function, say $w(x, h) = \alpha(h) + \beta(h)x$, where we have two equations $2\alpha(h) = -(u(x_0) + u(x_1) - 2)/h^2$ and $\beta(h) = (2 - (u(x_{n+1}) - u(x_n))/h)/h^2$ to fix the constants. Hence, $w(x, h)$ converges to $-d^2u/dx^2(0)) + \frac{1}{12} d^3u/dx^3(1)x$ for $h \to 0$. Substituting this w in (4.74) will give a bounded function of x.

It is beyond the scope of this text to prove stability of discretizations in general, but one can quite easily observe that some discretization may be prone to instability. The discrete operator will, due to consistency, very much be acting the same on smooth functions as the original continuous operator does. The difference is expected for fast-oscillating wave-like functions. If for such wave-like functions the difference operator is nearly zero and much smaller than when applied to a smooth function of the same magnitude, while the continuous form is not, then this indicates an instable discretization (equivalent to a very small c in (3.24)). It occurs, for example, for discretizations where the stencil of the discretization does not connect the odd points to the even points of the grid, for example the central discretization for u_x as used in the convection–diffusion equation in Example 4.2. In that example we also discussed how to be careful with the boundary condition in order not to overspecify the convective term. Moreover, at the end of Section 4.1.2

it is shown that one can add artificial diffusion to stabilize the discretization. In the next section, we will shortly discuss how one can construct a stable discretization. *See example Ex. 4.5*

The discrete Poincaré inequality is also useful to estimate the decay rate of the energy in parabolic equations, as will be shown in the next example.

Example 4.8 (Energy decay for convection diffusion in a closed domain)

In Theorem 4.2 we gave an expression for the decay of the energy. Assume we have homogeneous Dirichlet conditions for ϕ at all walls, that is, $\phi_{0,j} + \phi_{1,j} = 0$ for the left wall and similarly for the other walls. Then the first term of the first sum in the right-hand side of (4.54) becomes, apart from a factor,

$$\phi_{1,j}\delta_x\phi_{\frac{1}{2},j} = 2\phi_{1,j}^2.$$

Hence, (4.54) becomes

$$||\phi||\frac{d}{dt}||\phi|| = -\nu\sum_{j=1}^{N} k_j \left(2\frac{\phi_{1,j}^2}{h_{\frac{1}{2}}} + 2\frac{\phi_{M,j}^2}{h_{M+\frac{1}{2}}} + \sum_{i=1}^{M-1} \frac{(\delta_x\phi_{i+\frac{1}{2},j})^2}{h_{i+\frac{1}{2}}} \right)$$

$$-\nu\sum_{i=1}^{M} h_i \left(2\frac{\phi_{i,1}^2}{k_{\frac{1}{2}}} + 2\frac{\phi_{i,N}^2}{k_{N+\frac{1}{2}}} + \sum_{j=1}^{N-1} \frac{(\delta_y\phi_{i,j+\frac{1}{2}})^2}{k_{j+\frac{1}{2}}} \right).$$

Using the Poincaré inequality, we find

$$||\phi||\frac{d}{dt}||\phi|| \leq -\nu\sum_{j=1}^{N} k_j/L^2 \sum_{i=1}^{M} \phi_{i,j}^2 h_i$$

$$-\nu\sum_{i=1}^{M} h_i/L^2 \sum_{j=1}^{N} \phi_{i,j}^2 k_j$$

$$= -\frac{2}{L^2}\nu\sum_{i=1}^{M}\sum_{j=1}^{N} \phi_{i,j}^2 h_i k_j$$

$$= -\frac{2\nu}{L^2}||\phi||^2.$$

Hence, $\frac{d}{dt}||\phi|| \leq -\frac{2\nu}{L^2}||\phi||$, showing how fast the energy will go to zero.

Similarly we can give a bound for the decay of the kinetic energy of an incompressible fluid in a closed container without any forcing, starting from Theorem 4.3.

4.3.2 Construction of Stable Finite-Difference and Finite-Volume Discretizations

In the last section, we have seen that it is important that a bi-linear form related to a discretized second-order PDE satisfies a discrete Poincaré inequality. In order to achieve this, we could also work the other way around, starting from a continuous bi-linear form, discretize that one, and rework it to the discretization of the PDE.

Consider the 1D equation

$$-\frac{d}{dx}\left(v(x)\frac{d}{dx}u\right) = f(x) \tag{4.76}$$

on [0,1] with $v(x) > c > 0$. This leads to a well-posed problem if we prescribe u at the boundaries, say $u(0) = u(1) = 0$, because the associated bi-linear form is given by $a(v,u) = \int_0^1 \frac{dv}{dx}v(x)\frac{du}{dx}dx$ and is coercive; see Section 3.5.2. Also, the associated operator will be self-adjoint. Now we discretize this bi-linear form as follows:

$$\hat{a}(\mathbf{V},\mathbf{U})) \equiv \sum_{j=0}^{n} \delta V_{j+\frac{1}{2}}v\left(x_{j+\frac{1}{2}}\right)\delta U_{j+\frac{1}{2}}/h$$

$$= -V_0 v\left(x_{-\frac{1}{2}}\right)\delta U_{-\frac{1}{2}}/h + V_{n+1}v\left(x_{n+\frac{3}{2}}\right)\delta U_{n+\frac{3}{2}}/h - \sum_{j=0}^{n+1} V_j\delta(v(x_j)\delta U_j)/h$$

with $h = 1/(n+1)$ and where we used Lemma 4.1 to get all the operators acting on $\mathbf{U}$. Applying the Dirichlet boundary conditions, we get

$$\hat{a}(\mathbf{V},\mathbf{U})) = -\sum_{j=1}^{n} V_j\delta(v(x_j)\delta U_j)/h.$$

Similarly, one can discretize the linear form associated to (4.76):

$$F(\mathbf{V}) = \sum_{j=1}^{n} V_j f(x_j)h.$$

Considering the problem of finding $\mathbf{U}$ such that $\hat{a}(\mathbf{V},\mathbf{U})) = F(\mathbf{V})$ for arbitrary $\mathbf{V}$ leads to the desired discretization:

$$-\delta(v(x_j)\delta U_j)/h = f(x_j)h.$$

Next to this difference notation, we could also use a matrix notation for the bi-linear form:

$$\hat{a}(\mathbf{V},\mathbf{U}) = h(D\mathbf{V}, D\mathbf{U})$$

with D defined by $(Du)_j = (U_j - U_{j-1})/h$ for $j = 2, \cdots, n$, $(Du)_1 = U_1/h$, and $(Du)_{n+1} = -u_n/h$, hence $D \in \mathbb{R}^{n+1,n}$ and $F(\mathbf{V}) = (\mathbf{V},\mathbf{f})h$ with $\mathbf{f} = (f(x_1),\ldots,f(x_n))^T$. This leads immediately to a matrix-vector form of the discretized equation

$$D^T D\mathbf{U} = \mathbf{f}.$$

Let us next consider the 2D equation

$$- \nabla \cdot (A \nabla u) = f \tag{4.77}$$

on the unit square, that is, $[0, 1] \times [0, 1]$. Again we assume Dirichlet boundary conditions. The associated 2D bi-linear form is given by

$$a(v, u) = \int_0^1 \int_0^1 v \nabla \cdot A \nabla u \, dx dy = \int_0^1 \int_0^1 (\nabla v, A \nabla u) \, dx dy. \tag{4.78}$$

So, the coercivity is expected to be satisfied if A is positive definite. The discretized counterpart of this bi-linear form on a uniform grid is, with $x_0 = y_0 = 0$ and $x_{n+1} = y_{n+1} = 1, h = 1/(n + 1)$,

$$\hat{a}(\mathbf{V}, \mathbf{U}) = h \left(\begin{bmatrix} D_x \\ D_y \end{bmatrix} \mathbf{V}, \hat{A} \begin{bmatrix} D_x \\ D_y \end{bmatrix} \mathbf{U} \right)$$

where $D_x \equiv I \otimes D$ and $D_y \equiv D \otimes I$ are matrices with D as in the previous 1D case and $\otimes$ denotes the Kronecker product. Moreover, $\mathbf{U}$ and $\mathbf{V}$ are vectors where the grid U_{ij} are stored in lexicographical ordering. Next to this we have $F(\mathbf{V}) = (\mathbf{V}, \mathbf{f})h$, where $\mathbf{f}$ contains the evaluations of f at the grid points ordered in the same lexicographical ordering. This bi-linear form is clearly positive definite if $\hat{A}$ is positive definite. Note that $\hat{A}$ is not a square matrix of order 2 as A is. In general, it will also not be diagonal. Also, here we get the equation immediately in matrix-vector form:

$$\begin{bmatrix} D_x \\ D_y \end{bmatrix}^T \hat{A} \begin{bmatrix} D_x \\ D_y \end{bmatrix} \mathbf{U} = \mathbf{f}. \tag{4.79}$$

See example Ex. 4.6.

In general, positive definiteness is satisfied inside a domain with a standard discretization. However, at the boundaries we may experience difficulties. *Considering the positivity of the associated bi-linear form will help us to find the correct discretization of the boundary conditions.* The preceding is a simple example, but if a Poisson equation is defined in a cylinder, a sphere, or a thin layer on a sphere, then it is essential not to drop small terms from the equations just because they are provably small. One easily can lose positive definiteness from the continuous form of these equations. So, the recommendation is to write the operator in a form such that the self-adjointness is apparent. One should discretize this form to inherit positive definiteness into the discretization. *See example Ex. 4.7*

In some PDEs the self-adjoint part is not apparent. For example, using the product rule on (4.76), one finds that

$$- \left(\frac{d}{dx} v(x) \right) \frac{d}{dx} u - v(x) \frac{d^2}{dx^2} u = f(x) \tag{4.80}$$

gives a convection–diffusion equation. One can always try to find the adjoint operator by using the fact that every operator can be split in a self-adjoint and a skew-adjoint operator. So, one determines the adjoint operator from the preceding (see Section 3.5.1) and applies this rule, which after some algebraic manipulations leads back to (4.76). *See example Ex. 4.8*

4.4 Finite-Element Methods for Fluid Flows

In this section, a concise description of the finite-element approach for fluid flows is presented. We start off explaining the general concepts of the discretization for the Poisson equation, followed by that on the convection–diffusion, Stokes, and Navier–Stokes equations.

4.4.1 Discretization

The starting point for the finite-element method is a weak form such as (3.14). Let us state it here as follows. We are looking for a solution u in a linear space $\mathcal{V}$ such that for all $v \in \mathcal{V}$ it holds that

$$a(v,u) = F(v). \tag{4.81}$$

The space is such that all elements of it satisfy the essential boundary condition (in homogeneous form). We would like to find a solution of the form $\hat{u} = \sum_{j=1}^{N} c_j \phi_j(x)$, where all $\phi_i(x)$, $i = 1, \ldots, N$ span an N-dimensional sub-space of $\mathcal{V}$. This sub-space is the search space and written as $\mathcal{V}_N$. We now want (4.81) to hold on $\mathcal{V}_N$ (Galerkin approach), which gives the linear system

$$A\mathbf{c} = \mathbf{b},$$

where $A_{ij} = a(\phi_i, \phi_j)$ and $b_i = F(\phi_i)$. *See example Ex. 4.9*

The only thing left to do now is to define the basis functions. The most common choice is to use interpolation polynomials as basis functions. We could use polynomials that perform an interpolation over the whole domain called *global polynomial interpolation* or use *piecewise polynomial interpolation*. The former needs high-degree polynomials when we have many interpolation points and may suffer from the *Runge phenomenon* (resulting in oscillatory behaviour) if sharp gradients are present in the real solution. Another problem is to find interpolating polynomials on irregularly shaped domains. If no sharp gradients are present in the solution and the domain is built up by a few rectangles, then this may do a good job and it is favourable to use orthogonal polynomials as a basis (see Trefethen (2000)). However, if we expect strong gradients and the domain is quite irregular, piecewise interpolating polynomials are much more flexible. For the construction of piecewise polynomials, we partition the domain in smaller parts (the pieces) and define a low-order interpolation on each part; next, we require at the interfaces of the parts some form of continuity depending on the order of the PDE.

Now, assume we have defined the piecewise interpolating polynomials we want to use, for example a linear approximation on each part of the domain; then this spans a space $\mathcal{V}_N$. The next step is to find a suitable basis for these polynomials which also spans the space $\mathcal{V}_N$. The nice thing is that there exists a basis in which the basis functions have a very local support, and these are the building blocks of the finite-element approach. Let us consider the 1D case. Say we partition the domain $[0,1]$ into N intervals (not necessarily equal). Now we require the piecewise interpolating function to be linear on each element and that it is continous from one element to a neighbouring element (Fig. 4.8a). As basis function we take a function that is 1 at some interface of two neighbouring elements and zero at all other interfaces (Fig. 4.8b), so $\phi_i(x_j) = \delta_{ij}$ where we have used the Kronecker δ. Finally, we take from these basis functions the restriction to one element which yields the *element basis functions* (Fig. 4.8c). In practice one is computing all the contributions to the matrix and right-hand side per element, leading to an *element matrix* and *element load vector*. In the next step, all the information is *assembled* in the system matrix called the *stiffness matrix* and the *load vector*. Generally speaking, a finite element is defined by a piece of the domain and the interpolation used on that piece, where for the latter we should think of the element basis functions.

Example 4.9 *Consider the problem $-d^2u/dx^2 = 1$ with $u(0) = u(1) = 0$. For this problem, we want to determine the stiffness matrix and the load vector on an equidistant mesh with elements of length $h = 1/N$, using linear basis functions. We take $x_0 = 0$ and $x_N = 1$ and $x_i = ih$ for $i = 1, \ldots, N - 1$. On element i the element matrix and load vector are given by*

$$\begin{bmatrix} (\phi_i', \phi_i')_i & (\phi_i', \phi_{i+1}')_i \\ (\phi_{i+1}', \phi_i')_i & (\phi_{i+1}', \phi_{i+1}')_i \end{bmatrix}, \quad \begin{bmatrix} (\phi_i, 1)_i \\ (\phi_{i+1}, 1)_i \end{bmatrix},$$

respectively, where $(.,.)_i$ denotes the restriction of $(.,.)$ to the ith element, so $(f, g)_i = \int_{x_i}^{x_{i+1}} f(x)g(x)dx$. Note that on interval (x_i, x_{i+1}) $\phi_i' = -1/h$ and $\phi_{i+1}' = 1/h$ (see Fig. 4.8d). So, the element matrix and the element load vector (the right-hand side) become

$$\frac{1}{h}\begin{bmatrix} 1 & -1 \\ -1 & 1 \end{bmatrix}, \quad \frac{h}{2}\begin{bmatrix} 1 \\ 1 \end{bmatrix}.$$

Since $a_{ii} = (\phi_i', \phi_i') = (\phi_i', \phi_i')_{i-1} + (\phi_i', \phi_i')_i$ we have an overlap in the 2×2 blocks if we assemble. The final system is

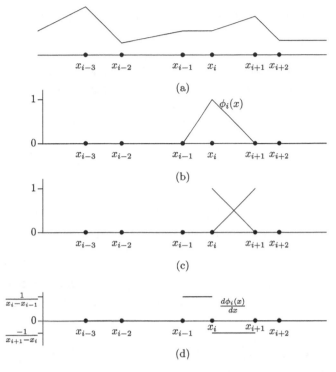

Figure 4.8 (a) Partitioning and piecewise linear polynomial, (b) basis function, (c) element basis functions, (d) derivative of basis function.

$$\frac{1}{h}\begin{bmatrix} 2 & -1 & & & \\ -1 & 2 & -1 & & \\ & \ddots & \ddots & \ddots & \\ & & -1 & 2 & -1 \\ & & & -1 & 2 \end{bmatrix}\begin{bmatrix} c_1 \\ c_2 \\ . \\ c_{N-2} \\ c_{N-1} \end{bmatrix} = h\begin{bmatrix} 1 \\ 1 \\ . \\ 1 \\ 1 \end{bmatrix}.$$

Observe that the system is the same as the one obtained from the finite-difference method (for $f \equiv 1$) if we divide by h on both sides. See example Exs. 4.10 and 4.11.

Having made the discretization, one likes to know the accuracy of it, in particular how the error decreases with a typical element size (h in the example). This proceeds in two steps. First we show that for a problem satisfying the Lax–Milgram theorem, one can bound the error of the solution on a sub-space $\mathcal{W}$ of $\mathcal{V}$ by the error of its least-squares approximation in that space. This is made explicit in Cea's lemma. This lemma also says that when the exact solution is in $\mathcal{W}$, it will be found. Secondly, we make use of the fact that the least-squares error achieves its mimimum over all functions in $\mathcal{W}$ and will in particular be less than the error for the interpolation of the exact solution in $\mathcal{W}$. This interpolation error can be expressed in h.

Theorem 4.6 (Cea's lemma) *Let $\mathcal{W} \subset \mathcal{V}$ be a Hilbert space. Let on $\mathcal{V}$ the bi-linear form of the weak problem (4.81) have a coercivity constant $c > 0$ (see (3.24)), and a bound (3.25a) with constant M. If $u_{\mathcal{W}}$ is the solution of the weak problem in $\mathcal{W}$ and u of that in $\mathcal{V}$, then $||u - u_{\mathcal{W}}||_{\mathcal{V}} \leq \frac{M}{c}||u - v||_{\mathcal{V}}$ for any $v \in \mathcal{W}$, in particular the least-squares approximation of u in $\mathcal{W}$.*

Proof Starting from coercivity (3.24), we have

$$c||u - u_{\mathcal{W}}||_{\mathcal{V}}^2 \leq a(u - u_{\mathcal{W}}, u - u_{\mathcal{W}}) = a(u - v + v - u_{\mathcal{W}}, u - u_{\mathcal{W}})$$
$$= a(u - v, u - u_{\mathcal{W}}) + a(v - u_{\mathcal{W}}, u - u_{\mathcal{W}}), \qquad (4.82)$$

where we used linearity in the last step. Since for each $w \in \mathcal{W}$ we have both $a(w, u) = F(w)$ and $a(w, u_{\mathcal{W}}) = F(w)$, we find by subtraction $a(w, u - u_{\mathcal{W}}) = 0$ for any $w \in \mathcal{W}$. If $v \in \mathcal{W}$, then we can choose $w = v - u_{\mathcal{W}}$, and using this in (4.82), we find

$$c||u - u_{\mathcal{W}}||_{\mathcal{V}}^2 \leq a(u - v, u - u_{\mathcal{W}}) \leq M||u - v||_{\mathcal{V}}||u - u_{\mathcal{W}}||_{\mathcal{V}},$$

where in the last step we used (3.25a). Dividing this expression by $||u - u_{\mathcal{W}}||_{\mathcal{V}}$ and c leads to the assertion. ∎

Corollary 4.1 *With the same assumptions as in Theorem 4.6, it holds that $||u - u_{\mathcal{W}}||_{\mathcal{V}} \leq \frac{M}{c}||u - \Pi_{\mathcal{W}}u||_{\mathcal{V}}$ where $\Pi_{\mathcal{W}}$ denotes the interpolation of u in $\mathcal{W}$.*

The interpolation of a function $u(x)$ on the space of piecewise linear polynomials as considered in Example 4.9 uses only $u(x_i)$, that is, $u(x)$ at the grid points. In between, we approximate $u(x)$ by linear polynomials.

Note that interpolation of a function is a linear operation, so $\Pi_{\mathcal{W}}$ is a linear operator.

In the following example we focus on bounding $||u - \Pi_{\mathcal{W}}u||_{\mathcal{V}}$ for smooth functions u for the simplest interpolation one can think of.

Example 4.10 *Given a function $f(x)$ and a space $\mathcal{V}_N$ of piecewise constant functions on [0,1] on a mesh with grid size $h = 1/N$. In this case, the least-squares error of $f(x)$ being approximated by a piecewise constant function can be bounded by the interpolation error of a piecewise constant interpolation of $f(x)$ denoted by $\Pi_{\mathcal{V}_N}f(x)$.*
Since the interpolation is also in $\mathcal{V}_N$, we have

$$\min_{g \in \mathcal{V}_N} ||f - g|| \leq ||f - \Pi_{\mathcal{V}_N}f||$$

and we just have to bound

$$||f - \Pi_{\mathcal{V}_N} f|| = \sqrt{\sum_{i=1}^{N} \int_{(i-1)h}^{ih} (f(x) - \Pi_{\mathcal{V}_N} f(x))^2 dx}.$$

We first consider the error on one interval. Recall that for constant interpolation we take the midpoint as the interpolation point and that the error is given by

$$f(x) - \Pi_{\mathcal{V}_N} f(x) = \left(x - \left(i - \frac{1}{2} \right) h \right) f'(\zeta(x)),$$

where $\zeta(x)$ is a point on the interval $[(i-1)h, ih]$, depending on the value of x. Hence,

$$\int_{(i-1)h}^{ih} (f(x) - \Pi_{\mathcal{V}_N} f(x))^2 dx = \int_{(i-1)h}^{ih} \left(x - \left(i - \frac{1}{2} \right) h \right)^2 (f'(\zeta(x)))^2 dx$$

$$\leq \max_{x \in [(i-1)h, ih]} (f'(x))^2 \int_{(i-1)h}^{ih} \left(x - \left(i - \frac{1}{2} \right) h \right)^2 dx$$

$$= \frac{1}{12} h^3 \max_{x \in [(i-1)h, ih]} (f'(x))^2.$$

So, we have

$$\sqrt{\sum_{i=1}^{N} \int_{(i-1)h}^{ih} (f(x) - \Pi_{\mathcal{V}_N} f(x))^2 dx} \leq \sqrt{\frac{1}{12} h^3 \sum_{i=1}^{N} \max_{x \in [(i-1)h, ih]} (f'(x))^2}$$

$$\leq \sqrt{\frac{1}{12}} h \sqrt{\max_{x \in [0,1]} (f'(x))^2} = \sqrt{\frac{1}{12}} h \max_{x \in [0,1]} |f'(x)|,$$

where in the one but last step we used that $Nh = 1$. From the preceding we also see that asymptotically the error bound tends to

$$\sqrt{\frac{1}{12}} h \sqrt{\int_0^1 f'(x)^2 dx}.$$

This informs us that the approximation shows first-order convergence in h.

Example 4.9 suggests that for functions in H^1, defined by (3.22), we have $O(h)$ convergence in the H^0 norm, that is, $||u||_{H^0} = \sqrt{(u,u)}$ with the inner product $(u, v) = \int_0^1 uv dx$ for a problem defined on $(0,1)$. One can prove that for a generalization to arbitrary piecewise interpolating polynomials of degree p for a function $u(x)$ in H^{p+1} we have $O(h^{p+1})$ convergence in the H^0 norm, and $O(h^{p+1-k})$ in the H^k norm (see Quarteroni *et al.* (2007, section 12.4.5) for more details). Here we introduced the generalization H^k of the space H^1, which is simply the space of functions for which the Sobolev norm $||u||_{H^k}$ is finite, the norm being defined by *See example Ex. 4.12*

$$\|u\|_{H^k} \equiv \sqrt{\sum_{i=0}^{k} \|\frac{d^i u}{dx^i}\|_{H^0}^2}.$$ (4.83)

So, in the finite-element discretization of the one-dimensional Poisson equation in the example we will have an error bound

$$\|u - u_{\mathcal{V}_N}\|_{H^1} \leq \frac{M}{c} \hat{c}(u)h$$ (4.84)

where $\hat{c}$ is a constant that depends on the exact solution u. One can also prove that, in accordance with the preceding, $\|u - u_{\mathcal{V}_N}\| = \|u - u_{\mathcal{V}_N}\|_{H^0} = O(h^2)$ (Hughes, 2000).

In general, the order of accuracy that can be obtained by piecewise interpolating functions depends on the following.

1. The order of the differential equation q.
2. The degree of piecewise interpolating polynomial p and the continuity of its derivatives m.
3. The order of the derivative of the unknown function k one is considering, that is, u ($k = 0$) or du/dx ($k = 1$) or d^2u/dx^2 ($k = 2$), etc.

In general, the order of accuracy in the approximation of $\frac{d^k u}{dx^k}$ in the $\|\cdot\|_{H^0}$ norm is $p + 1 - k$, so it is independent of q and m provided they satisfy the two conditions $2m \geq q$ and $p + 1 - q \geq 0$. So, for second-order PDEs ($q = 2$) using piecewise linear interpolation ($m = 1$, $p = 1$), we have that the function value ($k = 0$) is approximated to second order in h. Hence, if the mesh size is halved in all directions, then the error decreases by about a factor of four.

In the preceding discussion we assumed an exact integration to compute the coefficients of the stiffness matrix. It is, however, enough if the integration is exact for polynomials of degree $2p - 2$ for $m = 1$.

The introduction of the finite-element approach so far makes clear that there are two ways to increase the accuracy. The first is by decreasing the mesh size called *h-refinement* and the second is by increasing the order p of the polynomials called *p-refinement*.

The handling of constraints can be nicely done by using Lagrange multipliers as taught in calculus courses. Here it is introduced in the finite-dimensional case. Suppose we want to minimize

$$J(\mathbf{x}) = \frac{1}{2}(\mathbf{x}, A\mathbf{x}) - (\mathbf{b}, \mathbf{x}),$$ (4.85)

with A Symmetric and Positive Definite (SPD) (see Definition B.9), subject to the constraint $B^T \mathbf{x} = \mathbf{c}$. We can bring this constraint within the minimization using Lagrange multipliers. We now have to find the stationary point of

$$\hat{J}(\mathbf{x}, \mu) = \frac{1}{2}(\mathbf{x}, A\mathbf{x}) - (\mathbf{b}, \mathbf{x}) + (\mu, B^T\mathbf{x} - \mathbf{c}).$$

Taking the derivative with respect to μ and the components of $\mathbf{x}$, and setting the result to zero, leads to the following linear equations to be solved:

$$A\mathbf{x} - \mathbf{b} + B\mu = 0,$$
$$B^T\mathbf{x} = \mathbf{c}.$$

If A is not symmetric but positive definite, then we still can use the last equation to incorporate a constraint. *See example Ex. 4.13*

So far, not much has been written about the Petrov–Galerkin approach, where the test space is different from the search space. For that case we list here some common choices.

1. **Least squares** Here $\mathcal{W}_n = A\mathcal{V}_n$, which is equivalent to the minimization over the search space $\mathcal{V}_n$ of $||\mathcal{A}u_n - \mathbf{f}||_2^2$. This is an approach which always works, but the matrix of the resulting linear system may have a rather high condition number, that is, the number that quantifies the sensitivity of the solution to errors in the data (see Burden and Faires (2001, section 7.4)), which may give problems with reaching the required accuracy or with the convergence of the iterative method. So, this approach is a kind of last resort if other approaches fail. An advantage is that it leads to a SPD matrix. (In fact, it can also be viewed as the Galerkin method applied to $A^*Au = A^*\mathbf{f}$.)
2. **Collocation** In the collocation approach one requires that $\mathcal{A}u_n - \mathbf{f}$ should be zero in n judiciously chosen points in the domain. This is equivalent to choosing Dirac delta functions $\delta(x - x_j)$ as test functions.
3. **Finite volumes** If we choose the test functions to be indicator functions for which the ith function is one on control volume Ω_i (see Fig. 4.4), then we get an integral of the unknown function over the control volume. This choice leads to a method which is very close to the finite-volume approach.

The fill of the matrix resulting from a finite-element discretization will depend on the overlap of basis functions. In the extreme case that the *support*, that is, the part of the domain where the function is non-zero, of every $\phi_i(x)$ extends over the whole domain, the matrix of the linear problem will in general be full (unless we take eigenfunctions of the problem as basis functions). An example of this approach is the *pseudo-spectral method*. In this method orthogonal polynomials (discussed in section 8.2 of Burden and Faires (2001) or section 10.1 in Quarteroni and colleagues (2007)) are used as basis functions. We note that these basis functions allow for a high accuracy for smooth solutions.

Additional Material

- Pseudo-spectral methods are discussed in Trefethen (2000) and Quarteroni and colleagues (2007, section 12.3). Their application in bifurcation analysis of fluid flows can be found in Kuhlmann and Albensoeder (2014).
- A more elaborate discussion on finite elements can be found in Braess (2007), Ern and Guermond (2002), Hughes (2000), and Brenner and Scott (2008).

4.4.2 Navier–Stokes using FEM

We start off with the convection–diffusion equation since it is an essential part of the Navier–Stokes equation. Consider the one-dimensional convection–diffusion equation (3.38) with boundary conditions $u(0) = 0$ and $\mu du/dx(1) = g$, assuming a constant $\bar{u} \geq 0$. Then the bi-linear form is $a(v, u) = (v, \bar{u} du/dx - \mu d^2 u/dx^2) + \mu v(1) du/dx(1)$. Obviously the bi-linear form is coercive because $a(u, u) = \bar{u} u(1)^2 + \mu ||du/dx||^2 \geq \mu c_P ||u||^2$, where c_P is the Poincaré constant. However, if μ tends to zero, then the coercivity becomes weak and the solution is more sensitive to perturbations, that is, in Theorem 3.2 the constant c tends to zero. In such a case, still an inf-sup condition (3.34) with a c independent of μ may hold. We consider this condition on the linear space $\mathcal{V} = \{u \in C^2[0, 1] \mid u(0) = 0, \ du/dx(1) = 0\}$ where $a(v, u) = (v, \bar{u} du/dx - \mu d^2 u/dx^2)$. Then

$$
\inf_{u \in \mathcal{V}} \sup_{v \in \mathcal{V}} \frac{a(v, u)}{||v|| \, ||u||} = \inf_{u \in \mathcal{V}} \sup_{v \in \mathcal{V}} \frac{(v, \mathcal{A}u)}{||v|| \, ||u||}
$$

$$
= \inf_{u \in \mathcal{V}} \frac{(\mathcal{A}u, \mathcal{A}u)}{||\mathcal{A}u|| \, ||u||} = \inf_{u \in \mathcal{V}} \frac{||\mathcal{A}u||}{||u||}.
$$

Using this last expression, we can find a value for c using that for any $v \in \mathcal{V}$

$$
(\mathcal{A}v, \mathcal{A}v) = \left(\bar{u} \frac{dv}{dx} - \mu \frac{d^2 v}{dx^2}, \bar{u} \frac{dv}{dx} - \mu \frac{d^2 v}{dx^2} \right)
$$

$$
= \left(\bar{u} \frac{dv}{dx} - \mu \frac{d^2 v}{dx^2}, \bar{u} \frac{dv}{dx} - \mu \frac{d^2 v}{dx^2} \right)
$$

$$
= \left(\bar{u} \frac{dv}{dx}, \bar{u} \frac{dv}{dx} \right) - 2 \left(\mu \frac{d^2 v}{dx^2}, \bar{u} \frac{dv}{dx} \right) + \left(\mu \frac{d^2 v}{dx^2}, \mu \frac{d^2 v}{dx^2} \right)
$$

$$\geq \left(\bar{u}\frac{dv}{dx}, \bar{u}\frac{dv}{dx}\right) - 2\mu\bar{u}\left(\frac{dv}{dx}\right)^2 \Big|_0^1 = \left(\bar{u}\frac{dv}{dx}, \bar{u}\frac{dv}{dx}\right) + 2\mu\bar{u}\left(\frac{dv}{dx}(0)\right)^2$$

$$\geq \left(\bar{u}\frac{dv}{dx}, \bar{u}\frac{dv}{dx}\right) \geq \bar{u}^2 c_P(v, v).$$

So, this means that one can pick c in the inf-sup condition (3.34) as $\bar{u}\sqrt{c_P}$, and this entails that the continuous problem itself is stable if μ tends to zero.

Contrary to the coercivity, the inf-sup condition is not necessarily inherited on sub-spaces. For example, consider the space of continuous piecewise linear functions as we did for the Poisson problem. Then the element matrix related to the term $(v, du/dx)$ in the bi-linear form $a(v, u)$ is

$$\begin{bmatrix} (\phi_i, \phi_i')_i & (\phi_i, \phi_{i+1}')_i \\ (\phi_{i+1}, \phi_i')_i & (\phi_{i+1}, \phi_{i+1}')_i \end{bmatrix} = \frac{1}{2}\begin{bmatrix} -1 & 1 \\ -1 & 1 \end{bmatrix},$$

which leads to a matrix C of the shape

$$C = \frac{1}{2}\begin{bmatrix} 0 & 1 & & & \\ -1 & 0 & 1 & & \\ & \ddots & \ddots & \ddots & \\ & & -1 & 0 & 1 \\ & & & -1 & 1 \end{bmatrix}.$$

Observe that the application of this matrix to the vector $\mathbf{s} = [1, -1, 1, \cdots,]$ has only a non-zero entry at the first and last positions, so $(\mathbf{s}, C\mathbf{s}) = \frac{1}{2}(-1 + 2) = \frac{1}{2}$. To check the coercivity and inf-sup condition we also need (u, u) on the sub-space. The associated element matrix is

$$\begin{bmatrix} (\phi_i, \phi_i)_i & (\phi_i, \phi_{i+1})_i \\ (\phi_{i+1}, \phi_i)_i & (\phi_{i+1}, \phi_{i+1})_i \end{bmatrix} = \frac{1}{6}\begin{bmatrix} 2 & 1 \\ 1 & 2 \end{bmatrix},$$

which leads to the mass matrix M of the shape

$$M = \frac{1}{6}\begin{bmatrix} 4 & 1 & & & \\ 1 & 4 & 1 & & \\ & \ddots & \ddots & \ddots & \\ & & 1 & 4 & 1 \\ & & & 1 & 2 \end{bmatrix}.$$

It can be showed that $(M - \frac{1}{3}I)\mathbf{s}$ is non-zero everywhere except at the first and last positions and that $(\mathbf{s}, (M - \frac{1}{3}I)\mathbf{s}) = 0$, hence $(\mathbf{s}, M\mathbf{s}) = \frac{1}{3}(\mathbf{s}, \mathbf{s}) = \frac{1}{3}n$. In the absence of the diffusion term ($\mu = 0$), on the space of piecewise linear functions, the coercivity constant is less than

$$\min_v \frac{(\mathbf{v}, C\mathbf{v})}{(\mathbf{v}, M\mathbf{v})} \leq \frac{(\mathbf{s}, C\mathbf{s})}{(\mathbf{s}, M\mathbf{s})} = \frac{3}{2n} = \frac{3h}{2}.$$

Hence, we see that the coercivity constant for the matrix C will depend on h and go to zero for h to zero. In this case, the inf-sup condition will not help, since we have

$$\min_{\mathbf{v}} \frac{(C\mathbf{v}, C\mathbf{v})}{(\mathbf{v}, M\mathbf{v})} \leq \frac{(C\mathbf{s}, C\mathbf{s})}{(\mathbf{s}, M\mathbf{s})} = \frac{15}{2n} = \frac{15h}{2}.$$

Hence, also the inf-sup constant goes to zero with h. This means that without enough diffusion added we have an unstable discretization, resulting in wiggles in the solution.

One way out is to add artificial diffusion, similar to the finite difference and finite volume case as discussed at the end of Section 4.1.2. This means that we simply replace μ by $\mu + \alpha h$. The price for the improved stability, that is, a smooth solution, is that the discretization becomes first-order accurate. Therefore, we would like to keep α as small as possible. We know that the solution will be smooth if there is diagonal dominance; see Section 4.1.4. Moreover, we know from Example 4.9 what the diffusion matrix looks like, and adding C we see that we should choose $\alpha = \max(|\bar{u}|h/2 - \mu, 0) = \mu/2 \max(|P| - 2, 0)$, where P is the mesh-Péclet number (see Section 4.2). For large P this may have a big impact if the solution contains large gradients. In such a case one can try to counteract the effect by the following observation. Now, instead of adding $\alpha d^2 u/dx^2$ we could also add $\alpha d/dx$ times the whole equation to the original equation, which is equivalent to solving the equation

$$\left(1 + \frac{\alpha h}{\bar{u}} \frac{d}{dx}\right)\left(\bar{u}\frac{du}{dx} - \mu\frac{d^2u}{dx^2} - f\right) = 0.$$

In the weak form this is

$$\left(v - \frac{\alpha h}{\bar{u}}\frac{dv}{dx}, \bar{u}\frac{du}{dx} - \mu\frac{d^2u}{dx^2} - f\right) + \mu v(1)\left(\frac{du}{dx}(1) - g\right) = 0,$$

which is a Petrov–Galerkin approach to solve the equation. Applying this in full will require higher smoothness of the test functions. So, one can also decide to do it only in part. For instance, if the viscosity is relatively small with respect to the convection, one could just use

$$\left(v - \frac{\alpha h}{\bar{u}}\frac{dv}{dx}, \bar{u}\frac{du}{dx} - f\right) - \left(v, \mu\frac{d^2u}{dx^2}\right) + \mu v(1)\left(\frac{du}{dx}(1) - g\right) = 0, \qquad (4.86)$$

for which one can use our original space. For piecewise linear basis functions this is equivalent to what is called streamline upwind Petrov–Galerkin (SUPG) introduced by Brooks and Hughes (1982), because it will make the coefficient in upwind direction stronger at the expense of the coefficient in downwind direction.

Let us now consider the 2D case

$$- \bar{u}u_x - \bar{v}u_y + \mu \Delta u = f, \tag{4.87}$$

with $u = 0$ on Γ_D and $\mu(\nabla u, \mathbf{n}) = g$ on Γ_N.

The generalization of SUPG to the 2D case consists of using the test functions $w = v + \alpha h(\bar{u}v_x + \bar{v}v_y)$ for all terms except the diffusion term leading to the weak form

$$\hat{a}(v, u) = (v + \alpha h(\bar{u}v_x + \bar{v}v_y), \bar{u}u_x + \bar{v}u_y) - (v, \mu \Delta u) + \mu \int_{\Gamma_N} v(\nabla u, \mathbf{n})d\Gamma$$

$$= (v, \bar{u}u_x + \bar{v}u_y) + \alpha h(\bar{u}v_x + \bar{v}v_y, \bar{u}u_x + \bar{v}u_y) + \mu(\nabla v, \nabla u)$$

and

$$\hat{F}(v) = (v + \alpha h(\bar{u}v_x + \bar{v}v_y), f) + \int_{\Gamma_N} gv(1)d\Gamma.$$

In the 2D case the name SUPG is clearer in the sense that the artificial diffusion is only applied in streamline direction.

As a start towards the Navier–Stokes equations, we first discuss the discretization of the Stokes equation because the difficulty of the indeterminacy of the pressure is coming in. The Stokes equations are the ones given by (3.57), except that C is not present in the momentum equation, that is, in the first equation. The associated weak form is already given by (3.58a), on which we apply (3.36) to the $\mathbf{v}\Delta u$ term and the $\mathbf{v}\nabla p$ term leading to the bi-linear form

$$a(\mathbf{u}, p; \mathbf{v}) = ((I \otimes \mu \nabla)\mathbf{u}, (I \otimes \mu \nabla \mathbf{v})_\Omega + (p, \nabla \cdot \mathbf{v})_\Omega - (B(\mathbf{u}, p) - p, \mathbf{v})_\Gamma. \tag{4.88}$$

We add to this also a weak form of the second equation,

$$b(q, \mathbf{u}) \equiv (q, \nabla \cdot \mathbf{u})_\Omega = 0. \tag{4.89}$$

We can just add the latter equation to the previous with an arbitrary non-zero constant in between. So, we get

$$\hat{a}(\mathbf{u}, p; \mathbf{v}, q) = a(\mathbf{u}, p; \mathbf{v}) + \chi b(q, \mathbf{u}). \tag{4.90}$$

Now we consider coercivity and see that we get

$$\hat{a}(\mathbf{u}, p; \mathbf{u}, p) = a(\mathbf{u}, p; \mathbf{u}) + \chi b(p, \mathbf{u})$$

$$= ||(I \otimes \mu \nabla)\mathbf{u}||_\Omega^2 + (1 + \chi)(p, \nabla \cdot \mathbf{u})_\Omega + (B(\mathbf{u}, p) - p, \mathbf{u})_\Gamma. \tag{4.91}$$

Observe that there is no positive contribution of an inner product of p with itself over the domain. So, in order to get a non-negative expression we need to get rid of the cross terms containing both $\mathbf{u}$ and p, which occurs by taking $\chi = -1$. But even then, we only get coercivity for the velocities. This means that the problem will be well-posed for the velocity but not for the pressure. Therefore, we have to turn to the inf-sup condition. In Appendix C, it is shown that an inf-sup condition is satisfied if the Laplacian in the Stokes equation is coercive and bounded and that an inf-sup and boundedness condition holds for the pressure gradient operator.

This will be the case for the continuous equations, but since the inf-sup condition on the pressure gradient is in general not inherited on sub-spaces, we have to check it for any pair of sub-spaces we choose for $\mathbf{u}$ and p. And indeed if we use for both sub-spaces linear shape functions, then the property does not hold; in fact, this is exactly the same as in the discussion for $(v, du/dx)$. But in this case, the velocity is correct and the wiggles will appear in the pressure solution. So, the inf-sup condition for the pressure gradient couples the spaces possible for p to those used for the velocity. For instance, linear shape functions for the velocity and constant ones for the pressures are a good pair.

If one does not want such a coupling, then one can stabilize the pressure by changing the continuity equation to

$$\nabla \cdot \mathbf{u} - \epsilon \mathcal{D}p = 0,$$

where $(q, \mathcal{D}p)$ is coercive. This gives

$$b_\epsilon(q; \mathbf{u}, p) = b(q, \mathbf{u}) - \epsilon(q, \mathcal{D}p).$$

If we subtract this from (4.88), we end up with a bi-linear form which is coercive. Common choices are the identity operator, minus the Laplacian, or a biharmonic operator, that is, Δ^2. Of course, we are altering the equation by adding such a term, and hence it should be small and focussed towards suppressing wiggles. For that aim it is best to use a high-order differential operator since for a similar effect on the low frequencies it is suppressing high-frequency modes much harder than low-order differential operators will do. However, the drawback of higher-order differential operators is that the linear system becomes more and more difficult. So, it is very common to use the Laplace operator.

There is, however, an alternative to the preceding approach: just require that the pressure is perpendicular to the kernel of the gradient operator. Often the set of pressure nodes can be decoupled into a number of sets in which there is, in the discrete equation, no coupling between a pressure from one set to a pressure from another set. The pressures in each set are, however, well coupled and well-posed. In such a case, it is in general quite easy to determine the kernel of the discrete gradient operator. For instance, in a closed domain we will have the physical constant vector in the kernel and next to that the spurious vectors which have different constant values on each independent set of pressures. So, if there are two independent sets, then there is one spurious mode which has ones on one set and minus ones on the other. Such an approach can be used for both finite-element methods and finite-volume methods. We applied it in the ocean model THCM (Wubs *et al.*, 2006).

In any case, some prudence with stabilizing is justified; see Gresho and Lee (1981) for a relevant discussion. In the finite-volume world it is quite common to use an unstable discretization, a reason being that boundaries are easier to treat, for example the package ReFRESCO (see www.refresco.org). The challenge in such cases is not to excite unstable modes and to define postprocessing in accordance with the discretization.

For the Navier–Stokes equation we need the stabilization of the pressure precisely for the same reasons as for the Stokes equation. Moreover, since the convection is entering, we also need the stabilization of that part.

An overall generalization of the bi-linear form of the Navier–Stokes equation is

$$\hat{a}(\mathbf{u}, p; \mathbf{v}, q) = ((I \otimes \mu \nabla)\mathbf{u}, (I \otimes \mu \nabla \mathbf{v})_\Omega + ((I \otimes C)\mathbf{u}, \mathbf{v})_\Omega - b(p, \mathbf{v}) + b_\epsilon(q; \mathbf{u}, p)$$
$$+ (B(\mathbf{u}, p) - p, \mathbf{v})_\Gamma. \tag{4.92}$$

The simplest stabilization of the convection is adding a term $\alpha h((I \otimes C)\mathbf{u}, ((I \otimes C)\mathbf{v})$. Now one chooses, for example, the linear basis functions for both velocities and pressures and one is required to find $\mathbf{u}_h$ and p_h in these spaces such that

$$((I \otimes \mu \nabla)\mathbf{u}_h, (I \otimes \mu \nabla \mathbf{v}_h)_\Omega + ((I \otimes C)\mathbf{u}_h, \mathbf{v}_h)_\Omega + \alpha h((I \otimes C)\mathbf{u}_h, ((I \otimes C)\mathbf{v}_h)$$
$$- b(p_h, \mathbf{v}_h) + b_\epsilon(q_h; \mathbf{u}_h, p_h) + (B(\mathbf{u}_h, p_h) - p_h, \mathbf{v}_h)_\Gamma = 0 \tag{4.93a}$$
$$b_\epsilon(q_h; \mathbf{u}_h, p_h) = 0 \tag{4.93b}$$

for all $\mathbf{v}_h$ and q_h in the same spaces.

4.5 Time-Dependent PDEs

In this section we treat the discretization of time-dependent PDEs such as the hyperbolic and parabolic PDEs. We follow the method of lines in which first the space discretization is performed. Then the resulting system of ODEs is analyzed and based on that an appropriate time-integration method is chosen. We note, however, that in some textbooks discretization is performed immediately in both space and time, for example in Burden and Faires (2001, chapter 12). This approach gives of course a little bit more freedom which can, especially in the hyperbolic case, be an advantage to develop schemes that take good care of characteristics. A group of schemes in this direction is formed by the Lax, Lax–Wendroff, and MacCormack schemes.

4.5.1 Method of Lines

In this section, we will show how we transform a PDE into a system of ODEs by considering the 1D heat equation

$$\frac{\partial u}{\partial t} = a\frac{\partial^2 u}{\partial x^2} + f, \quad a > 0, \ t > 0, \ 0 < x < 1. \tag{4.94}$$

The initial condition is $u(x,0) = g(x)$, and for convenience we will assume homogeneous boundary conditions $u(0,t) = u(1,t) = 0$.

In the method of lines we first discretize in space. For that, we will treat the finite-difference and finite-element method subsequently.

Using a finite-difference method, we partition the x-interval in m equal parts, hence $h = 1/m$ and $x_i = ih$, $i = 0, 1, \ldots m$. According to (4.3), it holds for sufficient smooth u that

$$u_{xx}(x,t) = \frac{u(x+h,t) - 2u(x,t) + u(x-h,t)}{h^2} - \tfrac{1}{12}h^2 u_{xxxx}(x,t) + O(h^4).$$

Now, if we use the notation $u_j(t) = u(x_j, t)$, then we get, after substitution in (4.94) for [4] $x = x_j$, $j = 1, 2, \ldots, m-1$,

$$\frac{d}{dt}u_j(t) = a\frac{u_{j+1}(t) - 2u_j(t) + u_{j-1}(t)}{h^2} - \tfrac{1}{12}ah^2 u_{xxxx}(x_j, t) + O(h^4) + f_j(t).$$

After discarding the local discretization errors, we arrive at the system of ODEs

$$\frac{d}{dt}U_j = a\frac{U_{j-1} - 2U_j + U_{j+1}}{h^2} + f_j(t), \quad j = 1, 2, \ldots, m-1 \tag{4.95}$$

with initial condition $U_j(0) = g(x_j)$. Here the functions $U_j(t)$ approximate $u_j(t)$, hence they approximate the solution $u(x,t)$ along the lines $x = x_j$, $j = 1, 2, \ldots, m-1$. This is why this approach received the name *method of lines*.

We can also put the difference equations in matrix-vector form. For that we introduce the vector

$$\mathbf{U}(t) = [U_1(t), U_2(t), \ldots, U_{m-1}(t)]^T,$$

similarly a vector $\hat{f}$ and $\hat{g}$, and the $(m-1) \times (m-1)$ matrix

$$A = \frac{a}{h^2}\begin{bmatrix} -2 & 1 & & & \\ 1 & -2 & 1 & & \\ & \ddots & \ddots & \ddots & \\ & & 1 & -2 & 1 \\ & & & 1 & -2 \end{bmatrix}. \tag{4.96}$$

Now we can write the system of ODEs as

$$\frac{d}{dt}\mathbf{U} = A\mathbf{U} + \hat{\mathbf{f}} \tag{4.97}$$

with $\mathbf{U}(0) = \mathbf{g}$.

In using a finite-element method, the only difference with the approach we followed for the elliptic equations is that now the coefficients in the sum of the basis functions will depend on t. So, we write $u_N(x, t) = \sum_{j=1}^{N} c_j(t)\phi_j(x)$. In the Galerkin approach, we substitute this into (4.94) and test with the ϕ_i. This yields

$$\sum_{j=1}^{N} \left[\frac{d}{dt} c_j(t)(\phi_i, \phi_j) + ac_j(\phi_i', \phi_j') - (\phi_i, f) \right] = 0,$$

where $(u, v) \equiv \int_0^1 uvdx$. In matrix-vector form it can be written as

$$M\frac{d}{dt}\mathbf{c} = A\mathbf{c} + \hat{\mathbf{f}}, \tag{4.98}$$

where the *mass matrix* M is given by $M_{ij} = (\phi_i, \phi_j)$, and the stiffness matrix A by $A_{ij} = (\phi_i', \phi_j')$ and $\hat{f}_i = (\phi_i, f)$.

It is interesting to see that in both cases we have a similar ODE system of the shape

$$M\frac{d}{dt}\mathbf{c} = A\mathbf{c} + \hat{\mathbf{f}}, \quad \mathbf{c}(0) = \mathbf{c}_0, \tag{4.99}$$

where in the finite-difference case, M is just the identity. (Also the matrix A and load vector $\hat{\mathbf{f}}$ will differ.)

Note that, if in a system of PDEs one or more of the equations do not contain time derivatives, then these equations form a constraint to the solution. After space discretization they become an algebraic constraint, and then we arrive at so-called algebraic differential equations. For such equations special methods exist. In the previously mentioned methods it is best to apply the constraint immediately to the new time level. Time steps need not be constant, and this can be exploited by adapting the time step such that locally an approximation of the discretization error remains small (see Burden and Faires (2001), sections 5.5, 5.7). Usually, for example, in MATLAB's ODE solvers (ode23 etc.), one can set relative and absolute tolerances for this approximation.

Additional Material

- For more details on time integration methods and their application to computational fluid dynamics, see Peyret and Taylor (1983), Hirsch (1994), Wesseling (1999), and Zikanov (2010).

4.5.2 Stability Analysis

In order to make a judicious choice for a time-discretization method we must consider the stability problem for (4.99). In this case it is enough to consider what happens if we perturb the initial value. Let us denote the solution following from the perturbation by $\hat{\mathbf{c}}(t)$ which has initial condition $\hat{\mathbf{c}}(0) = \mathbf{c}_0 + \boldsymbol{\varepsilon}$. Since $\hat{\mathbf{c}}$ also satisfies the ODEs exactly, we can simply subtract the two systems to obtain a system for the difference $\mathbf{e}(t) = \hat{\mathbf{c}}(t) - \mathbf{c}(t)$, which assumes the form

$$M\frac{d}{dt}\mathbf{e} = A\mathbf{e}, \quad \mathbf{e}(0) = \boldsymbol{\varepsilon}. \tag{4.100}$$

If $\mathbf{e}(t) \to 0$ for $t \to \infty$, then we call (4.99) *stable*. One of the ways to investigate stability is to analyze what happens to the energy $||\mathbf{e}||_M^2 = (\mathbf{e}, M\mathbf{e})$ for $t \to \infty$. Since it also is the square of a norm, assuming M is positive definite, we have that if the energy goes to zero, then $\mathbf{e}$ also goes to zero. So, we are interested in showing that the time derivative of the energy is negative, that is,

$$\frac{d}{dt}||\mathbf{e}||_M^2 = \frac{d}{dt}(\mathbf{e}, M\mathbf{e}) = 2\left(\mathbf{e}, M\frac{d}{dt}\mathbf{e}\right) = 2(\mathbf{e}, A\mathbf{e}) < 0.$$

For (4.94) we have for the finite-difference/volume case a discrete Poincaré inequality, and for the finite-element discretization a Poincaré inequality is just inherited from the continuous case. So, for both finite-difference/volume and finite-element discretization there exists a c such that

$$(\mathbf{e}, A\mathbf{e}) \le -c(\mathbf{e}, M, \mathbf{e}), \tag{4.101}$$

and consequently the energy is a decaying function of t since

$$\frac{d}{dt}||\mathbf{e}||_M^2 \le -2c||\mathbf{e}||_M^2$$

or

$$\frac{d}{dt}||\mathbf{e}||_M \le -c||\mathbf{e}||_M$$

or

$$||\mathbf{e}(t)||_M \le \exp(-ct)||\mathbf{e}(0)||_M.$$

Instead of considering the energy, we can also perform an eigenvalue analysis of the system of ODEs. By finding suitable eigenvectors, one can transform the system of ODEs into a set of independent scalar equations. To find them, we employ separation of variables by setting

$$\mathbf{e}(t) = \exp(\lambda t)\mathbf{v}, \tag{4.102}$$

where **v** does not depend on time anymore. This leads to the generalized eigenvalue problem

$$\lambda M\mathbf{v} = A\mathbf{v}. \tag{4.103}$$

If M is non-singular, the eigenvalues are just the standard eigenvalues of $M^{-1}A$. Once we have found the eigenpairs (λ_i, v_i), we build the matrix $V = [v_1, \ldots, v_N]$, assuming that a complete set of eigenvectors exists, and use it to bring (4.100) to diagonal form, yielding the scalar equations

$$\frac{d}{dt}\hat{e}_i = \lambda_i \hat{e}_i \text{ for } i = 1, \ldots, N$$

with $\hat{\mathbf{e}} = V^{-1}\mathbf{e}$. The initial condition transforms into the condition $\hat{\mathbf{e}}(0) = V^{-1}\boldsymbol{\varepsilon}$. So, for studying the stability of systems of ODEs (4.98), it is enough to study this scalar equation where λ_i is running through the eigenvalues of $M^{-1}A$. The scalar equation informs us that any initial perturbation will vanish if the real part of each eigenvalue of $M^{-1}A$ is less than zero. So, $\hat{\mathbf{e}}$ goes to zero, and since V is non-singular also **e** goes to zero. We note that in the appendix there are some generalizations of bounds on growth, for example, Theorems B.30 and B.31, showing also the role of V.

Computing all the eigenvalues of $M^{-1}A$, that is, its *spectrum*, to verify that each of them is negative is an enormous task, but fortunately not necessary. In general, it is enough to know that all eigenvalues have a negative real part. In the context of ODEs coming from PDEs there are two main paths to show this: the matrix method and the Fourier method. In the former, we really find an upper bound for the eigenvalues, and in the latter we easily find approximate eigenvalues which can be scanned on the sign of the real part.

In the *matrix method*, one again can use the Poincaré inequality. Let (4.101) hold and suppose that **v** is an eigenvector; then we have respectively

$$(\mathbf{v}, A\mathbf{v}) \leq -c(\mathbf{v}, M\mathbf{v}),$$
$$(\mathbf{v}, \lambda M\mathbf{v}) \leq -c(\mathbf{v}, M\mathbf{v}),$$
$$\lambda \leq -c.$$

Since λ might be complex, this means that for each eigenvalue $\text{Re}(\lambda) \leq -c$, which provides us an upper bound.

In the *Fourier method*, we take the difference operator according to a typical row of the eigenvalue problem (4.103) as starting point and assume this equation to hold for all j disregarding the boundary conditions. Then the Fourier component $u_j = \exp(ij\varphi)$, with φ in $[-\pi, \pi)$ (for uniqueness) and i the imaginary unit, is an eigen(grid)function of this problem and its eigenvalue can be straightforwardly derived. Let us do this for the matrix A in (4.96) with M being identity. A typical row of the eigenvalue problem (4.103) reads

$$a\frac{u_{j+1}(t) - 2u_j(t) + u_{j-1}(t)}{h^2} = \lambda u_j.$$

Now we substitute $u_j = \exp(ij\varphi)$ and find

$$a\frac{e^{i\varphi} - 2 + e^{-i\varphi}}{h^2} \exp(ij\varphi) = \lambda \exp(ij\varphi).$$

Using that $e^{i\varphi} + e^{-i\varphi} = 2\cos(\varphi)$ and $\cos(\varphi) = 1 - 2\sin^2(\varphi/2)$, we obtain

$$\lambda = -4a\frac{\sin^2(\varphi/2)}{h^2}. \tag{4.104}$$

Observe that the eigenvalue is real and in the interval $[-4a/h^2, 0]$. A zero eigenvalue only occurs if $\varphi = 0$, which corresponds to $u_j \equiv 1$. One can immediately check that the constant vector is not satisfying the first and last equations of the eigenvalue problem (4.103), so we can exclude $\varphi = 0$ from the eigenvalue and hence $[-4a/h^2, 0)$. So, the real part of all eigenvalues are less than zero. This approach is less rigorous than the matrix method, since we have neglected the boundary conditions. Nevertheless, in many cases it gives a good indication of the location of the eigenvalues in the complex plane.

In the previous discussion we have assumed a linear problem (4.99); in view of this book, this is a severe restriction. In fact, in general a flow problem can be described by (2.16) and after discretization by (2.51). This led to (2.58) for the study of the stability of the steady state $\bar{x}$ which is equivalent to (4.100), the only difference being that in the current case $\bar{x}$ may still be time dependent. Here, this does not lead to a time-dependent Jacobian matrix, since due to the linearity the Jacobian is just the matrix A. *See example Ex. 4.14*

Example 4.11 *Consider the discretized Ginzburg–Landau equation (4.35), that is,*

$$M\frac{d}{dt}\mathbf{A} = (\gamma_1 M + \gamma_2 D)\mathbf{A} - M\gamma_3 |\mathbf{A}|^2 \mathbf{A} + \mathbf{g}, \tag{4.105}$$

with $\mathbf{A}(0) = \mathbf{a}_0$. Recall that M is a diagonal matrix. Let $\mathbf{B}$ satisfy the same equation with initial condition $\mathbf{B}(0) = \mathbf{a}_0 + \boldsymbol{\epsilon}$. Then the difference $\mathbf{e} = \mathbf{B} - \mathbf{A}$ satisfies the equation

$$M\frac{d}{dt}\mathbf{e} = (\gamma_1 M + \gamma_2 D)\mathbf{e} - M\gamma_3 \left((\mathbf{e} + \mathbf{A})|\mathbf{e} + \mathbf{A}|^2 - \mathbf{A}|\mathbf{A}|^2\right). \tag{4.106}$$

Defining the energy $(\mathbf{e}, M\mathbf{e})$, then we can show analogously to the analysis in Section 3.6.3 that

$$\left(\mathbf{e}, M(\mathbf{e} + \mathbf{A}) \, |\mathbf{e} + \mathbf{A}|^2 - \mathbf{A}|\mathbf{A}|^2 \right) = \left(||\mathbf{e}^2||_M^2 + 3(|\mathbf{e}|^2, \mathbf{e}\mathbf{A})_M + 2||\mathbf{e}\mathbf{A}||_M^2 + (\mathbf{e}^2, \mathbf{A}^2)_M \right.$$

$$\geq -\frac{5}{4} ||\mathbf{A}||_M^2 ||\mathbf{e}||_M^2.$$

So, for the right-hand side of (4.106),

$$(\mathbf{e}, (\gamma_1 M + \gamma_2 D)\mathbf{e} - M\gamma_3 \left((\mathbf{e} + \mathbf{A}) \, |\mathbf{e} + \mathbf{A}|^2 - \mathbf{A} \, |\mathbf{A}|^2 \right)) \leq (\gamma_1 - \gamma_2 c_P$$

$$+ \frac{5}{4} Re(\gamma_3) ||\mathbf{A}||_M^2 \bigg) \, ||\mathbf{e}||_M^2,$$

where c_P is the Poincaré constant. Hence, the energy will certainly go to zero if $\gamma_1 + \frac{5}{4} Re(\gamma_3)||\mathbf{A}||_M^2 < \gamma_2 c_P$, which makes the equation stable for that case.

We can also look at the eigenvalues of the Jacobian matrix involved. This follows from (4.106) for infinitesimal small $\mathbf{e}$. In that case, only the linear terms remain in the right-hand side:

$$M \frac{d}{dt} \mathbf{e} = \left((\gamma_1 I - \gamma_3 2|\mathbf{A}|^2)M + \gamma_2 D \right) \mathbf{e} - \gamma_3 \mathbf{A}^2 M \bar{\mathbf{e}}. \qquad (4.107)$$

By taking the adjoint of this equation we also get an equation for the adjoint of $\mathbf{e}$. The matrix occurring in these two equations is the sought Jacobian. Now, we need theorems that localize the eigenvalues of a matrix. One of techniques is to look at the Field of Values (see B.10). Actually this is related to the energy approach and we can as well apply the energy approach to the preceding equation. However, this means that we have

$$\left(\mathbf{e}, (\gamma_1 M + \gamma_2 D)\mathbf{e} - M\gamma_3 (2\mathbf{e}|\mathbf{A}|^2 + \mathbf{A}^2 \bar{\mathbf{e}}) \right) \leq \left(\mathbf{e}, (\gamma_1 M + \gamma_2 D - M\gamma_3|\mathbf{A}|^2)\mathbf{e} \right).$$

So, if $Re(\gamma_3) > 0$, then the non-linear term only adds to the stability, and therefore locally we have stability if $\gamma_1 < \gamma_2 c_P$; then certainly all eigenvalues are in the negative half plane.

Using the Fourier method, we find eigenvalues

$$\lambda(\varphi) = \gamma_1 - 4\gamma_2 \frac{\sin^2(\varphi/2)}{h^2}. \qquad (4.108)$$

So, we have a decay of the associated Fourier component (see (4.102)) of $\exp((\gamma_1 - 4\gamma_2 \frac{\sin^2(\varphi/2)}{h^2})t)$. This decay is stronger, the closer φ gets to π. This is typical for diffusion. Higher-frequency modes are damped much more strongly than low-frequency modes.

4.5.3 Generalization to Higher-Order Time Derivatives

Higher-order ODEs can be brought back to first-order ODEs. We provide an example for the wave equation, given by

$$\frac{\partial^2 u}{\partial t^2} = a \frac{\partial^2 u}{\partial x^2} + f, \quad a > 0, \ t > 0, \ 0 < x < 1. \qquad (4.109)$$

The initial conditions are $u(x, 0) = \phi(x)$, $u_t(x, 0) = \psi(x)$, and for convenience we will assume homogeneous boundary conditions $u(0, t) = u(1, t) = 0$. We first perform a space discretization exactly as in the parabolic case and arrive at a system of ODEs given by

$$\frac{d^2}{dt^2}\mathbf{U} = A\mathbf{U} + \hat{f} \qquad (4.110)$$

with $\mathbf{U}(0) = \hat{\phi}$ and $\frac{d}{dt}\mathbf{U}(0) = \hat{\psi}$ where $\hat{\phi}$ and $\hat{\psi}$ are the vectors that originate from ϕ and ψ in the grid points. The system of ODEs is written as the set of first-order equations given by

$$\frac{d}{dt}\mathbf{U} = \mathbf{W},$$

$$\frac{d}{dt}\mathbf{W} = A\mathbf{U} + \hat{f}, \qquad (4.111)$$

with $\mathbf{U}(0) = \hat{\phi}$ and $\mathbf{W}(0) = \hat{\psi}$. It can also be written in matrix-vector form

$$\frac{d}{dt}\begin{bmatrix} \mathbf{U} \\ \mathbf{W} \end{bmatrix} = J\begin{bmatrix} \mathbf{U} \\ \mathbf{W} \end{bmatrix} + \begin{bmatrix} 0 \\ \hat{f} \end{bmatrix}$$

where

$$J = \begin{bmatrix} 0 & I \\ A & 0 \end{bmatrix}. \qquad (4.112)$$

Here we have to analyze the properties of the matrix J in order to select an appropriate method for the time integration. *See example Ex. 4.15*

4.5.4 Time Integrators

In this section we show a number of time integrators that can solve a system of ODEs of the form $du/dt = f(t, u)$ with an initial condition given. In all cases we will define a grid in time direction with step size Δt.

In the previous section we studied the stability of the system of ODEs (4.98). In this section we will do that for the time-discretized system of ODEs, in fact the fully discretized PDEs. We would like that the system of time-discretized ODEs are also stable if the system of ODEs is stable. In a numerical method for time integration we are just making some approximation to the time derivative, and therefore the reduction to scalar equations can be done in precisely the same way as in Section 4.5.2, using the same V. In fact, the actions of discretization and diagonalization commute (see Fig. 4.9). This means that for stability of the numerical method we just can look at the scalar equation

$$\frac{d}{dt}u = \lambda u, \qquad (4.113)$$

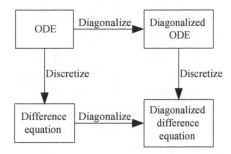

Figure 4.9 Discretization and diagonalization commute.

which is called the *test equation*, and where λ will run through the eigenvalues of the eigenvalue problem (4.103). When we apply a time integration method to the test equation, we will be able to derive the so-called *region of absolute stability* in the complex plane; we will show these for specific cases in what follows. If $\lambda \Delta t$, Δt being the time step used in the method, is in the region of absolute stability for all eigenvalues λ of $M^{-1}A$, then any initial perturbation will go to zero, that is, our desired property holds.

The *forward Euler method* is defined by

$$u^{(n+1)} = u^{(n)} + \Delta t f(t_n, u^{(n)}), \tag{4.114}$$

where $u^{(0)}$ is given and $t_n = t_0 + n\Delta t$. Using Taylor series, one can show that this method is first-order accurate. As indicated earlier, for stability we can apply it to (4.113), which gives

$$u^{(n+1)} = u^{(n)} + \Delta t \lambda u^{(n)} = (1 + \Delta t \lambda) u^{(n)}. \tag{4.115}$$

Here $1 + \Delta t \lambda$ is called the *amplification factor* and indicated by $\rho(\lambda \Delta t)$. Note that (4.115) is a recurrence and hence

$$u^{(n)} = (1 + \Delta t \lambda)^n u^{(0)}.$$

Since for stability we want the initial perturbation to damp out, we desire that $u^{(n)}$ tends to zero if n tends to infinity. This is the case only if the magnitude of the amplification factor is less than one:

$$|1 + \Delta t \lambda| < 1. \tag{4.116}$$

By defining $z \equiv \Delta t \lambda$ we find that the region of absolute stability of the Euler method is the domain in the complex domain for which an element z satisfies

$$|1 + z| < 1.$$

For equality we find in the complex plane a circle with center -1 and radius 1. The inequality holds for all z in this circle. For the finite-difference discretization (4.95) of the diffusion problem we found the Fourier eigenvalues (4.104) and argued that the eigenvalues are in the interval $[-4a/h^2, 0)$. Similar to the argument to exclude 0 as an eigenvalue, we can also exclude $-4a/h^2$ from being an eigenvalue. This value will occur if we set $\varphi = -\pi$ in (4.104), and that is related to $u_j = \exp(-ij\pi) = (-1)^j$. Using this in (4.103) shows once again that it does not satisfy the equations at the first and last rows, where it does on any other row. So, if $-\Delta t 4a/h^2$ is on the boundary of the region of absolute stability, we will have a stable intergration. Since the eigenvalues are real, this boundary is at -2. This leads to the following restriction on the time step:

$$\Delta t \leq \frac{h^2}{2a}. \tag{4.117}$$

Hence, we get a restriction which can become quite severe for small h.

Note that we need more information here on the location of the spectrum than needed for the stability of the system of ODEs in Section 4.5.2. In this case, one also would like to have a lower bound on the spectrum. This follows quite easily with the Fourier method. For the matrix method we can often make use of a bound on the spectral radius, that is, $\max_\lambda |\lambda|$, where λ runs through the eigenvalues of A, is less than $||A||_\infty$, which is simply $\max_i \sum_{j=1}^{N} |a_{ij}|$, that is, the maximum of the absolute row sums of A; see also Theorem B.1.

For example, consider the matrix A in (4.96). We see that the matrix is symmetric, so all its eigenvalues are real. Moreover, we know already from Section 4.5.2 that all eigenvalues of $-A$ are positive, and obviously they are real due to the symmetry of A. Next we observe that the infinity norm of $-A$ is $4a/h^2$, so the eigenvalues of A are on the interval $[-4a/h^2, 0)$, or. using the same argument as for the Fourier method, on $(-4a/h^2, 0)$, which leads to the same interval and hence to the same time-step restriction. The forward Euler method is an *explicit method* since we can just fill in a known $u^{(n)}$ in f, do an addition, and we have $u^{(n+1)}$. This contrasts with an *implicit method*, where we will have the unknown $u^{(n+1)}$ in f, which in general leads to the solution of a non-linear system of equations. The big difference between explicit and implicit methods is that the time step for explicit methods is always bounded in a form similar to that for the Euler method, while for implicit methods this need not be the case. This may make it worthwhile to use an implicit method.

The *backward Euler method* is an implicit method which has the form

$$u^{(n+1)} = u^{(n)} + \Delta t f\left(t_{n+1}, u^{(n+1)}\right)$$

and is, like the Forward Euler method, first-order accurate. From the region of absolute stability we observe that this method has no restriction on the time step for our model problem. In fact, it will not have a restriction for any problem with eigenvalues from which the real part is negative. Methods with such a property are called *A-stable*. The backward Euler method is the simplest example of the *Backward Differentiation Formulas (BDFs)*, where the f is always and only evaluated at t_{n+1} and for the discretization of the time derivative values in the past are used, which explains the name. In general BDF(k) uses k values of w in the past and has order of accuracy k.

The *trapezoidal method* has the symmetric form

$$u^{(n+1)} = u^{(n)} + \frac{\Delta t}{2} \left[f(t_n, u_n) + f(t_{n+1}, u^{(n+1)}) \right],$$

which causes this method to be second-order accurate in time. The trapezoidal method is, in PDE context, often called the *Crank–Nicolson method*.

The *theta method* is a generalization of the trapezoidal method and has the form

$$u^{(n+1)} = u^{(n)} + \Delta t \left[(1 - \theta) f(t_n, u_n) + \theta f(t_{n+1}, u^{(n+1)}) \right].$$

Observe that for $\theta = 0, 1/2, 1$ we get the forward Euler, trapezoidal method, and the backward Euler, respectively. *See example Ex. 4.16*

By applying it to the test equation (4.113), we find the amplification factor

$$\frac{1 + (1 - \theta)\Delta t \lambda}{1 - \theta \Delta t \lambda} \tag{4.118}$$

The method is A-stable for $\theta \geq 1/2$, since equality is only reached when λ is purely imaginary. So, if $\text{Re}(\lambda) < 0$, it is less than one. Note that the amplification factor tends to one for $\Delta t \lambda \to \infty$ and $\theta = 1/2$. If λ is real and negative, this behaviour differs severely from the behaviour of the solution of the test equation (4.113). The former will give almost *no* damping of an initial perturbation, while the latter will. It is useful to add some damping to the trapezoidal method by making $\theta > 1/2$.

There exists also a variant of the theta method called the implicit theta method:

$$u^{(n+1)} = u^{(n)} + \Delta t \left[f(t_n + \theta \Delta t, (1 - \theta)u^{(n)} + \theta u^{(n+1)}) \right].$$

This method differs from the first one for non-linear problems only. Hence, the stability analysis is equal. *See example Ex. 4.17*

Example 4.12 *Now, let us apply the implicit theta method to the space-discretized Ginzburg–Landau equation (4.35). We will get*

$$MA^{(n+1)} = MA^{(n)} + \Delta t[(\gamma_1 M + \gamma_2 D)((1-\theta)A^{(n)} + \theta A^{(n+1)})$$
$$- M\gamma_3|(1-\theta)A^{(n)} + \theta A^{(n+1)}|^2((1-\theta)A^{(n)} + \theta A^{(n+1)})] + \Delta t\mathbf{g},$$

with $A^{(0)} = \mathbf{f}$.

In Exercise 4.15 you have been asked to show that the eigenvalues of J defined in (4.112) are all purely imaginary. Hence, the (forward) Euler method is not fit for this problem; however, the backward Euler and the Trapezoidal method can be used. Two examples of explicit methods that can be used are the classical Runge–Kutta 4 method and the explicit midpoint method (see Burden and Faires (2001), Section 5).

4.5.5 Von Neumann Analysis

Instead of the method-of-lines approach, one can also discretize both time and space derivatives at the same time and next analyze its stability properties. So, if we apply forward Euler to (4.95), we get

$$U_j^{(n+1)} = U_j^{(n)} + \Delta t a \frac{U_{j-1}^{(n)} - 2U_j^{(n)} + U_{j+1}^{(n)}}{h^2} + f_j(t^{(n)}), \quad j = 1, 2, \ldots, m-1, \tag{4.119}$$

for $n = 0, 1, 2, \cdots$. The initial condition is $U_j^0 = \phi(x_j)$, and the boundary conditions are $U_0^{(n)} = U_m^{(n)} = 0$. As before, we consider stability due to perturbations in the initial condition and hence we can look at the homogeneous equation. Now we substitute $U_j^{(n)} = \rho^n \exp(ij\varphi)$ and obtain

$$\rho^{n+1} \exp(ij\varphi) = \rho^n \exp(ij\varphi) + \Delta t a \rho^n \frac{\exp(i(j-1)\varphi) - 2\exp(ij\varphi) + \exp(i(j+1)\varphi)}{h^2}.$$

After simplification we find that $\rho(\varphi) = 1 - \Delta t 4a \frac{\sin^2(\varphi/2)}{h^2}$, which is also called the *amplification factor*. Before, we called $1 + \Delta t \lambda$ the amplification factor from the forward Euler method. By just substituting the Fourier eigenvalue of the Jacobian of (4.95), that is, $\lambda = 4a \sin^2(\varphi/2)/h^2$, we get the same result. In the Von Neumann analysis we require that the amplification factor should be less than or equal to one in absolute sense for all φ. We see that in this case $1 - \Delta t 4a/h^2 \le \rho \le 1$, so the lower bound should not be less than -1. This results in the same requirement as (4.117).

Example 4.13 *For the Ginzburg–Landau equation, we can use the computed Fourier eigenvalue (4.108) and substitute it in the amplification factor (4.118) of the theta method, leading to the amplification factor of the Von Neumann analysis:*

$$\rho(\varphi) = \frac{1 + (1-\theta)\Delta t \left(\gamma_1 - 4\gamma_2 \frac{\sin^2(\varphi/2)}{h^2}\right)}{1 - \theta\Delta t \left(\gamma_1 - 4\gamma_2 \frac{\sin^2(\varphi/2)}{h^2}\right)}.$$

Observe that for small φ, $\rho(\varphi) \approx \exp(\Delta t(\gamma_1 - \gamma_2\varphi^2)/h^2)$. We see that it is less than one if $\gamma_1 - \gamma_2\varphi^2/h^2 < 0$. However, for $\varphi \approx \pi$ and $\Delta t(\gamma_1 - \gamma_2\varphi^2/h^2)$ large (and negative), $\rho(\varphi) = (1-\theta)/\theta$. So, for $\theta = 1/2$ there is no damping. Note that the Fourier component $\exp(ij\pi) = (-1)^j$ is the highest oscillating mode that can be represented on the grid. It is unnatural that such a high-frequency component would not damp out. Recall that the solution of the test equation (4.113) is $u(t) = u(0)\exp(\lambda t)$. Substituting the Fourier eigenvalue, we find $u(t) = u(0)\exp\left(\left(\gamma_1 - 4\gamma_2\frac{\sin^2(\varphi/2)}{h^2}\right)t\right)$, which will go to zero rapidly for $\varphi \approx \pi$ and $\gamma_1 - 4\gamma_2/h^2 << 0$. So, if one really uses large time steps, then the trapezoidal method and implicit midpoint may have too little damping for high-frequency components in the solution, and either one should take a smaller time step or increase θ.

4.6 Summary

- The aim of discretization is to approximate an infinite-dimensional problem by a finite-dimensional one where a balance has to be sought between accuracy and computation time. There are essentially two starting points: (i) approximating the PDE on a grid and (ii) expressing the unknown function in a series of basis functions with unknown coefficient. The finite-difference and finite-volume methods are examples of the former, and the finite-element and spectral methods of the latter.
- Finite differences are found by combining nearby function values on the grid in such a way that they approximate a derivative. By using Taylor series, the truncation error of such an approximation can be derived, and it is of order p if it is $O(h^p)$, where h is a typical mesh size. The order can be increased by using function values at more grid points. In the finite-volume method, the domain is partitioned in volumes and on each volume one starts from the integral form of the PDE. In conservation laws such an integral can be cast into a boundary integral by Gauss' theorem. By discretizing the contributions at an interface between two volumes, exactly the same one obtains discrete conservation. Herewith mass, momentum and energy can be preserved exactly.
- For convergence, (i) a discretization of a PDE should become a better and better approximation of the PDE (consistency) when the mesh size is refined, and (ii) the discretization should be stable.

- Essential for numerical stability of the discretization is to use as close as possible neighbours for approximating differences. To achieve this, it is in many cases handy to locate the unknowns in systems of PDEs differently, leading to *staggered grid* discretizations. In some cases, there are conflicts on how to choose the staggering. In such a case, the most important balancing terms should determine the choice. One can, and does, also use unstable discretizations, but then one should be careful not to excite spurious modes in the solution space, for example by a judicious boundary treatment, and possibly add artificial diffusion to counteract instability.
- Since finite-differences and finite-volume discretizations do not inherit the coercivity (important to prove stability) of a weak form, one has to prove the associated theorems for finite differences for each type of discretization.
- For time-dependent PDEs it is convenient to first discretize in space. For a specific semi-discretization one estimates the time steps to be taken for accuracy reasons and this determines whether to use explicit methods, which have a restriction on the time step for stability, or implicit methods, which have in general no restriction on the time step or a very mild one. Here one should take into account that implicit methods are usually an order of magnitude more expensive than explicit methods due to the (non-)linear systems that need to be solved.
- For a stable PDE, we have that a perturbation in the initial condition has a limited growth for all time. In a numerical time integrator, this property is mimicked in the requirement of absolute stability.

4.7 Exercises

Exercise 4.1 *Consider the linear convection–diffusion equation*

$$-\frac{d^2u}{dx^2} - 3\frac{du}{dx} = f(x) \text{ on } [0,1] \text{ with } u(0) = u(1) = 1.$$

a. *On an equidistant grid, give the finite-difference discretization of the equation and the corresponding linear system. Take the Dirichlet conditions at the grid points.*
b. *Let $g(\xi) = a + b\exp(\xi)$. Determine a and b such that g is a map from $[0,1]$ to itself.*
c. *Determine the discretization of the convection–diffusion equation on the non-uniform grid defined by the function in the previous part. Check that for the choice $g(\xi) = \xi$ we get back the discretization found in a. What will the linear system look like for the non-uniform case?*

d. *On the non-equidistant grid, which difference equations and changes to the linear system result from (i) a Neumann condition $du/dx = 2$ and from (ii) a Robin condition $u + 2du/dx = 5$ at the left boundary?*

e. *Suppose the left boundary is put between two grid points, so the fictive point x_0 is at $g(-h/2)$ and $x_1 = g(h/2)$. Use the values at these two points for the Dirichlet, Neumann, and Robin condition, respectively, to derive the difference equation at the left boundary. (This positioning is commonly used in CFD.)*

f. *Suppose that the convective term is absent, will the matrix related to the discretization on the non-uniform grid be symmetric then? And, if the diffusive term is absent, will the matrix be skew symmetric then?*

Exercise 4.2

a. *Show that the Poisson equation and the Cauchy–Riemann equation are of the form $div(\mathcal{M}u) = f$ and give $\mathcal{M}$. Give also the condition on the flux in these cases.*

b. *Consider the equation $u_t = div(\mathcal{M}u)$ and assume $\int_\Gamma (\mathcal{M}u, n)d\Gamma = 0$. Show that the integral $\int_\Omega u d\Omega$ is independent of t.*

Exercise 4.3 *Make second-order accurate finite-difference and finite-volume discretizations (except for a) on a uniform grid of the problems below. Consider for each case boundary conditions at the grid points and boundary conditions in the middle between the grid points. Moreover, give the position of the grid points and the mesh size.*

a. $-\cos(x)\frac{d^2u}{dx^2} - 3x\frac{du}{dx} + \tanh(x/10)u = \exp(x)$ *on* $[0, 1]$ *with* $u(0) = 1$, $u(1) = -2$.

b. $-\frac{d}{dx}\left(\cos(x)\frac{du}{dx}\right) - 3\frac{d(xu)}{dx} + \tanh(x/10)u = \exp(x)$ *on* $[0, 1]$ *with* $u(0) = 1$, $u(1) = -2$.

c. $-\frac{\partial}{\partial x}(\bar{u}u) - \frac{\partial}{\partial y}(\bar{v}u) + \frac{\partial}{\partial x}(\mu\frac{\partial u}{\partial x}) - \frac{1}{2}\frac{\partial}{\partial x}\left(\mu\frac{\partial u}{\partial y}\right) - \frac{1}{2}\frac{\partial}{\partial y}\left(\mu\frac{\partial u}{\partial x}\right) + \frac{\partial}{\partial y}\left(\mu\frac{\partial u}{\partial y}\right) = 0$, *where* $\bar{u}(x, y) \equiv 1$, $\bar{v}(x, y) = 1 + x$, *and* $\mu(x, y) = 1 + xy$. *The boundary conditions are* $u(0, y) = 1$, $u(x, 0) = \cos(x)$, *and* $\frac{\partial u}{\partial x}(1, y) = \frac{\partial u}{\partial y}(x, 1) = 0$.

d.

$$-\frac{\partial}{\partial x}(u^2) - \frac{\partial}{\partial y}(vu) + \mu\Delta u = 0,$$

$$-\frac{\partial}{\partial x}(uv) - \frac{\partial}{\partial y}(v^2) + \mu\Delta v = 0,$$

with boundary conditions $u(0, y) = 1$, $u(x, 0) = \cos(x)$, *and* $\frac{\partial u}{\partial x}(1, y) = \frac{\partial u}{\partial y}(x, 1) = 0$, $v(0, y) = 1 - x$, $v(x, 0) = \sin(x)$, *and* $\frac{\partial v}{\partial x}(1, y) = \frac{\partial v}{\partial y}(x, 1) = 0$.

Exercise 4.4 *Consider the discretization by finite volumes of the equation in Example 4.1 on a non-equidistant grid (part d).*

a. *Determine its truncation error.*

b. *Determine its bi-linear form.*

c. Which constant will hold in the Poincaré inequality for this problem?
d. Show that the global discretization error is determined by the order in the truncation error.

Exercise 4.5 *Consider the central discretization for du/dx and apply it to the grid function $f_i = (-1)^i$. Make the grid function continuous differentiable using a sine function that interpolates the grid function. Take the derivative of this sine function and observe the difference with the discrete result. What does this difference mean for the stability of the discretization?*

Exercise 4.6 *Suppose $A = [2, 1; 1, 2]$. What will be the shape of the matrix $\hat{A}$ in (4.79)?*

Exercise 4.7 *Find, for example on the Internet, the Navier–Stokes equations in cylindrical coordinates. Consider the diffusion operator in these equations and show that in these coordinates also the operator is self-adjoint and determine the quadratic form that can be used as a starting point for the discretization.*

Exercise 4.8 *Determine the adjoints of the operators in (4.80) (assuming Dirichlet boundary conditions) and show that the skew-adjoint part will be zero and the self-adjoint part the operator found in (4.76).*

Exercise 4.9

a. *Derive the linear system following from substituting ϕ_i for v and $\hat{u} = \sum_{j=1}^{N} c_j \phi_j(x)$ for u in (4.81) (see Example 3.4).*
b. *Show that, if $a(*, *)$ satisfies (3.24) for all v in $\mathcal{V}$, then the matrix derived in the previous part is positive definite, and if $a(u, v) = a(v, u)$ for arbitrary u and v in $\mathcal{V}$, then it is symmetric.*
c. *Show that, if the $c_j, j = 1, \ldots, N$ are known, we can compute the function values of the interpolating $\hat{u} = \sum_{j=1}^{N} c_j \phi_j(x)$ polynomial easily, using the element basis functions.*

Exercise 4.10

a. *For a 1D domain, draw the element basis functions for quadratic interpolation on each element. In this case, next to the endpoints, the midpoint also is used in the interpolation. Determine also the element matrices for the same case as in Example 4.9.*
b. *A cubic interpolation polynomial can be fixed by giving at the endpoints of an interval both the function value and its derivative. Give the equations that define the four element basis functions.*
c. *Consider a triangulation of a 2D domain, that is, the domain is partitioned in triangles. Assume that we want to do a linear approximation on each triangle determined by the function value in its three corners. This is in fact a plane through these function values. Make a sketch of the associated basis function.*

d. *Try to find the description of the quadratic and cubic 2D triangular elements, for example on the Internet, and sketch the location of interpolation points. In which way is this a generalization of the corresponding 1D elements?*

Exercise 4.11 *Create the linear systems that arise from the positive definite weak forms occurring in parts a, b, and c of Example 3.15 for both shape functions of degree 1 and 2.*

Exercise 4.12 *Consider on the interval [0,1] a grid with grid size h with $h = 1/N$, where on each grid interval a polynomial of degree p is defined and where the coefficients are chosen such that it is continuous on the whole interval [0,1]. These piecewise polynomials define a sub-space V.*

a. *Show that the best approximation in the H^0 norm of a function $f(x)$ can be bounded by the error of the piecewise interpolation polynomial of f in the sub-space and that the error behaves as $O(h^{p+1})$.*
b. *Repeat the previous part for the Sobolev norm defined in Section 3.5.2 which contains derivatives up to order k, leading to the conclusion that now the error behaves like $O(h^{p+1-k})$.*

Exercise 4.13 *Determine the equations that result from a minimization of the following expression over u, v, where both u and v satisfy homogeneous boundary conditions at all boundaries:*

$$\frac{1}{2}(u, -\Delta u) - (f_1, u) + \frac{1}{2}(v, -\Delta v) - (f_2, v)$$

subject to the constraint $u_x + v_y = 0$.

Exercise 4.14

a. *Consider the problem in Exercise 3.2 on the interval [0,1]. Give the systems of ODEs that arise if we use for the space discretization, respectively, (4.4), (4.6), and (4.5), where for the equidistant grid $x_0 = 0$ and $x_m = 1$. Now there will also be an unknown in x_m since there is no boundary condition at $x = 1$. Use for all three cases (4.6) at x_m. What will be the solution with the last discretization for all time? What should be the conclusion?*
b. *Try the difference/Fourier method on the three discretizations in part a. Which of the three has positive eigenvalues? What will this mean if the initial solution has a small perturbation? So, what is the conclusion?*

Exercise 4.15

a. Suppose that in (4.111) A is similar to a diagonal matrix D through the similarity transformation matrix Q (see Section B.1). Show that the matrix J is brought to a bidiagonal
   form using the transformation matrix diag(Q, Q), that is, a block diagonal matrix with
   two diagonal blocks both equal to Q. Show that the result matrix is reducible (see Def.
   B.7) and put it in the associated form. Show that the eigenvalues of that matrix are
   purely imaginary.
b. Re-writing a second-order ODE to a first-order system of ODEs is not unique. Suppose
   we set $\frac{d}{dt}\mathbf{U} = B\mathbf{W}$ where B is non-singular and of the same order as A. How does the
   first-order system look now? By at least two choices we can turn the new matrix into
   a skew-symmetric matrix. The first is by taking $B = \sqrt{-A}$, and the second by taking a
   Cholesky decomposition of $-A$ (see Section 6.4). Show this.

Exercise 4.16 Give the difference equation that emerges after the application of the theta
method to (4.99).

Exercise 4.17

a. Show that the Forward Euler method is not suited for problems with purely imaginary
   eigenvalues.
b. For the Backward Euler method, show that the stability analysis leads to $|1 - z| > 1$
   and draw its region of absolute stability.
c. Search for plots of the regions of stability of the BDF(k) methods on the Internet. For
   which values of k are they A-stable and for which values can they solve our model
   parabolic problem without a restriction on the time step?
d. Show that the stability analysis of the trapezoidal method leads to $|2 - z| > |2 + z|$ and
   that this method also is A-stable.
e. Show that the amplification factor of the trapezoidal method is tending to minus one for
   z tending to infinity.

5

Numerical Bifurcation Analysis

Moving continuously through phase space
to uncover the structure of attractors

Almost any computational fluid dynamical problem, including the ones discussed in Chapter 1, starts from the Navier–Stokes equations possibly supplemented by conservation equations for scalar quantities. The resulting problem can be written in the general form

$$\mathcal{M}\frac{\partial \mathbf{u}}{\partial t} = \mathcal{F}(\mathbf{u}, \lambda, t), \tag{5.1}$$

where $\mathbf{u}$ includes the velocity field, pressure, and possibly additional scalar quantities; λ is a scalar parameter. The quantity $\mathcal{M}$ is the so-called mass matrix, needed to take equations into account which contain no time derivative, such as the continuity equation for incompressible flows. Appropriate boundary conditions are added to these equations to make the problem well-posed, such as discussed in Chapter 3. Next, the equations in (5.1) are discretized using finite-element, finite-difference, finite-volume, or spectral methods such as those discussed in Chapter 4.

The discretized system of equations in (5.1) can then be written as

$$M\frac{d\mathbf{x}}{dt} + L(\lambda)\mathbf{x} + N(\mathbf{x}, \lambda) = \mathbf{F}(\lambda), \tag{5.2}$$

where the vector $\mathbf{x} \in \mathbb{R}^d$ denotes the values of $\mathbf{u}$ on a grid (when using finite differences) or the coefficients in the expansion in basis functions (when using a finite-element or spectral method). In addition, L is the discretized linear operator, N the discretized non-linear operator, and $\mathbf{F}$ is usually the forcing. In general, L, N, and $\mathbf{F}$ depend on the parameter λ. Note that M is typically not invertible because of the lack of time derivatives in some equations. We may also implicitly include the boundary conditions in the formulation (5.2). For instance, Dirichlet conditions correspond to M and N being zero on the boundary of the domain and L being the identity operator.

The more common approach in investigating solutions of (5.2) for specific model parameters is to solve the initial value problem, that is, the discretized model is integrated forward in time from a given initial condition and the long-time behaviour of quantities of interest are studied. To determine transition behaviour and critical conditions, parameters are subsequently changed and the transient and asymptotic behaviour of the model solutions is studied. In this way, transitions between different types of equilibrium behaviour or to different types of time-dependent behaviour, such as those discussed in Chapter 2, can be found.

However, often the primary interest is in changes in time-asymptotic behaviour when parameters are changed, and another class of numerical methods can be used, which focuses directly on the computation of the asymptotic flow states in the models. As we have seen, these states may be steady states, periodic orbits, quasi-periodic orbits, or more complicated states, usually referred to as equilibrium solutions of the model. The methods of numerical bifurcation theory, in particular continuation techniques, consist of efficient numerical schemes to determine equilibrium solutions when varying parameters. It enables one to obtain meaningful and generic information on the local dynamics of the model (5.2) for a large range of parameter values. Although time integration of the model may ultimately be needed to detect more complicated bifurcations as well as statistical properties of the flow, continuation methods are able to determine the first bifurcations from the branches of steady states and periodic orbits in an efficient way. When interested in steady states, continuation also avoids potentially long-time integrations for many parameter values.

Continuation methodology was originally developed (Keller, 1977; Doedel, 1980) for tackling bifurcation problems in models consisting of a small number of ordinary differential equations, hence with a small number of degrees of freedom d. Over the last decades, the methods have been extended and applied to a number of flow problems governed by systems of discretized partial differential equations such as Navier–Stokes (and Boussinesq) equations with a large number of degrees of freedom, and this development forms the core of this book. In this chapter, we follow mainly the presentation in Doedel and colleagues (1991), but with a slightly different notation to be consistent with the remainder of the book. An important part of the computational work is the solution of large linear systems of equations and large-dimensional generalized eigenvalue problems. The success of the methods to solve the latter problems mainly determines the dimension d of the dynamical system which can be handled. Whereas for small-dimensional dynamical systems robust direct techniques can be used (Chapter 6), one must turn to sophisticated (and less robust) iterative techniques for large-dimensional systems (Chapters 7 and 8).

5.1 Continuation of Steady States

To determine steady solutions of (5.2), we need to solve the set of non-linear algebraic equations

$$\Phi(\mathbf{x}, \lambda) = L(\lambda)\mathbf{x} + N(\mathbf{x}, \lambda) - \mathbf{F}(\lambda) = 0, \tag{5.3}$$

where $\Phi \colon \mathbb{R}^{d+1} \to \mathbb{R}^d$.

The aim of continuation methods is to determine solutions $\mathbf{x}$ of (5.3) when varying the parameter λ. The key theorem in this approach is the implicit function theorem (Guckenheimer and Holmes, 1990; Krantz and Parks, 2002) which states that when there is a solution $(\mathbf{x}_0, \lambda_0)$ of (5.3) for which the Jacobian $\Phi_\mathbf{x}^0 = \Phi_\mathbf{x}(\mathbf{x}_0, \lambda_0)$ is non-singular and $\Phi_\mathbf{x}$ and Φ are both smooth at $\mathbf{x}_0$, then there exists a unique, smooth solution family $\mathbf{x}(\lambda)$ for values of λ near λ_0, for which $\mathbf{x}(\lambda_0) = \mathbf{x}_0$. A solution for which the Jacobian $\Phi_\mathbf{x}$ is non-singular is also called a regular (or isolated) solution.

Now let $\mathbf{X} = (\mathbf{x}, \lambda) \in \mathbb{R}^{d+1}$; then, Equation (5.3) can be written as $\Phi(\mathbf{X}) = 0$. A solution $\mathbf{X}_0$ is called a regular solution if the matrix $\Phi_X^0 = \Phi_\mathbf{X}(\mathbf{X}_0)$ (with d rows and $d + 1$ columns) has rank d. Because

$$\Phi_\mathbf{X}^0 = (\Phi_\mathbf{x}^0 \ \Phi_\lambda^0), \tag{5.4}$$

where $\Phi_\lambda^0 = \Phi_\lambda(\mathbf{x}_0, \lambda_0)$, there are two cases for which this matrix has rank d: (i) $\Phi_\mathbf{x}^0$ is non-singular or (ii) the dimension of the null space of $\Phi_\mathbf{x}^0$ is one and the vector Φ_λ^0 is not in the span of the d columns of $\Phi_\mathbf{x}^0$ (that is, not in the range of $\Phi_\mathbf{x}^0$).

When $\mathbf{X}_0$ is a regular solution of $\Phi(\mathbf{X}) = 0$, then there locally exists a unique, one-dimensional solution family $\mathbf{X}(s)$ with $\mathbf{X}(0) = \mathbf{X}_0$, parameterized by s. This solution family is usually referred to as a solution branch. In case (i), when $\Phi_\mathbf{x}^0$ is non-singular, this is guaranteed by the implicit function theorem, that is, $\mathbf{x}(\lambda)$. In case (ii), the rows of $\Phi_\mathbf{X}^0$ can be interchanged to see that the solution can locally be parameterized by one of the components of $\mathbf{x}$. The latter case occurs at a saddle-node bifurcation, where the solution branch cannot be parameterized by λ itself, as $\Phi_\mathbf{x}^0$ is singular.

The main aim of a continuation method (Fig. 5.1a) is to compute a next point $\mathbf{X}_1 = (\mathbf{x}_1, \lambda_1)$ on a solution branch, when a regular solution $\mathbf{X}_0 = (\mathbf{x}_0, \lambda_0)$ is available plus a tangent $\dot{\mathbf{x}}_0 = d\mathbf{x}/d\lambda$ at $(\mathbf{x}_0, \lambda_0)$. As will be shown in much more detail in Chapter 6, the equation $\Phi(\mathbf{X}_1) = 0$ will be solved by a Newton–Raphson method where we will use as an initial guess

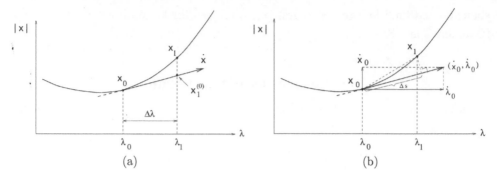

Figure 5.1 (a) Sketch of a simple parameter continuation method, where the solution is represented as $\mathbf{x}(\lambda)$. (b) Sketch of the pseudo-arclength continuation method, where the solution is represented as $(\mathbf{x}(s), \lambda(s))$, with s the arclength parameter.

$$\mathbf{x}_1^0 = \mathbf{x}_0 + \Delta\lambda\dot{\mathbf{x}}_0$$

with $\Delta\lambda = \lambda_1 - \lambda_0$. When the Jacobian $\Phi_\mathbf{x}(\mathbf{x}_1, \lambda_1)$ is non-singular, then Newton–Raphson is guaranteed to converge. Hence, we find a solution branch $\Phi(\mathbf{x}(\lambda), \lambda) = 0$ and at each point the new tangent can be determined by differentiating this expression to λ giving at $(\mathbf{x}_1, \lambda_1)$

$$\Phi_\mathbf{x}(\mathbf{x}_1, \lambda_1)\dot{\mathbf{x}}_1 + \Phi_\lambda(\mathbf{x}_1, \lambda_1) = 0. \tag{5.5}$$

However, when $\Phi_\mathbf{x}$ is singular, this does not work, so one cannot continue around a saddle-node bifurcation.

A continuation method that solves this problem is the pseudo-arclength method. Here, a branch of solutions is parameterized by an arclength parameter s. A branch $\mathbf{X} = (\mathbf{x}(s), \lambda(s))$, $s \in [s_a, s_b]$ is a smooth one-parameter family of solutions of (5.3). Since an extra degree of freedom is introduced by the arclength s, a normalization condition of the form

$$\Sigma(\mathbf{x}(s), \lambda(s), s) = 0 \tag{5.6}$$

is needed to close the system of equations. We thus end up solving a system of non-linear algebraic equations of dimension $d + 1$ for the $d + 1$ unknowns $(\mathbf{x}(s), \lambda(s))$.

The tangent space of the curve at $s = s_0$ is spanned by the vector $\dot{\mathbf{X}}(s_0) = (\dot{\mathbf{x}}(s_0), \dot{\lambda}(s_0))^T$ (Fig. 5.1b), where the dot now indicates differentiation to s. In the pseudo-arclength method (Keller, 1977), the function Σ is chosen as

$$\Sigma(\mathbf{x}, \lambda, s) = \dot{\mathbf{x}}_0^T(\mathbf{x} - \mathbf{x}_0) + \dot{\lambda}_0(\lambda - \lambda_0) - (s - s_0) \tag{5.7}$$

which is added to Equation (5.3), that is, $\Phi(\mathbf{x}(s), \lambda(s)) = 0$. Differentiation of both Equations (5.3) and (5.7) to s gives

$$\Phi_{\mathbf{x}}\dot{\mathbf{x}} + \Phi_{\lambda}\dot{\lambda} = 0, \tag{5.8a}$$
$$\dot{\mathbf{x}}_0^T\dot{\mathbf{x}} + \dot{\lambda}_0\dot{\lambda} = 1, \tag{5.8b}$$

from which the new tangent $(\dot{\mathbf{x}}, \dot{\lambda})$ can be determined.

The advantage of this method is that the Jacobian of the extended system (5.3)–(5.7) is non-singular at saddle-node bifurcations, whereas the Jacobian $\Phi_{\mathbf{x}}$ is. To show this, the pseudo-arclength system is written as

$$\Phi(\mathbf{X}) = 0, \tag{5.9a}$$
$$\dot{\mathbf{X}}_0^T(\mathbf{X} - \mathbf{X}_0) - (s - s_0) = 0. \tag{5.9b}$$

The Jacobian of this system is the matrix $G = (\Phi_{\mathbf{X}} \dot{\mathbf{X}}_0^T)^T$. From (5.8) it follows that the null space of $\Phi_{\mathbf{X}}$ at $\mathbf{X}_0$ is spanned by $\dot{\mathbf{X}}_0$. Now suppose that G is singular at a regular point; then there exists a vector $\mathbf{z} \neq 0$ such that $\Phi_{\mathbf{X}}\mathbf{z} = 0$ and hence $\mathbf{z} = \alpha\dot{\mathbf{X}}_0$. In addition, the vector $\mathbf{z}$ has to satisfy $\dot{\mathbf{X}}_0^T\mathbf{z} = 0$, from which it follows that $\alpha = 0$ and hence $\mathbf{z} = 0$. Therefore, G is non-singular and one can easily follow a branch around a saddle-node bifurcation (Keller, 1977).

In many applications, a trivial state can be easily found; for example, for zero forcing, a motionless solution may exist (examples are given in Chapter 1); let this solution be indicated by $(\mathbf{x}_0, \lambda_0)$, found at $s = s_0$. Often, also an initial tangent can be computed as, for example, a solution may exist for all values of a parameter. When a tangent is not available, a more general way of computing the initial tangent is the following. By differentiating $\Phi(\mathbf{X}(s)) = 0$ to s, we find again

$$\Phi_{\mathbf{X}}\dot{\mathbf{X}}(s) = \begin{pmatrix} \frac{\partial\Phi_1}{\partial x_1} & \cdots & \frac{\partial\Phi_1}{\partial x_d} & \frac{\partial\Phi_1}{\partial\lambda} \\ \frac{\partial\Phi_d}{\partial x_1} & \cdots & \frac{\partial\Phi_d}{\partial x_d} & \frac{\partial\Phi_d}{\partial\lambda} \end{pmatrix} \dot{\mathbf{X}}(s) = 0. \tag{5.10}$$

If $(\mathbf{x}_0, \lambda_0)$ is a regular point, $\Phi_{\mathbf{X}} = (\Phi_{\mathbf{x}} \ \Phi_{\lambda})$ has rank d. In this case, it can be triangulated into the form

$$\begin{pmatrix} * & * & * & * \\ 0 & * & * & * \\ 0 & 0 & * & * \end{pmatrix}. \tag{5.11}$$

where this matrix (an $*$ indicates a possible non-zero element) is shown for $d = 3$. The last row cannot be entirely zero, and therefore the (permuted) tangent vector $\dot{\mathbf{X}}_0 = (\dot{\mathbf{x}}_0, \dot{\lambda}_0)$ can be computed by solving

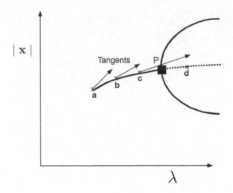

Figure 5.2 Example of computations of steady states $\mathbf{x}_a, \ldots, \mathbf{x}_d$ versus a parameter λ on a branch passing through a pitchfork bifurcation P.

$$
\begin{pmatrix}
* & * & * & * \\
0 & * & * & * \\
0 & 0 & * & * \\
0 & 0 & 0 & 1
\end{pmatrix}
\dot{\mathbf{X}}_0 =
\begin{pmatrix}
0 \\
0 \\
0 \\
1
\end{pmatrix},
\tag{5.12}
$$

and usually its length is normalized, such that $\|\dot{\mathbf{X}}_0\| = 1$.

5.2 Detection of Bifurcations

Suppose we have computed the points **a**, **b**, **c**, and **d** on a branch of steady solutions as indicated in Fig. 5.2 by varying a parameter λ. Assume, as in Fig. 5.2, that a pitchfork P occurs on this branch. How can one now determine the value of λ at this pitchfork?

5.2.1 *Detection of Simple Bifurcation Points*

As we have seen in Chapter 2, a bifurcation point of (5.3) is a point where the Jacobian Φ_x is singular. Hence, to determine simple bifurcation points (trans-critical, pitchfork, and saddle-node bifurcations), a test function τ is needed which changes sign as a bifurcation point (such as in Fig. 5.2) is crossed.

Using pseudo-arclength continuation, saddle-node bifurcations can be easily detected by taking $\tau = \dot{\lambda}$ along a solution branch, where the dot indicates differentiation to the arclength parameter s. For trans-critical and pitchfork bifurcations, one possible choice is $\tau = \det(\Phi_x)$, the determinant of the Jacobian matrix. The most efficient way to compute $\det(\Phi_x)$ is to perform an LU factorization of the Jacobian matrix such that $\det(\Phi_x) = \det(L)\det(U)$. In this case, $\det(\Phi_x)$ can be found from the diagonal elements of L and U.

However, making an *LU* factorization is expensive to compute for many large-dimensional problems, and other alternatives must be considered.

In Seydel (1994), a family of test functions τ_{pq} is based on a matrix Φ_x^{pq}, which is the Jacobian matrix Φ_x in which the pth row is replaced by the qth unit vector. If the linear system

$$\Phi_x^{pq}\mathbf{v} = \mathbf{e}_p \tag{5.13}$$

is solved for $\mathbf{v}$, where $\mathbf{e}_p$ is the pth unit vector, then it can be shown (Seydel, 1994) that

$$\tau_{pq} = \mathbf{e}_p^T \Phi_x \mathbf{v} \tag{5.14}$$

changes sign when Φ_x is singular. In principle, the choices of p and q are arbitrary as long as Φ_x^{pq} is non-singular. Of course, for any solution method, it is advantageous that Φ_x^{pq} and Φ_x have the same structure. The preceding is mathematically close to one step of the inverse iteration procedure as described in Section 7.4. *See example Ex. 5.1*

Once a change in sign is found in one of the scalar quantities (e.g., $\dot{\lambda}$, $\det(\Phi_x)$, τ_{pq}) between two points along a branch, say s_a and s_b, a secant process can be used to locate the zero of each function numerically. In more detail, let either function be indicated by $f(s)$; then a zero of $f(s)$ is determined by

$$s_{l+1} = s_l - f(s_l)\frac{s_l - s_{l-1}}{f(s_l) - f(s_{l-1})}, \tag{5.15a}$$

$$s_0 = s_a \quad ; \quad s_1 = s_b. \tag{5.15b}$$

When $s_a \neq 0$, the stopping criterion on the iteration can be chosen as $|s_{l+1} - s_l|/s_a < \varepsilon$, where ε must be chosen to achieve the desired accuracy. In some cases, a larger ε must be taken because the matrix Φ_x may become nearly singular. It is recommended to check a posteriori that the value of $f(s)$ is substantially smaller than the value of this function at both s_a and s_b.

5.2.2 Detection of Hopf Bifurcations

Suppose a stationary solution $\mathbf{x}_0$ of (5.3) at a certain value of λ has been determined. Then its linear stability is investigated by considering perturbations $\mathbf{x} = \mathbf{x}_0 + \tilde{\mathbf{x}}$. Substituted into the general equations (5.2) and omitting quadratic terms in the perturbations quantities, one gets

$$M\frac{\partial \tilde{\mathbf{x}}}{\partial t} = \Phi_x(\mathbf{x}_0, \lambda)\tilde{\mathbf{x}}, \tag{5.16}$$

where Φ_x is again the Jacobian matrix of Φ.

These equations admit solutions of the form $\tilde{\mathbf{x}} = \hat{\mathbf{x}}\, e^{\sigma t}$ and this leads to a generalized matrix eigenvalue problem of the form

$$A(\lambda)\hat{\mathbf{x}} = \sigma B(\lambda)\hat{\mathbf{x}} \qquad (5.17)$$

with $A = \Phi_{\mathbf{x}}(\mathbf{x}_0)$ and $B = M$ in general depending on the parameter λ and with the matrix B in general being singular.

Let A and B be two $d \times d$ matrices; then the set of matrices of the form $A - \sigma B$ is called a pensil. The eigenvalues of the pensil belong to the set $\sigma(A, B)$ defined by

$$\sigma(A, B) = \{z \in \mathbb{C} \mid \det(A - zB) = 0\}. \qquad (5.18)$$

When B is non-singular, the generalized eigenvalue problem is equivalent to an ordinary eigenvalue problem:

$$B^{-1}A\mathbf{x} = \sigma\mathbf{x}. \qquad (5.19)$$

However, when B is singular, complications arise. The set $\sigma(A, B)$ may consist of a finite number of eigenvalues, no eigenvalue may exist, or infinitely many may occur (Golub and van Loan, 1996). We will discuss solution methods for these eigenvalue problems in Chapters 6, 7, and 8. *See example Ex. 5.2*

Although more sophisticated methods exist (Kuznetsov, 1995; Meerbergen and Spence, 2010), Hopf bifurcation points are usually determined by solving the linear stability problem (5.17). In this case, a complex conjugate pair of eigenvalues $\sigma = \sigma_R \pm i\,\sigma_I$ crosses the imaginary axis at $\lambda = \lambda_H$ transversally, that is, with

$$\frac{d\sigma_R}{d\lambda}(\lambda_H) \neq 0.$$

To detect such a Hopf bifurcation, the test function $\tau = \sigma_R(\lambda)$ has to be calculated. To precisely locate the Hopf bifurcation, one can use again the secant method (5.15).

5.3 Branch Switching

If, for example, $\det(\Phi_{\mathbf{x}})$ changes sign but $\dot{\lambda}$ does not, a simple bifurcation point (trans-critical or pitchfork) is detected. Subsequently, a branch switch process can be started to locate solutions on the nearby branch.

5.3.1 Simple Bifurcation Points

In Fig. 5.3, this situation is sketched near a pitchfork bifurcation. Let Φ_x be the Jacobian matrix at the bifurcation point $(\mathbf{x}_0, \lambda_0)$ just after the secant iteration (5.15) has converged for $s = s_*$. Furthermore, let the tangent along the already known

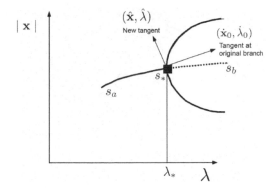

Figure 5.3 Example of branch switching near a pitchfork bifurcation.

branch in $s = s_*$ be indicated by $(\dot{\mathbf{x}}_0, \dot{\lambda}_0)$. To determine a point on the new branch, first the null vector $\mathbf{v}$ of $\boldsymbol{\Phi}_\mathbf{x}$ is calculated, for example by inverse iteration (Section 7.4). Next, a null vector $(\hat{\mathbf{x}}, \hat{\lambda})$ of the extended Jacobian matrix is constructed which is orthogonal to $(\dot{\mathbf{x}}_0, \dot{\lambda}_0)$ by solving

$$\begin{pmatrix} \boldsymbol{\Phi}_\mathbf{x} & \boldsymbol{\Phi}_\lambda \\ \dot{\mathbf{x}}_0^T & \dot{\lambda}_0 \end{pmatrix} \begin{pmatrix} \hat{\mathbf{x}} \\ \hat{\lambda} \end{pmatrix} = \begin{pmatrix} \mathbf{0} \\ 0 \end{pmatrix}. \tag{5.20}$$

The solution of this problem is

$$\hat{\lambda} = \frac{-\dot{\mathbf{x}}_0^T \mathbf{v}}{\dot{\lambda}_0 - \dot{\mathbf{x}}_0^T \mathbf{z}} \; ; \; \hat{\mathbf{x}} = \mathbf{v} - \hat{\lambda}\mathbf{z},$$

where $\mathbf{z}$ is the solution of $\boldsymbol{\Phi}_\mathbf{x}\mathbf{z} = \boldsymbol{\Phi}_\lambda$.

To determine a point on the new branch (Fig. 5.3), we start in the direction of the new tangent, that is,

$$\mathbf{x}^1 = \mathbf{x}_0 \pm \Delta s \, \hat{\mathbf{x}} \; ; \; \lambda^1 = \lambda_0 + \Delta s \, \hat{\lambda}. \tag{5.21}$$

The $\pm$ indicates that points can be found on either side of the known branch. This solution will serve as initial condition for the Newton–Raphson method. When a point on a new branch is found, the pseudo-arclength procedure is again used to compute additional points on this branch.

If one already anticipates a pitchfork bifurcation, one can also determine the new branch by a technique which makes use of the so-called imperfections of the bifurcations. Slight perturbations from symmetry in a particular model will also destroy the pitchfork bifurcation. Consider, for example, such a perturbation in the dynamical system

$$\frac{dx}{dt} = \epsilon + \lambda x - x^3 \tag{5.22}$$

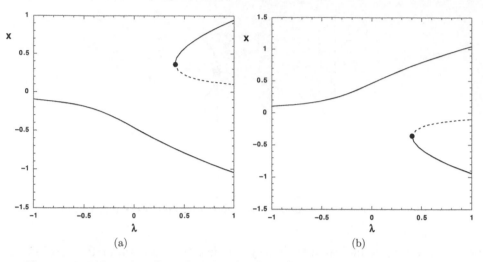

Figure 5.4 Bifurcation diagrams of (5.22) for different values of ϵ with (a) $\epsilon =$ -0.1 and (b) $\epsilon = 0.1$. A drawn (dashed) curve indicates stable (unstable) steady solutions.

for some (small) ϵ. When $\epsilon = 0$, a pitchfork bifurcation occurs at $(x = 0, \lambda = 0)$, but for $\epsilon \neq 0$, the pitchfork bifurcation is no longer present because the reflection symmetry no longer exists. Bifurcation diagrams are sketched for both negative and positive ϵ in Fig. 5.4 and show different re-connections of the branches. In both cases, there is only one solution for negative λ, and three solutions exist for large positive λ.

Another example is the imperfection of the trans-critical bifurcation given by

$$\frac{dx}{dt} = \epsilon + \lambda x + x^2. \tag{5.23}$$

For $\epsilon = 0$, a trans-critical bifurcation occurs at $(x = 0, \lambda = 0)$, but for $\epsilon \neq 0$, this bifurcation is no longer present. The different re-connections for this case are shown in Fig. 5.5 for both positive and negative ϵ. For $\epsilon < 0$, the stable and unstable branches connect up, whereas for $\epsilon > 0$ an unstable and a stable branch connect to give two saddle-node bifurcations. In the latter case, a window in λ opens where no steady solutions exist.

Suppose, in the case of a pitchfork bifurcation, that two points A and B on a branch are computed where some test function τ changes sign (Fig. 5.6a). From imperfection theory, we know that introducing an additional parameter μ which breaks the symmetry (for example, introducing some asymmetric component in the forcing), the pitchfork no longer exists for small μ. Hence, one continues a few steps into the parameter μ from point A up to $\mu = \varepsilon$. Then, a point C on the bifurcation diagram as in Fig. 5.6b is obtained. Next, the parameter λ is increased

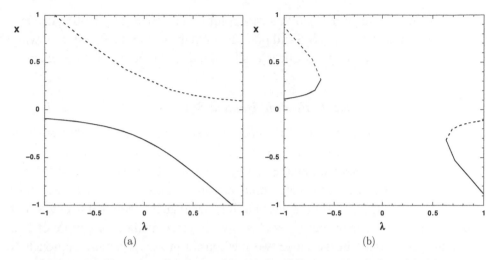

Figure 5.5 Bifurcation diagrams of (5.23) for different values of ϵ with (a) $\epsilon = -0.1$ and (b) $\epsilon = 0.1$. A drawn (dashed) curve indicates stable (unstable) steady solutions.

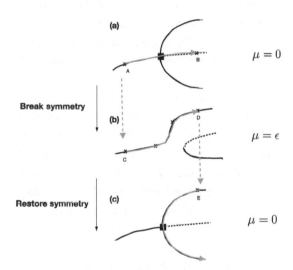

Figure 5.6 Example of how knowledge of imperfections can be used to locate bifurcation points. The control parameter is λ. (a) Symmetric situation with computed points A and B, where a sign switch in one of the test functions τ has been detected. (b) The imperfect pitchfork bifurcation is created by adding artificial asymmetry into the set of equations using a parameter μ. Point A is first followed up to B to detect the pitchfork. Next, point A is followed up to point C in μ. As a next step, one continues the branch from C to D for a value of λ approximately up to the value at B. (c) Finally, symmetry is restored, point D is followed up to E, and the pitchfork can be found as the point where λ changes sign. Continuation directions are indicated by the grey curves.

up to the value of λ at point B; in this way point D is reached (Fig. 5.6b). As a last step, μ is continued back to zero and point E is obtained (Fig. 5.6c). By following the branch back in λ, the pitchfork is easily found as the point where $\dot{\lambda}$ changes sign.

5.3.2 Finding Isolated Branches

In many applications there exist branches of steady-state solution that are disconnected from the branch containing a starting solution, the so-called isolated branches. One can already anticipate that in a dynamical system in which there is no symmetry, it is likely that isolated branches are present. There are at least four methods to compute isolated branches, but it is never guaranteed that one will find all branches with either of these methods. Two of those methods are more or less 'trial and error', while in the latter two methods, a more systematic approach is followed.

(i) *Transient integration*

In this approach, a set of initial conditions is chosen and a transient computation is started. If one is lucky, one of the initial conditions is in the attraction basin of a steady state on the isolated branch, and once found (Fig. 5.7a), one can continue tracing this branch using the pseudo-arclength continuation method.

(ii) *Isolated non-linear algebraic system solver search*

One can also start a Newton–Raphson solver uncoupled from the pseudo-arclength continuation from several chosen starting points. Since the convergence of these solvers is only good when one is near the steady state, this method may not work very well; but again, if no special considerations are made, one has to be very lucky to find a point on an isolated branch (Fig. 5.7b). One can make use of so-called deflation techniques (Farrell *et al.*, 2015), where the right-hand side $\Phi(\mathbf{x}, \lambda)$ is multiplied with function $H(x; \mathbf{x}^*)$ where $\mathbf{x}^*$ is an already computed solution of $\Phi(\mathbf{x}, \lambda) = 0$. This function H is chosen such that the original solution will not be reached in a Newton process on $H(x; \mathbf{x}^*)\Phi(\mathbf{x}, \lambda) = 0$. A recent application of this deflation technique on the 2D Rayleigh–Bénard problem can be found in Boullé and colleagues (2022).

(iii) *Two-parameter continuation*

In many cases, a second parameter can be varied such that the isolated branch targeted connects to an already known branch. An important example is where there are values of the second parameter for which the dynamical system has a particular symmetry and pitchfork bifurcations are present. Once the connection is present, the isolated branch can be computed by restoring the second parameter to its original value (Fig. 5.7c).

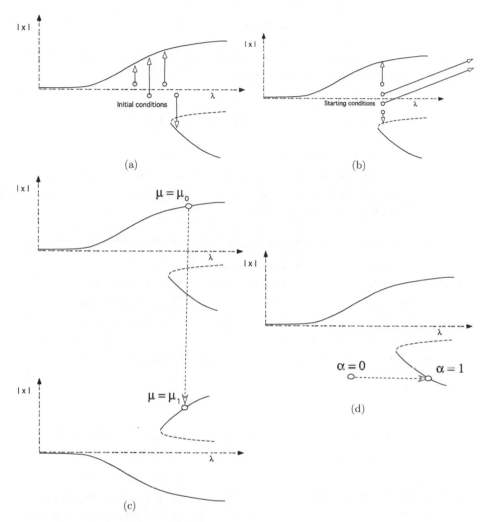

Figure 5.7 Illustrations of the computation of isolated branches using four different methods. (a) Transient integration; the open circles indicate the initial conditions and the arrows the direction of the trajectories. Note that only stable steady states can be reached. (b) Isolated non-linear algebraic system solver search; the open circles indicate the starting points. The two large arrows indicate a possible divergence of this process. (c) Two-parameter continuation; a pitchfork occurs for $\mu_0 < \mu < \mu_1$. (d) Residue continuation, where α is the 'homotopy' parameter.

(iv) *Residue continuation*

This method is a special case of a two-parameter continuation where one starts with a guess of the solution on the isolated branch, say indicated by $\mathbf{x}_G$, at some value of a parameter λ. Because this is no steady solution, it follows that

$$\Phi(\mathbf{x}_G, \lambda) = \mathbf{r}_G \neq 0,$$

where $\mathbf{r}_G$ is the non-zero residue. One now defines a second (so-called 'homotopy') parameter α and considers

$$\Phi(\mathbf{x}, \lambda) - (1 - \alpha)\mathbf{r}_G = 0.$$

For $\alpha = 0$, the solution is given by $\mathbf{x}_G$ (by construction), and hence this is the starting point of the pseudo-arclength continuation with parameter α. By tracing the steady solution branch from $\alpha = 0$ to $\alpha = 1$, one may eventually find an isolated branch (Fig. 5.7d).

Additional Material

- There are quite a few books on continuation methods, some focusing on more mathematical aspects (Govaerts, 2000; Allgower and Georg, 2003; Krauskopf *et al.*, 2007). Others focus more on the application to general ordinary (Seydel, 1994) and partial differential (Uecker, 2021) equations or on a specific type of problem, e.g. reaction–diffusion equations (Mei, 2000).

5.4 Continuation of Bifurcation Points in a Secondary Parameter

Often in applications, one wants to determine so-called regime diagrams where paths of bifurcation points are plotted in a two-parameter plane. One can do this by varying the second parameter, say μ, and compute bifurcation diagrams in the primary parameter λ for each value of μ. An example is shown in Fig. 5.8, where the paths of two saddle-node bifurcations eventually intersect in a so-called cusp bifurcation. However, it can be much more efficient to directly compute the path of a bifurcation point versus μ, which we will discuss in this section for the saddle-node bifurcations and Hopf bifurcations.

5.4.1 Saddle-Node Bifurcations

In what follows we indicate the null space of a matrix A by $\mathcal{N}_s(A)$ and its range by $\mathcal{R}_a(A)$. A regular solution $\mathbf{X}_0 = (\mathbf{x}_0, \lambda_0)$ is called a simple fold when dim $\mathcal{N}_s(\Phi_\mathbf{x}^0) = 1$ and $\Phi_\lambda^0 \notin \mathcal{R}_a(\Phi_\mathbf{x}^0)$, where again $\Phi_\mathbf{x}^0 = \Phi_\mathbf{x}(\mathbf{x}_0, \lambda_0)$ and $\Phi_\lambda^0 = \Phi_\lambda(\mathbf{x}_0, \lambda_0)$. By differentiating $\Phi(\mathbf{x}(s), \lambda(s)) = 0$ to s, where s is the arclength parameter, and evaluating this at $\mathbf{X}_0$, we obtain

$$\Phi_\mathbf{x}^0 \dot{\mathbf{x}}_0 = -\Phi_\lambda^0 \dot{\lambda}_0. \tag{5.24}$$

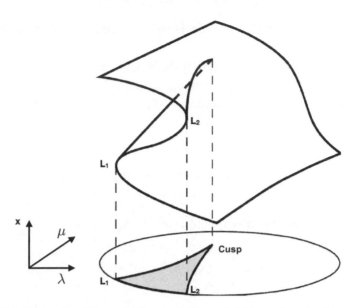

Figure 5.8 Example of the path of two saddle-node bifurcations (in the parameter λ) versus a second parameter μ. When these paths intersect, a cusp bifurcation occurs.

At a simple saddle-node bifurcation, $\Phi_\lambda^0 \notin \mathcal{R}_a(\Phi_x^0)$ and hence $\dot{\lambda}_0 = 0$. This property was already used in the detection of the saddle-node bifurcation based on the shape of the branches and its normal form. As a consequence, $\Phi_x^0 \dot{x}_0 = 0$ and hence $\mathcal{N}_s(\Phi_x^0) = \text{Span}\{\dot{x}_0\}$.

When (5.24) is differentiated again to s and evaluated at X_0, we obtain

$$(\Phi_{xx}^0 \dot{x}_0)\dot{x}_0 + \Phi_x^0 \ddot{x}_0 + 2\Phi_{x\lambda}^0 \dot{x}_0 \dot{\lambda}_0 + \Phi_{\lambda\lambda}^0 (\dot{\lambda}_0)^2 + \Phi_\lambda^0 \ddot{\lambda}_0 = 0. \qquad (5.25)$$

Now, because Φ_x^0 is singular with a one-dimensional kernel, its adjoint $(\Phi_x^0)^*$ also is singular. Indicate the kernel of this adjoint matrix by the vector ψ^*, where the $*$ indicates the conjugate transpose. This vector is orthogonal to the every vector y in $\mathcal{R}_a(\Phi_x^0)$ since

$$< \psi^*, \Phi_x^0 y > = < (\Phi_x^0)^* \psi^*, y > = 0,$$

where, for clarity, the standard inner product $<\ ,\ >$ in $\mathbb{R}^d$ is used. Left-multiplying (5.25) by ψ^* and using that $\dot{\lambda}_0 = 0$ at the simple saddle-node bifurcation, gives

$$\psi^* \Phi_\lambda^0 \ddot{\lambda}_0 + \psi^* (\Phi_{xx}^0 \dot{x}_0)\dot{x}_0 = 0,$$

and since $\Phi_\lambda^0 \notin \mathcal{R}(\Phi_x^0)$, we find for the curvature $\ddot{\lambda}_0$

$$\ddot{\lambda}_0 = -\frac{\psi^* (\Phi_{xx}^0 \dot{x}_0)\dot{x}_0}{\psi^* \Phi_\lambda^0} \qquad (5.26)$$

When $\ddot{\lambda}_0 \neq 0$, then the saddle-node is called a simple quadratic saddle-node bifurcation.

Suppose that a simple quadratic saddle-node has been detected, giving a solution $(\mathbf{x}_0, \lambda_0, \mathbf{v}_0, \mu_0)$ at a value μ_0 of a second parameter μ, where $\mathcal{N}(\Phi_x^0) = \mathbf{v}_0$. The quadratic saddle-node is hence determined (for each value of μ) by the system of equations

$$\Phi(\mathbf{x}, \lambda, \mu) = 0, \tag{5.27a}$$

$$\Phi_{\mathbf{x}}(\mathbf{x}, \lambda, \mu)\mathbf{v} = 0, \tag{5.27b}$$

$$\mathbf{v}_0^T \mathbf{v} - 1 = 0. \tag{5.27c}$$

The Jacobian J of this extended system at $(\mathbf{x}_0, \lambda_0, \mathbf{v}_0, \mu_0)$ is

$$J = \begin{pmatrix} \Phi_{\mathbf{x}}^0 & 0 & \Phi_\lambda^0 \\ \Phi_{\mathbf{xx}}^0 \mathbf{v}_0 & \Phi_{\mathbf{x}}^0 & \Phi_{\mathbf{x}\lambda}^0 \mathbf{v}_0 \\ 0 & \mathbf{v}_0^T & 0 \end{pmatrix}. \tag{5.28}$$

It turns out that this Jacobian is non-singular (Doedel *et al.*, 1991). The implicit function theorem now states that locally a unique smooth branch of saddle-node bifurcations in λ of (5.27) exists which can be parameterized by μ. *See example Ex. 5.3*

In practice, one also uses pseudo-arclength continuation, and the equation added is

$$\dot{\mathbf{x}}_0^T(\mathbf{x} - \mathbf{x}_0) + \dot{\mathbf{v}}_0^T(\mathbf{v} - \mathbf{v}_0) + \dot{\lambda}_0(\lambda - \lambda_0) + \dot{\mu}_0(\mu - \mu_0) - (s - s_0) = 0. \tag{5.29}$$

Equations (5.27) and (5.29) define an extended algebraic system of $2d + 2$ equations for the $2d + 2$ unknowns $(\mathbf{x}, \lambda, \mathbf{v}, \mu)$, which can be solved again by the Newton–Raphson method.

5.4.2 Hopf Bifurcations

At a Hopf bifurcation $(\mathbf{x}_0, \lambda_0)$, a nearby periodic orbit exists, and hence the linearized problem

$$M\frac{d\phi}{dt} = \Phi_{\mathbf{x}}(\mathbf{x}_0, \lambda_0)\phi \tag{5.30}$$

with $\phi(0) = \phi(p)$ has a solution for some value of the period p. If the linear stability problem (5.17) has only a simple pair of complex conjugate eigenvalues $\sigma = \pm i\sigma_I$ with imaginary part $\sigma_I = 2\pi/p$, then (5.30) has the solution

$$\phi(t) = \sin(\sigma_I t)\hat{\mathbf{x}}_R + \cos(\sigma_I t)\hat{\mathbf{x}}_I, \tag{5.31}$$

where the corresponding eigenvector of (5.17) is given by $\mathbf{v} = \hat{\mathbf{x}}_R + i\hat{\mathbf{x}}_I \in \mathbb{C}^d$ with $\|\mathbf{v}\| = 1$. Note that the Jacobian $\Phi_{\mathbf{x}}$ is non-singular at the Hopf bifurcation and hence the implicit function theory guarantees that nearby fixed-point solutions $\mathbf{x}$ can be parameterized by λ.

Giving a reference solution $(\mathbf{x}_0, \mathbf{v}_0, \sigma_I^0, \lambda_0, \mu_0)$, a Hopf bifurcation can be followed in a second parameter μ. The extended system is

$$\Phi(\mathbf{x}, \lambda, \mu) = 0, \tag{5.32a}$$

$$\Phi_\mathbf{x}(\mathbf{x}, \lambda, \mu)\mathbf{v} - i\sigma_I M\mathbf{v} = 0, \tag{5.32b}$$

$$\mathbf{v}^*\mathbf{v}_0 - 1 = 0. \tag{5.32c}$$

Consider now parameter continuation in μ, parameterizing the solution as $(\mathbf{x}(\mu), \mathbf{v}(\mu), \sigma_I(\mu), \lambda(\mu))$. The Jacobian of the extended system is

$$\begin{pmatrix} \Phi_\mathbf{x}^0 & 0 & 0 & \Phi_\lambda^0 \\ \Phi_{\mathbf{xx}}^0\mathbf{v}_0 & \Phi_\mathbf{x}^0 - i\sigma_I^0 M & -iM\mathbf{v}_0 & \Phi_{\mathbf{x}\lambda}^0\mathbf{v}_0 \\ 0 & \mathbf{v}_0^* & 0 & 0 \end{pmatrix}. \tag{5.33}$$

One can show (Doedel *et al.*, 1991) that this Jacobian matrix is non-singular when the transversality condition

$$\frac{d\sigma_R}{d\lambda}(\lambda_0) \neq 0$$

holds.

In practice, again pseudo-arclength continuation is used, and the additional equation

$$\dot{\mathbf{x}}_0^T(\mathbf{x}-\mathbf{x}_0)+(\mathbf{v}-\mathbf{v}_0)^*\dot{\mathbf{v}}_0+\dot{\sigma}_I^0(\sigma_I-\sigma_I^0)+\dot{\lambda}_0(\lambda-\lambda_0)+\dot{\mu}_0(\mu-\mu_0)-(s-s_0) = 0 \tag{5.34}$$

is added. Equations (5.32) and (5.34) form a system of $3d + 2$ equations for the $3d + 3$ unknowns $(\mathbf{x}, \mathbf{v}, \lambda, \sigma_i, \mu)$ and hence an additional condition is needed to compute a branch of Hopf bifurcations in μ.

Note that when $\phi(t)$ is a solution of (5.30), $\phi(t+\xi)$ is also a solution, so the phase of ϕ still has to be fixed along the branch of Hopf bifurcations. Another way to look at this is that the null space of (5.32b) is two-dimensional; apart from $\mathbf{v}_1 = \hat{\mathbf{x}}_R + i\hat{\mathbf{x}}_I$ $\mathbf{v}_2 = -\hat{\mathbf{x}}_I + i\hat{\mathbf{x}}_R$ also is an eigenvector. Let ϕ_0 be an already computed solution at a Hopf bifurcation point for a specified value of $\mu = \mu_0$. To fix the phase ξ of $\phi(t)$, it is required that

$$\delta(\xi) = \int_0^p (\phi(t + \xi) - \phi_0(t))^*(\phi(t + \xi) - \phi_0(t))dt, \tag{5.35}$$

where p is the period, is minimized with respect to ξ. Substitution of $\phi(t)$ and $\phi_0(t)$ and performing the minimization leads to *See example Ex. 5.4*

$$(\hat{\mathbf{x}}_I^0)^T \hat{\mathbf{x}}_R - (\hat{\mathbf{x}}_R^0)^T \hat{\mathbf{x}}_I = 0, \qquad (5.36)$$

which is the extra equation to be added to (5.32) and (5.34) to obtain a well-defined problem for the continuation of Hopf bifurcations in a second parameter μ.

5.5 Continuation of Periodic Orbits

Suppose we have located a Hopf bifurcation; then a nearby periodic solution exists. Hence, this solution $(\mathbf{x}(t), \lambda)$ satisfies (where M is again the mass matrix)

$$M\frac{d\mathbf{x}}{dt} = \Phi(\mathbf{x}, \lambda), \qquad (5.37a)$$
$$\mathbf{x}(0) = \mathbf{x}(p). \qquad (5.37b)$$

Using pseudo-arclength continuation methods, such periodic solutions can be computed in two ways, that is,

(i) as a solution of a boundary value problem in time. These methods require a time integration of (5.37) and will be discussed in Section 5.5.1.
(ii) as a solution of a fixed point of the Poincaré map. A set of algebraic equations is derived which can be solved in the same way as in fixed-point continuation (Section 5.5.2).

Additional Material

- There are several software packages to analyse the bifurcation diagrams of ordinary differential equations, such as AUTO (Doedel, 1980), see http://indy.cs.concordia.ca/auto/; and MATCONT (Dhooge et al., 2003), see https://sourceforge.net/projects/matcont/. The AUTO user manual and accompanying notes (Doedel et al., 2007) are highly recommended material to learn to use these methods. More recent packages in Python, such as PyDSTool, are available on GitHub: https://pydstool.github.io/PyDSTool/FrontPage.html.
- Software packages for large-dimensional dynamical systems, such as those arising from the discretization of partial differential equations, will be mentioned in Chapter 10.

5.5.1 Boundary Value Problem Methods

As the unknown period p is part of the problem, time is re-scaled with p such that the equations (5.37) become

$$M\frac{d\mathbf{x}}{dt} = p\Phi(\mathbf{x}, \lambda), \qquad (5.38a)$$

$$\mathbf{x}(0) \quad = \quad \mathbf{x}(1), \tag{5.38b}$$

and the state vector is now indicated by $(\mathbf{x}(t), p)$.

Suppose that $(\mathbf{x}_0(t), p_0, \lambda_0)$ is a previously computed periodic orbit for a value of λ_0; then the pseudo-arclength condition to compute a periodic solution at a parameter λ is

$$\int_0^1 (\mathbf{x}(t) - \mathbf{x}_0(t))^T \dot{\mathbf{x}}_0(t) dt + (p - p_0)\dot{p}_0 + (\lambda - \lambda_0)\dot{\lambda}_0 = s - s_0, \tag{5.39}$$

where the dot indicates again the derivative to s.

The periodic solution is not totally determined yet by (5.38) and (5.39) because the phase of the periodic orbit is still unknown. Note that if $\tilde{\mathbf{x}}(t)$ is a periodic solution, then also $\tilde{\mathbf{x}}(t + \xi)$ is a solution, where ξ is a constant phase shift. A phase condition can be derived by minimizing the distance δ between the solutions at λ^0 and at λ, that is, to minimize

$$\delta(\xi) = \int_0^1 \|\tilde{\mathbf{x}}(t + \xi) - \mathbf{x}_0(t)\|^2 \, dt. \tag{5.40}$$

Differentiation to ξ gives

$$\frac{\partial \delta}{\partial \xi} = 0 \to \int_0^1 (\tilde{\mathbf{x}}(t + \xi) - \mathbf{x}_0(t))^T \frac{d\tilde{\mathbf{x}}(t + \xi)}{dt} dt = 0. \tag{5.41}$$

The first product cancels because of periodicity, and the second term can be re-written using partial integration and the vector identity $\mathbf{a}^T \mathbf{b} = \mathbf{b}^T \mathbf{a}$. With the choice $\mathbf{x}(t) = \tilde{\mathbf{x}}(t + \xi)$, the phase condition becomes

$$\int_0^1 \mathbf{x}^T(t) \frac{d\mathbf{x}_0(t)}{dt} dt = 0. \tag{5.42}$$

So, the total system of equations to compute a branch of periodic solutions in λ is (5.38), (5.39), and (5.42).

A way to implement this numerically is to discretize time, using $t_i = i\Delta t, i = 0, \ldots, N$, with $N = 1/\Delta t$. Any time discretization of (5.38), (5.39), and (5.42) will lead to an algebraic system which can again be solved with the Newton–Raphson method. For example, a time-explicit Euler forward discretization gives

$$M\frac{\mathbf{x}^{n+1} - \mathbf{x}^n}{\Delta t} \quad = \quad p\Phi(\mathbf{x}^n, \lambda), \ n = 1, \ldots, N - 1 \tag{5.43a}$$

$$\mathbf{x}^0 \quad = \quad \mathbf{x}^N, \tag{5.43b}$$

which gives, combined with (5.39) and (5.42), $dN + 2$ equations for the $dN + 2$ unknowns $(\mathbf{x}^1, \ldots, \mathbf{x}^N, p, \lambda)$.

As the periodic orbit computation starts often at a Hopf bifurcation, already a starting solution is given by $\mathbf{x}_s(t) = \mathbf{x}_0 + \alpha\,\phi_0(t), p_s = p_0, \lambda_s = \lambda_0$, where ϕ_0 is given by (5.31), that is,

$$\phi_0(t) = \sin(2\pi t)\hat{\mathbf{x}}_R + \cos(2\pi t)\hat{\mathbf{x}}_I, \tag{5.44}$$

and $\mathbf{x}_0$ does not depend on time. A crude choice is $\alpha = \Delta s$. For the initial phase condition (5.42), we then find

$$\int_0^1 \mathbf{x}^T(t)\frac{d\mathbf{x}_s(t)}{dt}dt = 0 \rightarrow \int_0^1 \mathbf{x}^T(t)\frac{d\phi_0(t)}{dt}dt = 0, \tag{5.45}$$

and since $\dot{p}_0 = \dot{\lambda}_0 = 0$, the pseudo-arclength condition (5.39) reduces at the first step to

$$\int_0^1 (\mathbf{x}(t) - \mathbf{x}_0(t))^T \dot{\mathbf{x}}_s(t)dt - \Delta s = 0 \rightarrow \int_0^1 (\mathbf{x}(t) - \mathbf{x}_0(t))^T \phi_0(t)dt - \Delta s = 0. \tag{5.46}$$

As discussed in Section 2.5, the stability of the periodic orbits can be determined by computing the eigenvalues of the monodromy matrix. In general, this matrix has to be determined from the linearized system around the periodic solution (the fundamental solution), but in special cases, it can be obtained as a by-product of the continuation procedure (Doedel *et al.*, 1991). *See example Ex. 5.5*

5.5.2 *Fixed Points of the Poincaré Map*

The Poincaré map was already introduced in Chapter 2 through a Poincaré section Σ^+, here defined by

$$g(\mathbf{x}) = \mathbf{n}^T M(\mathbf{x} - \mathbf{x}_{\Sigma^+}), \tag{5.47}$$

where $\mathbf{n}$ is the normal to this section and M is again the mass matrix. The Poincaré map on this section is defined as

$$\mathcal{P}(\mathbf{x}, \lambda) = \big(\phi(t(\mathbf{x}), \mathbf{x}, \lambda) \in \Sigma^+\big), \tag{5.48}$$

where ϕ indicates the periodic solution and $t(\mathbf{x})$ is the time it takes to return to Σ^+ starting at $\mathbf{x} \in \Sigma^+$.

Fixed points of the Poincaré map correspond to periodic orbits and are determined by

$$\mathbf{x} - \mathcal{P}(\mathbf{x}, \lambda) = 0. \tag{5.49}$$

This is an algebraic system, and hence we can use pseudo-arclength continuation to find solutions while varying λ. The normalization condition for the tangent is written as

$$\Sigma(\mathbf{x}, \lambda) = \dot{\mathbf{x}}_0^T(\mathbf{x} - \mathbf{x}_0) + (\lambda - \lambda_0)\dot{\lambda} - \Delta s = 0, \tag{5.50}$$

where Δs is the step length.

The extended Jacobian matrix J can be written

$$J = \begin{pmatrix} I - \mathcal{P}_{\mathbf{x}} & -\mathcal{P}_{\lambda} \\ \dot{\mathbf{x}}_0^T & \dot{\lambda}_0 \end{pmatrix}. \tag{5.51}$$

The derivatives can be determined in terms of the original dynamical system (5.38) as $\phi(t, \mathbf{x}, \lambda) \in \Sigma^+$ and consequently

$$\mathbf{n}^T M(\phi(t(\mathbf{x}), \mathbf{x}, \lambda) - \mathbf{x}_{\Sigma^+}) = 0. \tag{5.52}$$

Differentiation of (5.52) to $\mathbf{x}$ gives

$$\mathbf{n}^T M(\phi_t \frac{\partial t}{\partial \mathbf{x}} + \phi_{\mathbf{x}}) = 0, \tag{5.53}$$

and this yields

$$\frac{\partial t}{\partial \mathbf{x}} = -\frac{\mathbf{n}^T M \phi_{\mathbf{x}}}{\mathbf{n}^T M \phi_t}. \tag{5.54}$$

Because $M\phi_t = \Phi(\phi, \lambda)$, the derivatives of the Poincaré map are given by

$$\mathcal{P}_{\mathbf{x}} = \phi_{\mathbf{x}} - \frac{\mathbf{n}^T M \phi_{\mathbf{x}}}{\mathbf{n}^T \Phi(\phi, \lambda)} \phi_t, \tag{5.55a}$$

$$\mathcal{P}_{\lambda} = \phi_{\lambda}, \tag{5.55b}$$

and similar to the implementation of fixed point continuation, we only need to solve systems of equations with $\mathcal{P}_{\mathbf{x}}$.

The Floquet multipliers can also be obtained from the Poincaré map $\mathcal{P}$. For a periodic orbit, a point $\mathbf{x}_*$ on the Poincaré section is mapped to itself after one period p, that is,

$$\mathcal{P}(\mathbf{x}_*) = \mathbf{x}_*. \tag{5.56}$$

The eigenvalues $\rho_j, j = 1, \ldots d - 1$ of the Jacobian matrix $\mathcal{P}_{\mathbf{x}}$ at $\mathbf{x} = \mathbf{x}_*$ determine the stability of the periodic orbit. In fact, these are the eigenvalues of the monodromy matrix (see Section 2.5), except (in case of autonomous systems) the unit eigenvalue.

5.6 Summary

- The pseudo-arclength continuation method is an efficient method to follow steady states and periodic orbits in parameter space. Its particular advantage is that branches of steady states can be followed around saddle-node bifurcations.
- Detection of bifurcation points on branches of steady states can be accomplished by several test functions, and then the precise value of the parameter at such a bifurcation is determined by the secant method.
- There are two techniques to compute new branches arising from pitchfork and trans-critical bifurcations: branch switching and continuation through imperfections.
- Instead of computing bifurcation diagrams in a first parameter λ for different values of a second parameter μ, there are techniques where the paths of a saddle-node bifurcation and of a Hopf bifurcation can be determined directly in the (λ, μ) plane.
- Periodic orbits arising from a Hopf bifurcation can be directly computed with pseudo-arclength continuation. The stability of these periodic orbits can be evaluated by computing the eigenvalues of the monodromy matrix or by constructing the Poincaré map and determining the eigenvalues of the Jacobian of this map.

5.7 Exercises

Exercise 5.1 *In Section 5.2.1, several test functions were suggested to detect simple bifurcations.*

a. Show that the test function τ_{pq} in (5.14) changes sign when a simple bifurcation (saddle-node, trans-critical, or pitchfork bifurcation) is crossed in a one-parameter continuation.

b. Show that the procedure of computing this test function is mathematically equivalent to one step of the inverse iteration algorithm used to determine a one-dimensional kernel of a singular matrix.

Exercise 5.2 *In several applications, one may want to compute curves of eigenvalues, versus a parameter λ, of the eigenvalue problem*

$$A(\lambda)\mathbf{x} = \sigma B(\lambda)\mathbf{x}.$$

a. In which case can this tracing method be much more useful compared to computing eigenvalues for discrete values of λ?

b. Design a pseudo-arclength continuation scheme to compute eigenvalues versus λ.

Exercise 5.3 *In this exercise we will show that the Jacobian matrix (5.28) is non-singular. Assume that it is singular; then there exists a vector $z = (z_1, z_2, z_3)^T$, z_3 being a scalar such that $Jz = 0$.*

a. Show that $z_3 = 0$ and hence $z_1 = c_1 v_0$.

b. Show next that for a quadratic saddle node, according to (5.26), we must have $c_1 = 0$.

c. Show that $z_2 = c_2 v_0$ and that $c_2 = 0$.

Exercise 5.4 *Follow the same procedure as in Section 5.5 to show that the minimization of (5.35) leads to the condition (5.36).*

Exercise 5.5 *When $x(t)$ is a periodic solution of an autonomous dynamical system, then, for every τ, $x(t + \tau)$ is also a periodic solution. For small τ, define $y(t) = x(t + \tau) - x(t)$; then $y(0)$ is an initial disturbance along the orbit of the periodic solution X.*

a. Show that for every integer k,

$$y(t + kp) = y(t).$$

b. Show that for autonomous systems, there is always one Floquet multiplier $\rho = 1$.

6

Matrix-Based Techniques

*Having an explicit Jacobian
is more valuable than having many right-hand sides.*

In Chapter 5 we have seen that, when a pseudo-arclength continuation method is applied, the actual computational work is to determine the solution of non-linear algebraic systems and generalized eigenvalue problems. In this chapter, we discuss the numerical bifurcation analysis that can be carried out when the application code supplies a Jacobian matrix. Actually, a significant number of simulation codes for complex flow simulations compute a Jacobian matrix (Dijkstra *et al.*, 2014), and such code development is typically motivated by investigations of flow behaviour that occurs on long time scales (or only for steady states). Having a Jacobian matrix, and not merely an algorithm for applying the Jacobian operator as in the matrix-free techniques discussed in Chapter 9, can lead to more robust and efficient techniques to solve the non-linear algebraic systems and generalized eigenvalue problems arising from continuation methods.

For modest sized problems, that is, with dimension d up to 10^4–10^5, direct solvers can be used, as will be discussed in this chapter. For large-scale applications where direct solvers are not practical, iterative methods (such as Krylov methods) are required, and these will be discussed in Chapters 7 and 8. Having an explicit Jacobian matrix also leads to a more efficient application of the Jacobian operator for Krylov iteration methods. Thus for stiff linear algebra problems for which effective preconditioners and many Krylov iterations are needed to solve the linear system, the cost of computing a Jacobian matrix is usually worthwhile for efficiency and even necessary for robustness.

6.1 Evaluating the Jacobian Matrix

Suppose we want to solve the algebraic system

$$\Phi(\mathbf{x}) = 0, \tag{6.1}$$

where the state vector $\mathbf{x} \in \mathbb{R}^d$, $\Phi: \mathbb{R}^d \to \mathbb{R}^d$ and we have already an approximation $\mathbf{x}_0$. Then we expand the $\Phi(\mathbf{x})$ around this point into a Taylor series:

$$\Phi(\mathbf{x}) = \Phi(\mathbf{x}_0) + \Phi_{\mathbf{x}}(\mathbf{x}_0)(\mathbf{x} - \mathbf{x}_0) + \mathcal{O}(|\mathbf{x} - \mathbf{x}_0|^2) \tag{6.2}$$

where $\Phi_{\mathbf{x}}$ is the Jacobian matrix, with (for $i = 1, \ldots, d; j = 1, \ldots, d$):

$$(\Phi_{\mathbf{x}})_{i,j} = \frac{\partial \Phi_i}{\partial x_j}. \tag{6.3}$$

We can divide the approaches to calculate this Jacobian into analytic methods which are exact derivatives up to floating point precision, and numerical methods which are generated with finite-difference approximations to the derivatives.

Additional Material

- With Automatic Differentiation, a code for the analytic derivatives is created, either by source transformation using a pre-processor, for example, ADIFOR (www.mcs.anl.gov/research/projects/adifor) and OpenAD (www.mcs.anl.gov/OpenAD), or through the operator overloading approach for C++ codes (http://trilinos.sandia.gov/packages/sacado).

Analytic Jacobian matrices can be programmed 'by hand' only in relatively simple cases, including the canonical problems as discussed in Chapter 1. They can also be computed with Automatic Differentiation, a symbolic approach for generating derivatives. Automatic Differentiation can lead to a considerable decrease in code development time over hand-coding the analytic Jacobian. If the equations (6.1) are fixed, this might be a modest gain, but it becomes a tremendous benefit when part of the research effort involves changes to the form of the equations (such as variation of source terms, solution-dependent properties, or equations of state). The numerical efficiency depends on many factors, but an Automatic Differentiation-generated Jacobian could take much (even up to 50 per cent) longer to compute than a 'hand-coded' Jacobian matrix. For large-scale problems where the linear solver time dominates, this difference decreases in importance. *See example Ex. 6.1*

Numerical methods to compute Jacobian matrices perform a finite-difference approximation by repeated calculation of the right-hand side. The numerical Jacobian matrix at $\mathbf{x}_0$ can be constructed, using the approximation of a derivative, according to

$$\frac{\partial \Phi_i}{\partial x_j}(\mathbf{x}_0) \approx \frac{\Phi_i(\mathbf{x}_0 + \epsilon \mathbf{e}_j) - \Phi_i(\mathbf{x}_0)}{\epsilon} \tag{6.4}$$

for $i = 1, \ldots, d; j = 1, \ldots, d$; and small ϵ; here $\mathbf{e}_j$ is the unit vector in the j-direction. In order to use numerical differentiation for large problems, the sparsity pattern of the Jacobian matrix must be exploited. In general it is not possible to evaluate a certain entry of the right-hand-side function. Usually there is a vector input and a vector output. So, it is crucial to limit the number of calls to the right-hand-side function; one should try to get as many derivatives as possible from one call to the right-hand-side function. For instance, if the Jacobian matrix is just a diagonal matrix, then due to the independence only two calls are needed (for $\mathbf{x}_0$ and $\mathbf{x}_0 + \epsilon \mathbf{1}$) to calculate all the derivatives.

In general, unknowns can be sub-divided by a colouring algorithm in such a way that a small number of right-hand-side evaluations yields all of the coefficients (Coleman *et al.*, 1984). Let, for each column j of $\Phi_{\mathbf{x}}$, the non-zero row elements be in the set R_j. Then the algorithm finds a partition $C_1, C_2, \ldots, C_p$ of the columns of $\Phi_{\mathbf{x}}$ such that no pair of columns in the same C_q shares a non-zero element on the same row. So, if columns j and k are both in C_q, then $R_j \cap R_k = \emptyset$. If we define $R_0^q = \{1, \ldots, d\}/\cup_{j \in C_q} R_j$, that is, the rows that have no contribution at all in the set C_q. Then we have that

$$
\frac{\left(\Phi(\mathbf{x} + \epsilon \sum_{j \in C_q} \mathbf{e}_j) - \Phi(\mathbf{x})\right)}{\epsilon} = \frac{\left(\sum_{j \in C_q} \Phi(\mathbf{x} + \epsilon \mathbf{e}_j)_{R_j} + \Phi(\mathbf{x} + \epsilon \sum_{j \in C_q} \mathbf{e}_j)_{R_0^q} - \Phi(\mathbf{x})\right)}{\epsilon}
$$

$$
= \frac{\left(\sum_{j \in C_q} \Phi(\mathbf{x} + \epsilon \mathbf{e}_j)_{R_j} + \Phi(\mathbf{x})_{R_0^q} - \Phi(\mathbf{x})\right)}{\epsilon}
$$

$$
= \frac{\sum_{j \in C_q} \left(\Phi(\mathbf{x} + \epsilon \mathbf{e}_j) - \Phi(\mathbf{x})\right)_{R_j}}{\epsilon}.
$$

So the evaluation will give us for each column j in C_q the entries of the approximate Jacobian on the rows i in R_j. Note that one can take the sets R_j bigger than where the jth column of Φ_x has non-zeroes, which goes at the expense of a bigger p.

A more efficient approach, particularly suited to finite-element calculations, is to perform the finite-difference calculation (6.4) at a local (element) level. If an element has q degrees of freedom, then the dense element stiffness matrix can be computed by $q + 1$ residual evaluations for this element. This is slightly more invasive in the code but tends to be much easier to implement and requires far fewer residual evaluations compared to the colouring approach.

One should be aware that numerical differentiation comes with an approximation error and a round-off error. For (6.4) the approximation error will behave proportionally to ϵ and the round-off error to u/ϵ, where u is the unit round. The unit round is the smallest positive real u such that $1 \hat{+} u > 1$ when rounding (indicated by the hat on the +) is employed. Since the round-off error increases with a decreasing ϵ, there is an optimal ϵ for which the highest accuracy of the approximate derivative is obtained. This accuracy depends on u; for example, for single, standard, and double precision this is $u = 10^{-8}$, 10^{-16}, and 10^{-32}, respectively. For completeness, we mention that numerical differentiation for first derivatives of real functions can be drastically improved if one uses complex arithmetic; for details and software see, Martins and colleagues (2003) and Shampine (2007).

6.2 Fixed Point Computation

From the previous chapter, we recall that the non-linear algebraic system of equations to be solved in a pseudo-arclength fixed-point continuation method, that is, (5.3) and (5.7), are given by

$$\Phi(\mathbf{x}, \lambda) = 0 \tag{6.5a}$$

$$\dot{\mathbf{x}}_0^T(\mathbf{x} - \mathbf{x}_0) + \dot{\lambda}_0(\lambda - \lambda_0) - (s - s_0) = 0. \tag{6.5b}$$

This is essentially a root-finding problem for a large-dimensional, non-linear algebraic system. Note that we can just define $\mathbf{X}$ in terms of the current unknowns by $\mathbf{X} = (\mathbf{x}, \lambda)$ and extend Φ by the equation (6.5b) yielding again an equation of the type (6.1) with state vector $\mathbf{X}$.

Root-finding methods are extensively treated in introductory courses/books on Numerical Analysis (Burden and Faires, 2001, chapters 2 and 10). The generic non-linear algebraic system solver applied is the Newton–Raphson method, or simply the Newton method. This method for systems is derived using Taylor series for vector functions. Omitting the last term in the right-hand side of (6.2), this equation becomes just a linear approximation of $\Phi(\mathbf{x})$ around $\mathbf{x}_0$. In fact it is the tangential hyperplane touching $\Phi(\mathbf{x})$ at $\mathbf{x}_0$. The zero of this linear approximation is the next approximation of the zero of Φ. This leads to the following algorithm. Given a first Newton guess $\mathbf{x}^1$, perform the following iteration, with Newton iterate k:

  (i). Evaluate $\Phi(\mathbf{x}^k)$,
  (ii). Build an (approximate) Jacobian matrix $\Phi_\mathbf{x}(\mathbf{x}^k)$ of Φ,
  (iii). Solve the linear system $\Phi_\mathbf{x}(\mathbf{x}^k)\Delta\mathbf{x}^k = -\Phi(\mathbf{x}^k)$,
  (iv). Update $\mathbf{x}^k$ as follows: $\mathbf{x}^{k+1} = \mathbf{x}^k + \alpha^k\Delta\mathbf{x}^k$.

The Newton–Raphson method is graphically illustrated in Fig. 6.1 in one dimension for a function $f(x) = 0$, where $x^{k+1} = x^k - f(x^k)/f'(x^k)$, and the prime indicates differentiation to x.

In step (i), one should be aware of the fact that the attainable accuracy of a Newton method is determined by the accuracy by which one can evaluate Φ. Building a Jacobian matrix in step (ii) may be expensive, therefore often the Jacobian is usually kept fixed for a number of Newton steps until the convergence deteriorates. Another way is to evaluate the Jacobian only on a sub-space. In step (iii) usually a large linear system has to be solved, for which the building of a factorization for

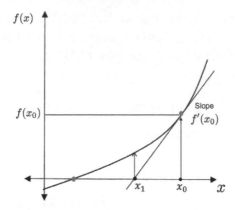

Figure 6.1 Sketch of the Newton–Raphson method for a one-dimensional problem $f(x) = 0$, with a function $f : \mathbb{R} \to \mathbb{R}$.

the Jacobian matrix is quite expensive. One could decide here to re-use the factorization of the Jacobian as long as the convergence does not deteriorate. In step (iv), a damping parameter α^k is applied. By choosing this parameter judiciously one can assure that $||\Phi(\mathbf{x}_k)||$ is becoming smaller in each step, which is not necessarily the case for $\alpha^k = 1$ in the standard Newton method. Hence, by introducing an α^k one can turn a locally convergent method into a globally convergent one.

Additional Material

• To dive more into the properties of the Newton–Raphson method and its convergence properties, see Kelley (2007) and Argyros (2008).

To solve the equations (6.5), an Euler predictor/Newton corrector algorithm is applied. Let the steady state which is already known be indicated by $\mathbf{x}_0$; then a good guess for the first Newton iterate to the next steady state is the Euler predictor given by

$$\mathbf{x}^1 = \mathbf{x}_0 + \Delta s \ \dot{\mathbf{x}}_0 \tag{6.6a}$$

$$\lambda^1 = \lambda_0 + \Delta s \ \dot{\lambda}_0 \tag{6.6b}$$

where the dot indicates differentiation to s. Now in the Newton method, $\mathbf{x}^k$ and λ^k are updated by (where the damping factor α^k is taken unity for convenience)

$$\mathbf{x}^{k+1} = \mathbf{x}^k + \Delta\mathbf{x}^{k+1} \tag{6.7a}$$

$$\lambda^{k+1} = \lambda^k + \Delta\lambda^{k+1} \tag{6.7b}$$

where $(\Delta \mathbf{x}^{k+1}, \Delta \lambda^{k+1})$ are solved from the correction equation

$$
\begin{pmatrix} \Phi_x(\mathbf{x}^k, \lambda^k) \Phi_\lambda(\mathbf{x}^k, \lambda^k) \\ \dot{\mathbf{x}}_0^T \dot{\lambda}_0 \end{pmatrix} \begin{pmatrix} \Delta \mathbf{x}^{k+1} \\ \Delta \lambda^{k+1} \end{pmatrix}
$$
$$
= \begin{pmatrix} -\Phi(\mathbf{x}^k, \lambda^k) \\ \Delta s - \dot{\mathbf{x}}_0^T(\mathbf{x}^k - \mathbf{x}_0) - \dot{\lambda}_0(\lambda^k - \lambda_0) \end{pmatrix}. \tag{6.8}
$$

Hence, within each iteration, a linear system of equations has to be solved. If the Newton–Raphson process has converged up to a desired accuracy, a new steady state has been found.

The system of equations (6.8) is an example of a so-called *bordered* linear system, which is of the form

$$
\begin{bmatrix} J & V \\ W^T & C \end{bmatrix} \begin{bmatrix} \mathbf{x} \\ \mathbf{c} \end{bmatrix} = \begin{bmatrix} \mathbf{r}_x \\ \mathbf{r}_c \end{bmatrix}, \tag{6.9}
$$

where $J = \Phi_x$ is the Jacobian matrix, and V and W contain an equal number of vectors. A standard approach to solve (6.9) is to carry out the block-factorization

$$
\begin{bmatrix} J & V \\ W^T & C \end{bmatrix} = \begin{bmatrix} J & 0 \\ W^T & I \end{bmatrix} \begin{bmatrix} I & J^{-1}V \\ 0 & C - W^T J^{-1} V \end{bmatrix}. \tag{6.10}
$$

Hence, pre-multiplying (6.9) by the inverse of the lower block triangular factor in the right-hand side of (6.10) gives

$$
\begin{bmatrix} I & J^{-1}V \\ 0 & C - W^T(J^{-1}V) \end{bmatrix} \begin{bmatrix} \mathbf{x} \\ \mathbf{c} \end{bmatrix} = \begin{bmatrix} J^{-1}\mathbf{r}_x \\ \mathbf{r}_c - W^T(J^{-1}\mathbf{r}_x) \end{bmatrix}. \tag{6.11}
$$

This shows that we have to solve two systems of equations with matrix J and right-hand sides $\mathbf{r}_x$ and V, respectively. Next, a small system for $\mathbf{c}$ has to be solved, followed by computing $\mathbf{x}$. The advantage of this is that we can re-use the factorization of J in the process. Note that the formulation also allows V and W to consist of multiple columns, though both should have the same amount of them.

For the solution of (6.8) this results in two steps in which only linear systems with Φ_x are solved. Let $\mathbf{r} = -\Phi(\mathbf{x}^k, \lambda^k)$ and $r_{d+1} = \Delta s - \dot{\mathbf{x}}_0^T(\mathbf{x}^k - \mathbf{x}_0) - \dot{\lambda}_0(\lambda^k - \lambda_0)$; then if $\mathbf{z}_1$ and $\mathbf{z}_2$ are solved from

$$
\Phi_x(\mathbf{x}^k, \lambda^k)\mathbf{z}_1 = \mathbf{r} \tag{6.12a}
$$
$$
\Phi_x(\mathbf{x}^k, \lambda^k)\mathbf{z}_2 = \Phi_\lambda(\mathbf{x}^k, \lambda^k), \tag{6.12b}
$$

then the solution $(\Delta \mathbf{x}^{k+1}, \Delta \lambda^{k+1})$ is found from

$$
\Delta \lambda^{k+1} = \frac{r_{d+1} - \dot{\mathbf{x}}_0^T \mathbf{z}_1}{\dot{\lambda}_0 - \dot{\mathbf{x}}_0^T \mathbf{z}_2} \tag{6.13a}
$$
$$
\Delta \mathbf{x}^{k+1} = \mathbf{z}_1 - \Delta \lambda^{k+1} \mathbf{z}_2. \tag{6.13b}
$$

This approach is fine as long as J is non-singular. However, at simple bifurcation (e.g., saddle-node bifurcations) points the matrix Φ_x becomes singular, and applying its inverse leads to an unstable algorithm. This is not a problem when using pseudo-arclength continuation to pass around a saddle-node bifurcation point, since the stepping algorithm will not land on the bifurcation. However, this does become an issue when attempting to converge to the bifurcation point. For the case of saddle-node bifurcations, there are essentially two approaches to deal with this situation. The first is to integrate the border rows and columns into the system, and the second is to stabilize the block factorization near the bifurcation point (Dijkstra *et al.*, 2014). Also iterative refinement, discussed in Section 6.4, makes it possible to converge to a bifurcation point (Govaerts, 1991).

6.3 Periodic Orbit Computation

Suppose that a previously computed periodic solution, now indicated by $(\bar{\mathbf{x}}(t), \bar{p})$, has been computed for a value of $\bar{\lambda}$. It was explained already in Section 5.5.1 how to obtain such a starting solution when a periodic orbit is arising from a Hopf bifurcation. In a pseudo-arclength method using a boundary value solution approach, the following equations have to be solved (see (5.38) and (5.39)):

$$M\frac{d\mathbf{x}}{dt} = p\Phi(\mathbf{x}, \lambda), \tag{6.14a}$$

$$\mathbf{x}(0) = \mathbf{x}(1), \tag{6.14b}$$

$$\int_0^1 (\mathbf{x}(t) - \bar{\mathbf{x}}(t))^T \dot{\bar{\mathbf{x}}}(t)dt + (p - \bar{p})\dot{\bar{p}} + (\lambda - \bar{\lambda})\dot{\bar{\lambda}} = \Delta s \tag{6.14c}$$

$$\int_0^1 \mathbf{x}^T(t)\frac{d\bar{\mathbf{x}}(t)}{dt}dt = 0. \tag{6.14d}$$

In this case, we also use the Euler estimate:

$$\mathbf{x}(t) = \bar{\mathbf{x}}(t) + \Delta s\dot{\bar{\mathbf{x}}}(t) \tag{6.15a}$$

$$p = \bar{p} + \Delta s\dot{\bar{p}} \tag{6.15b}$$

$$\lambda = \bar{\lambda} + \Delta s\dot{\bar{\lambda}}. \tag{6.15c}$$

There are two approaches to solve the system of equations (6.14), which will be discussed below.

6.3.1 *Extended System Methods*

In a first method, the time interval $[0, 1]$ is discretized as $0 = t_0, t_1, \ldots, t_N = 1$ and the resulting algebraic system is solved by Newton's method; for the time-discretization there are many options. For example, suppose that we use the backward Euler method with time step Δt. This results in a system of equations

$$M\mathbf{x}_j \;=\; M\mathbf{x}_{j-1} + p\Delta t\Phi(\mathbf{x}_j,\lambda), \text{for } j = 1,\ldots,N-1, \tag{6.16a}$$

$$M\mathbf{x}_0 \;=\; M\mathbf{x}_{N-1} + p\Delta t\Phi(\mathbf{x}_0,\lambda). \tag{6.16b}$$

The arclength condition becomes

$$\sum_{j=0}^{N-1}(\mathbf{x}_j - \bar{\mathbf{x}}_j)^T\dot{\bar{\mathbf{x}}}_j + (p - \bar{p})\dot{\bar{p}} + (\lambda - \bar{\lambda})\dot{\bar{\lambda}} = \Delta s \tag{6.17}$$

and the phase condition

$$\sum_{j=0}^{N-1}\mathbf{x}_j^T\frac{d}{dt}\bar{\mathbf{x}}_j = 0. \tag{6.18}$$

The part of the Jacobian related to (6.16) has the shape

$$\begin{pmatrix} M - p\Delta t\Phi_{\mathbf{x}}(\mathbf{x}_0,\lambda) & & & -M \\ -M & M - p\Delta t\Phi_{\mathbf{x}}(\mathbf{x}_1,\lambda) & & \\ & & \ddots & \ddots & \\ & & & -M & M - p\Delta t\Phi_{\mathbf{x}}(\mathbf{x}_{N-1},\lambda) \end{pmatrix}. \tag{6.19}$$

This will be extended by two full rows due to arclength and phase condition and two full columns for the dependencies on λ and p, respectively. Hence, this leads to a tremendously large system if the dimension of the system of discretized equations, that is, d, is large. A way to keep the number of time steps low is to use a higher-order time integration method. This goes at the expense of sub-time steps in this high-order method, increasing again the dimension of the system. Hence, a balance has to be sought here.

For example, in AUTO (Doedel, 1980), a collocation method is used with the Lagrangian basis polynomials $(w_{j,i}(t), j = 0,\ldots,N-1, i = 0,1,\ldots,m)$, defined by

$$w_{j,i}(t) = \prod_{k=0, k\neq i}\frac{t - t_{j+k/m}}{t_{j+i/m} - t_{j+k/m}} \;;\; t_{j+i/m} = t_j + \frac{i}{m}\Delta t_j \tag{6.20}$$

with $\Delta t_j = t_{j+1} - t_j$. The collocation method now consists of finding the unknowns $\hat{\mathbf{x}}_{j+i/m}$ in the representation

$$\mathbf{x}_j(t) = \sum_{i=0}^{m}w_{j,i}(t)\hat{\mathbf{x}}_{j+i/m} \tag{6.21}$$

such that (6.14a) is satisfied at time points $z_{j,i}$, that is,

$$M\frac{d\mathbf{x}}{dt}(z_{j,i}) = p\Phi(\mathbf{x}(z_{j,i}),\lambda). \tag{6.22}$$

The $z_{j,i}$ in each interval $[t_{j-1}, t_j]$ are the zeroes of the mth-degree Legendre poly-nomial relative to that sub-interval. With the preceding choice of the basis, $\hat{\mathbf{x}}_j$ and $\hat{\mathbf{x}}_{j+i/m}$ are to approximate the solution $\mathbf{x}(t)$ at times t_j and $t_{j+i/m}$, respectively. *See example Ex. 6.2*

The arclength condition (6.14b) is approximated by

$$\sum_{j=0}^{N-1}\sum_{i=0}^{m}\omega_{j,i}(\hat{\mathbf{x}}_{j+i/m} - \bar{\mathbf{x}}_{j+i/m})^T \dot{\bar{\mathbf{x}}}_{j+i/m} + (p - \bar{p})\dot{\bar{p}} + (\lambda - \bar{\lambda})\dot{\bar{\lambda}} = \Delta s \qquad (6.23)$$

where the $\omega_{j,i}$ are the Lagrange quadrature coefficients. The complete set of discrete equations now consists of $mN + 2$ equations for the Nm unknowns $\hat{\mathbf{x}}_{j+i/m}$ plus λ and p.

An alternative to introducing a time stepper is to use a Fourier expansion, which fits nicely with (5.44). We pose that $\mathbf{x}(t) = \sum_{k=0}^{N}(\mathbf{w}_k)_R \sin(k2\pi t) + (\mathbf{w}_k)_I \cos(k2\pi t)$, substitute this in the equation, require that the equation is exactly satisfied at $2N$ points in time (collocation), or multiply the equations by $\sin(l2\pi t)$ and $\cos(l2\pi t)$, $l = 0, \ldots, N$, respectively, and integrate in time over the interval $[0,1]$ (i.e., a Galerkin projection in time). What remains is a set of $2N$ coupled equations for $(\mathbf{w}_k)_R$ and $(\mathbf{w}_k)_I$, $k = 0, \ldots, N$. Note that the phase present in (5.44) will here be consumed in the unknown vectors and the phase condition is again needed to make them unique. *See example Ex. 6.3*

6.3.2 Multiple Shooting Methods

The name of the method is a reference to shooting a bullet in such a way that you hit the target, which may be influenced by gravity, wind, and other factors. For the numerical method one very much does the same. There is a current position, which forms one boundary condition, and there is the target, which is another boundary condition. At the current position, we add a fictive condition, which makes it pos-sible to treat our problem as an initial value problem and to integrate it towards the target boundary. In this way, the shooting method is an alternative for solving the two-point boundary value problem. Solving the initial value problem, we observe how far we are off from the target and adapt the fictive condition until convergence to the target (Fig. 6.2).

The method to determine periodic orbits through fixed points of the Poincaré map, as discussed in Section 5.5.2, can be considered as a shooting method where the 'initial condition' near the orbit is adapted such that after a period p, the orbit intersects the same point on the Poincaré section. When a Newton–Raphson method is applied to solve a system with a Jacobian (5.51), the correction equation that needs to be solved is

$$\begin{pmatrix} I - \mathcal{P}_{\mathbf{x}}(\mathbf{x}^k, \lambda^k) & -\mathcal{P}_{\lambda}(\mathbf{x}^k, \lambda^k) \\ (\dot{\mathbf{x}}_0)^T & \dot{\lambda}_0 \end{pmatrix} \begin{pmatrix} \Delta\mathbf{x}^{k+1} \\ \Delta\lambda^{k+1} \end{pmatrix} = -\begin{pmatrix} \mathbf{x}^k - \mathcal{P}(\mathbf{x}^k, \lambda^k) \\ \Sigma(\mathbf{x}^k, \lambda^k) \end{pmatrix}. \qquad (6.24)$$

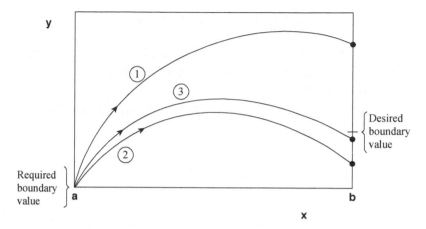

Figure 6.2 Sketch of the shooting method, where different initial conditions are used to generate the trajectories. These are chosen such that the boundary condition at the end point is satisfied.

As can be seen from (5.55a) and (5.55b), we need $\phi(t(\mathbf{x}^k), \mathbf{x}^k, \lambda^k)$, $\phi_\mathbf{x}(t(\mathbf{x}^k), \mathbf{x}^k, \lambda^k)$, and $\phi_\lambda(t(\mathbf{x}^k), \mathbf{x}^k, \lambda^k)$ to compute $\mathcal{P}_\lambda(\mathbf{x}^k, \lambda^k), \mathcal{P}_\mathbf{x}(\mathbf{x}^k, \lambda^k)$ and $\mathcal{P}_\lambda(\mathbf{x}^k, \lambda^k)$. Here, $\phi(t(\mathbf{x}^k), \mathbf{x}^k, \lambda^k)$ is the solution $\tilde\phi$ of

$$M\tilde\phi_t = \Phi(\tilde\phi, \lambda), \ \tilde\phi(0) = \mathbf{x}^k, \tag{6.25}$$

at the end time t such that $\tilde\phi(t)$ is in the Poincaré section, hence, $\phi(t(\mathbf{x}^k), \mathbf{x}^k, \lambda^k) = \tilde\phi(t) \in \Sigma^+$. For the other two, we can simply take the derivative of the equation (6.25) with respect to $\mathbf{x}$ and λ, respectively:

$$M(\tilde\phi_\mathbf{x})_t \ = \ \Phi_\mathbf{x}(\tilde\phi, \lambda^k)\tilde\phi_\mathbf{x}, \ \tilde\phi_\mathbf{x}(0) = I \tag{6.26a}$$

$$M(\tilde\phi_\lambda)_t \ = \ \Phi_\mathbf{x}(\tilde\phi, \lambda^k)\tilde\phi_\lambda + \Phi_\lambda(\tilde\phi, \lambda^k), \ \tilde\phi_\lambda(0, \mathbf{x}^k) = 0, \tag{6.26b}$$

and evaluate $\phi_\mathbf{x}(t(\mathbf{x}^k), \mathbf{x}^k, \lambda^k) = \tilde\phi_\mathbf{x}(t(\mathbf{x}^k))$ and $\phi_\lambda(t(\mathbf{x}^k), \mathbf{x}^k, \lambda^k) = \tilde\phi_\lambda(t(\mathbf{x}^k))$. The last three equations have to be solved simultaneously, because we do not want to store $\phi(t(\mathbf{x}^k), \mathbf{x}^k, \lambda^k)$. So, one has to perform a time integration to construct the Jacobian and right-hand side in (6.24)

The shooting method allows for parallelization if we factor the operator $\mathcal{P}$ by

$$\mathcal{P}(\mathbf{x}, \lambda) = P_N \circ P_{N-1} \circ \cdots P_1(\mathbf{x}, \lambda)$$

where $\mathbf{x}_{k+1} = P_k(\mathbf{x}_k, \lambda)$ and $\mathbf{x}_k$ is an approximation of $\mathbf{x}(t_k)$. Note that we use a subscript k to indicate this, while previously we used a superscript k for the Newton iterate. This means that we need N planes Σ_k which intersect transversally an orbit $\mathbf{x}^0(t)$. Assuming the period p of the orbit $\mathbf{x}^0(t)$ is known, we can just define $t_k = kp/N$ and define Σ_k by

$$< d\mathbf{x}^0/dt(t_k), M(\mathbf{y} - \mathbf{x}^0(t_k)) >= 0.$$

Each P_k maps a point $\mathbf{x}_k$ from Σ_k to a point $\mathbf{x}_{k+1}$ on Σ_{k+1} by a time integration of (6.14) with initial condition $\mathbf{x}_k$ and we just integrate in time until we reach a point in Σ_{k+1} which will be called $\mathbf{x}_{k+1}$.

These definitions can be used to define the multiple shooting method:

$$\mathbf{x}_{k+1}^{n+1} = P_k(\mathbf{x}_k^n, \lambda^n), \text{ for } k = 0, \ldots, N - 1, \ \mathbf{x}_0^{n+1} = \mathbf{x}_N^{n+1}.$$

So, if we introduce the vector of unknowns $\mathbf{X} = (\mathbf{x}_0, \mathbf{x}_1, \ldots, \mathbf{x}_{N-1})^T$ the iteration can be cast in the shape

$$\mathbf{X}^{n+1} = \hat{\mathcal{P}}(\mathbf{X}^n, \lambda^n) \tag{6.27}$$

of which we want to determine the fixed point. So we can also start from the equation $\mathbf{X} = \hat{\mathcal{P}}(\mathbf{X}, \lambda)$ together with the normalization condition for the tangent

$$\Sigma(\mathbf{X}, \lambda) = (\mathbf{X} - \mathbf{X}_0)^T \dot{\mathbf{X}}_0 + (\lambda - \lambda_0)\dot{\lambda} - \Delta s = 0 \tag{6.28}$$

and apply Newton's method to this. This leads to a similar correction equation as in (6.24), replacing $\mathcal{P}$ by $\hat{\mathcal{P}}$ and $\mathbf{x}$ by $\mathbf{X}$. Also, the solution process is the same. What is gained by this approach is that we can do the propagations on each sub-interval completely independent from those on other sub-intervals. Hence, it is embarrassingly parallel. So, if we have N processors available, then we can do one Newton iteration in one Nth of the time of that of the original shooting method. In Sanchez Umbria and Net (2013) a speedup of a factor 10 is reported for a fluid flow problem.

To start from a Hopf bifurcation, we can use (5.44), giving $d\mathbf{x}/dt(t_k) = \alpha\sigma_I(\cos(\sigma_I t_k)\hat{\mathbf{x}}_R - \sin(\sigma_I t_k)\hat{\mathbf{x}}_I)$ and $\mathbf{x}^0(t_k) = \mathbf{x}_0 + \alpha(\sin(\sigma_I t_k)\hat{\mathbf{x}}_R + \cos(\sigma_I t_k)\hat{\mathbf{x}}_I)$, where σ_I is the imaginary part of the eigenvalue at the Hopf bifurcation. (The real part is zero by definition.)

After the fixed point of (6.27) is found, we can compute the dominant eigenvalues of the monodromy matrix. Also here we can in parallel compute the matrices for each interval $[t_k, t_{k+1}]$

$$M(\tilde{\phi}_\mathbf{x})_t = \Phi_\mathbf{x}(\tilde{\phi}, \lambda)\tilde{\phi}_\mathbf{x}, \ \tilde{\phi}_\mathbf{x}(t_k) = I, \tag{6.29}$$

which gives us $\phi_\mathbf{x}(\mathbf{x}_k, t_{k+1} - t_k, \lambda)$. The product of all these matrices for $k = 0, \ldots, N - 1$ provides the full monondromy matrix. One can compute this matrix explicitly, but this is not needed when an iterative procedure is used, as discussed in the next two chapters. In this case, only the application of the monodromy matrix to a vector is needed, and this can be done by applying the matrices on each sub-interval one after another.

Another possibility is to compute the eigenvalues of $\hat{\mathcal{P}}_{\mathbf{X}}$. According to Theorem B.23, if λ is an eigenvalue of $\hat{\mathcal{P}}_{\mathbf{X}}$, then λ^N is an eigenvalue of the monodromy matrix. The standard way to compute all eigenvalues of a matrix is to compute its Schur form by the QR method (see Section 6.5). However, there exist special methods to create a block cyclic Schur factorization; for an overview see Watkins (2005).

Example 6.1 Ginzburg–Landau equation

Consider the Ginzburg–Landau equation as in Example 2.1. The discretization of the equation was already given in (4.35) where $\gamma_1 = \alpha + i\beta$. Next, we scale the time as earlier, leading to

$$M\frac{d}{dt}\mathbf{A} = p\big((\gamma_1 M + \gamma_2 D)\mathbf{A} - M\gamma_3|\mathbf{A}|^2\mathbf{A} + \mathbf{g}\big). \tag{6.30}$$

Since $|\mathbf{A}|^2 = \mathbf{A}\bar{\mathbf{A}}$ is not an analytical function, we have to split the system into a real and an imaginary part. Let $\mathbf{A} = \mathbf{A}_R + i\mathbf{A}_I$; then we get

$$M\frac{d}{dt}\mathbf{A}_R = p\big((\alpha M + \gamma_2 D)\mathbf{A}_R - \beta M\mathbf{A}_I - M\gamma_3(\mathbf{A}_R^2 + \mathbf{A}_I^2)\mathbf{A}_R + Re(\mathbf{g})\big), \tag{6.31a}$$

$$M\frac{d}{dt}\mathbf{A}_I = p\big((\alpha M + \gamma_2 D)\mathbf{A}_I + \beta M\mathbf{A}_R - M\gamma_3(\mathbf{A}_R^2 + \mathbf{A}_I^2)\mathbf{A}_I + Im(\mathbf{g})\big). \tag{6.31b}$$

See example Ex. 6.4
The Jacobian matrix $\Phi_{\mathbf{x}}$ of this system is

$$\big(I_2 \otimes (\alpha M + \gamma_2 D - \gamma_3 M|\mathbf{A}|^2)\big) + \begin{pmatrix} 0 & -1 \\ 1 & 0 \end{pmatrix} \otimes (\beta M) - 2\gamma_3 M \begin{pmatrix} \mathbf{A}_R \\ \mathbf{A}_I \end{pmatrix} \begin{pmatrix} \mathbf{A}_R & \mathbf{A}_I \end{pmatrix}. \tag{6.32}$$

Suppose we set boundary conditions $A(0,t) = 0$, $A(1,t) = 0.1(1+i)$ and parameters $\gamma_2 = \gamma_3 = 1$, $\gamma_1 = \alpha + i$, hence, $\beta = 1$. Furthermore, we use $m = 10$ (see Example 4.5), making $\mathbf{A}$ a vector of length 10, and the number of time steps is 20, hence $\Delta t = 1/20$. In Fig. 6.3 we depicted in several plots the behaviour of the periodic solution of the discretized system for α between 12 and 28.2 computed with the MATLAB/Octave package BifAnFF.[1] In Fig. 6.3a we show the 2-norm of the full periodic solution, indicated by the subscript x,t of the norm, so a vector containing all (20) time instances of the vector $\mathbf{A}$, as a function of α. We see that the norm goes from small to big with increasing α, which is in line with Fig. 6.3d. Since $\gamma_2 = 1$ we know that the first negative eigenvalue of $\gamma_2 D$ will be $-\pi^2$, so we expect a (Hopf) bifurcation near $\alpha = \pi^2$, which is in line with what we observe in Figs. 6.3a and 6.3d. From Example 2.1 we know that for $\gamma_2 = 0$ and $\beta = 1$ there is a period $p = 2\pi$. However, here γ_2 is not zero, and, moreover, we have a non-zero right boundary condition. This causes the increase of the period near the Hopf bifurcation. We have checked this by decreasing the magnitude of the boundary condition.

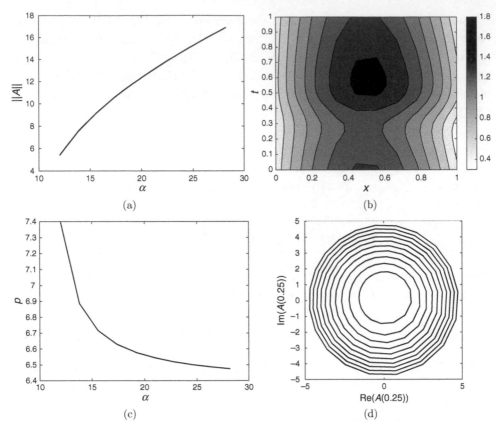

Figure 6.3 Continuation of orbits: (a) $||A||_{x,t}$ as function of α, (b) $|A(x,t)|$ for $\alpha =$ 12, (c) p as a function of α, (d) $A(0.25)$ as a function of time from $\alpha = 12$ (inner circle) to $\alpha = 28.2$ (outer circle), with step size 1.8. Parameters $\gamma_2 = \gamma_3 = 1$, $A(0,t) = 0$, $A(1,t) = 0.1(1+i)$, $\beta = 1$, $m = 10$, $\Delta t = 1/20$.

6.4 Robust Linear Systems Solvers

In general, there are two approaches to solve the linear systems arising from the Newton method. With *direct methods* the solution is solved to machine accuracy in one step, and with *iterative methods* the solution is approximated in a succession of steps, in each of which the accuracy is improved, until a user set tolerance on the accuracy is met.

For presenting the direct solvers, consider the basic problem

$$A\mathbf{x} = \mathbf{b} \tag{6.33}$$

where A is a matrix of order N. The usual way to solve this is using Gaussian elimination with pivoting (Burden and Faires, 2001, Sections 6.1–2). In this method a permutation matrix P, a lower triangular matrix L, and an upper triangular matrix

U are constructed such that $PA = LU$ (Burden and Faires, 2001, Section 6.5). The permutation matrix P shows which rows of A are permuted due to the pivoting. Once the decomposition is available, the solution $\mathbf{x}$ can be obtained by back substitution as follows, with $\mathbf{y} = U\mathbf{x}$:

$$Ly = P\mathbf{b} \to \mathbf{y} = L^{-1}P\mathbf{b} \to \mathbf{x} = U^{-1}L^{-1}P\mathbf{b}. \tag{6.34}$$

For large N, the computational work to construct this factorization is approximately $\frac{2}{3}N^3$. If A is symmetric, there is a factorization of the form LDL^T where L has ones on its diagonals and D is a diagonal matrix with 1×1 and 2×2 blocks on the diagonal. If A is also positive definite, we can make a Cholesky factorization LL^T (Burden and Faires, 2001, Section 6.6). It is known that for irreducible weakly diagonally dominant matrices, for example for M-matrices, there is no need for pivoting; see Theorems B.8 and B.35. Neither is this needed for symmetric positive definite matrices; see Definition B.9. The amount of computational work and *complexity*, that is, how the amount of work behaves as a function of the number of unknowns, depends on the ordering of the matrix. In order to keep the work low, we should try to keep the fill low. The *fill* are the elements occurring in the L and U which were not there in the original matrix. Hence, often *fill-reducing orderings* are applied (Quarteroni *et al.*, 2007; Saad, 2003).

A *symmetric reordering* of a matrix can be described in terms of a permutation matrix: PAP^T. A few symmetric orderings by which we can reduce the fill (Davis, 2006; Duff *et al.*, 2017; Davis *et al.*, 2016) are the reversed Cuthill–McKee, minimum degree, and nested dissection methods. The idea behind standard Cuthill–McKee is to minimize the bandwidth. A reversion of the ordering creates slightly better results in some cases. The minimum-degree ordering is based on the fact that a row with a low number of elements cannot produce much new fill. Since, the number of elements per row varies during the elimination, one tries to pick the row with the lowest number of elements as pivot row (Tinney and Walker, 1967; Duff *et al.*, 1986). Nested dissection is a divide-and-conquer strategy (George, 1973). In this technique, one simply starts off by splitting the domain in about two equal parts, which are separated by a number of unknowns that have a connection to both domains. An unknown in one of the domains does not have a direct connection to any unknown in the other domain, as it is only connected via an unknown on the separator. The unknowns on the separator are put last in the vector of unknowns. This process is repeated on each of the two domains recursively.

Example 6.2 Poisson equation
Consider the structure of the matrix due to the discretization of a scalar Poisson equation $\nabla^2 \psi = f$ on a rectangular two-dimensional domain Ω using Dirichlet boundary conditions. The sides of the domain are parallel to the x- and y-axes. The domain Ω is

Table 6.1 *Amount of work and number of non-zeroes in L, with $A = LL^T$ for various orderings.*

Numbering				Flops/1,000			nnz(L)/1,000		
	$N=$	100	400	1,600	6,400	100	400	1,600	6,400
Random		35	968	78,944	4,477,865	1.5	14	216	3,110
Rev. Cuthill–McKee		7	96	1,410	21,510	0.8	6	45	351
Nested dissection		7	78	804	7,637	0.8	5	28	153
Minimum degree		5	53	590	7,337	0.7	4	22	126

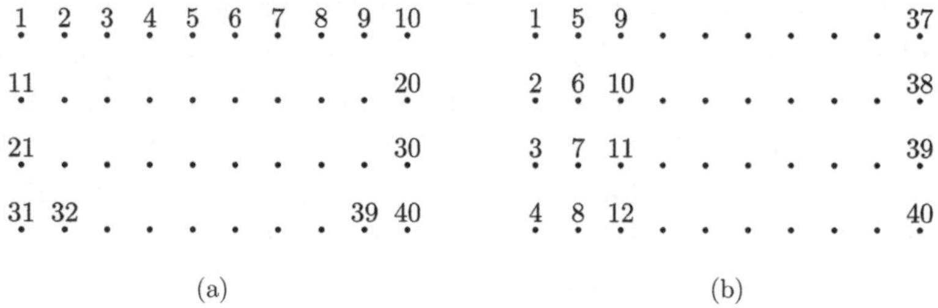

```
 1  2  3  4  5  6  7  8  9 10        1  5  9  .  .  .  .  .  . 37
11  .  .  .  .  .  .  .  . 20        2  6 10  .  .  .  .  .  . 38
21  .  .  .  .  .  .  .  . 30        3  7 11  .  .  .  .  .  . 39
31 32  .  .  .  .  .  . 39 40        4  8 12  .  .  .  .  .  . 40
              (a)                                (b)
```

Figure 6.4 (a) Row-wise ordering, (b) column-wise ordering.

covered by a uniform grid with, in both directions, an equal mesh width h and in x- and y-directions with m and n internal grid points, respectively. We number the internal grid points in lexicographical ordering as depicted for the case m = 10 and n = 4 in Fig. 6.4.

If the standard five-point difference molecule (4.13) is used, the state vector **x** *consists of values of ψ at all $n \times m$ grid points and the forcing* **b** *of values of f at the grid points. In the i-th row of A*x = b *there is a connection to x_{i-m} and x_{i+m}, but not to any x_j with $j < i - m$ or $j > i + m$. This means that the bandwidth of the matrix is $2m + 1$. It is easy to show that without pivoting there will be no fill outside the band. This bandwidth can be a lot less than that of a full matrix. The amount of work is now approximately $2m^2 N$ flops, with $N = nm$. If $m \gg n$, then it is advantageous to take the ordering as in Fig. 6.4a, because in that case the bandwidth is $2n + 1$. The original matrix has also a lot of zeros within the band, but during the elimination process this property is lost. See example Ex. 6.5*

To illustrate the influence of ordering on the amount of work and the number of non-zeroes in the factorization we consider in Table 6.1 a Cholesky factorization for the described Laplace problem for a number of grid resolutions m = n = 10, 20, 40, and 80.

These results show that it pays to use a fill-reducing ordering. The problems shown are still quite small. For bigger problems, nested dissection will eventually do a better job than minimum degree. In Table 6.2 we show how the complexity behaves for nested dissection on a Poisson problem.

Table 6.2 *Complexity of nested dissection on Poisson problem on a hypercube* $(N = n^d$ *unknowns).*

	1D		2D		3D		dD	
Factorization	n	N	n^3	$N\sqrt{N}$	n^6	N^2	$n^{3(d-1)}$	$N^{3(d-1)/d}$
Storage	n	N	$n^2 \log_2(n)$	$N \log_2(N)$	n^4	$N^{4/3}$	$n^{2(d-1)}$	$N^{2(d-1)/d}$

The preceding example shows that an LU-factorization is expensive with respect to computation time. The storage of the factorization is still moderate in the 2D and 3D cases. The solution time of $LU\mathbf{x} = \mathbf{b}$ is proportional to the storage and therefore comparatively cheap with respect to the factorization. Hence, if the matrix is fixed and the right-hand side changes, then reusing the factorization makes a direct method a very attractive alternative to iterative methods.

The structure in the sparsity pattern of a matrix is usually advantageous for fill-reducing orderings which will lead to less fill. However, structure may be destroyed by pivoting and lead to a lot of fill. Hence, in some cases, one accepts less stable orderings leading to a less accurate factorization. This is repaired by a few steps of *iterative refinement* (Burden and Faires, 2001, Section 7.4). Here the residual $\mathbf{r} = \mathbf{b} - A\tilde{\mathbf{x}}$ is computed, where $\tilde{\mathbf{x}}$ is the solution found from the direct solver and contaminated with round-off error. Next, a correction $\Delta\mathbf{x}$ is computed from $LU\Delta\mathbf{x} = \mathbf{r}$ and $\tilde{\mathbf{x}}$ is updated by adding $\Delta\mathbf{x}$. This can be repeated a few times until $|\mathbf{r}|$ is small enough. It is beneficial to use higher precision to compute the residual, but even if the same precision is used it does in general improve the accuracy (Quarteroni *et al.*, 2007).

Additional Material

- A few well-known direct solver codes are PARDISO (Bollhöfer *et al.*, 2019), see www.pardiso-project.org, UMFPACK (Davis, 2004), and MUMPS (Amestoy *et al.*, 2000), see http://mumps.enseeiht.fr.

Direct methods are robust, so they can be used to solve any problem with a non-singular matrix. Nowadays, these are quite well developed and usually available in mathematical kernels coming with compilers. Sometimes the user can choose the ordering, but usually it is set to nested dissection or (approximate) minimum degree. Some implementations, however, may simply be faster or better adapted to the hardware (cache, distributed memory). Therefore, if possible one could try a few and just pick the fastest one. In 2D, direct methods are usually faster than iterative methods for systems up to 10,000–100,000 unknowns. In 3D, the break-even point occurs for much smaller systems.

Example 6.3 (Ginzburg–Landau equation) *At this point we discuss which methods can be used for the linear systems we have met so far for the Ginzburg–Landau equations. For both the steady state and the time dependent case we have a Jacobian related to a one-dimensional problem: the matrix is tridiagonal and symmetric. If this tridiagonal matrix is diagonally dominant, then one can use a band solver and no pivoting is needed. If it is not diagonally dominant, then pivoting will be needed. This will destroy the structure of the matrix, and it is therefore appropriate to use a method which can deal with unstructured matrices. For a steady solution, diagonal dominance of the Jacobian is related to its stability. So loss of diagonal dominance occurs in the Jacobian if the Newton process is converging to an unstable solution. For time integration, the case is more subtle, since the time integrator adds a term to the diagonal. So even if we are doing time integration in the vicinity of an unstable steady state, the tridiagonal system may remain diagonal dominant if the time step is sufficiently small. To perform continuation, the matrix has a non-symmetric structure already and, moreover, diagonal dominance is not present. In this case, one needs a solver for unstructured matrices, since pivoting will happen. By the way, the Ginzburg–Landau problem needs complex arithmetic. Not all packages may have this available. If not, one can transform the system to real form (equating real and imaginary parts), but this is less efficient.*

6.5 Robust Eigenvalue Problem Solvers

Unlike the situation for linear systems solving, there are no truly direct numerical methods for the solution of an eigenvalue problem, in the sense that in general one cannot compute the eigenvalues (or eigenvectors) exactly in a finite number of floating-point operations. To compute all the eigenvalues from a matrix, one uses for the standard eigenvalue problem the QR method and for the generalized eigenvalue problem the QZ method. They also are a building block in computing few eigenvalues in case of large, sparse matrices, where they are used to compute the eigenvalues of a (Petrov)–Galerkin projection of such a matrix on a suitable sub-space, obtained in some clever way, for example by a Krylov sub-space method.

In Algorithm 1, the basic QR method is given. The matrix $A^{(n)}$ is decomposed at the nth step into an orthogonal matrix $Q^{(n)}$ (with $Q^T = Q^{-1}$) and an upper triangular matrix $R^{(n)}$. For this algorithm, it is easy to show that $A^{(n)}$ and A have the same eigenvalues. If the matrix A is real then, on convergence, $A^{(n)}$ becomes the real Schur normal form (see also (B.25)) of A. This is an upper triangular matrix having diagonal blocks of size 1, giving a real eigenvalue, and of size 2, giving complex

---

Algorithm 1 The QR method

Set $A^{(0)} = A$
for $n = 1, 2, \ldots$ **until convergence**
$\quad Q^{(n)} R^{(n)} = A^{(n)}$
$\quad A^{(n+1)} = R^{(n)} Q^{(n)}$
end

---

conjugate eigenvalues of A. A specific property of the algorithm is that the eigenvalues occur from large to small, in absolute value, on the diagonal. This is due to the fact that the QR method is related to the power method and its generalizations (as we will see in Chapter 7), in particular it is equivalent to orthogonal sub-space iteration if one iterates on the full space and starts with a space V consisting of the elementary basis vectors, that is, $V^{(0)} = I$.

As we will see in Chapter 7, the convergence of the mth eigenvalue in the QR method, assuming an ordering $|\lambda_1| \geq |\lambda_2| \geq \cdots \geq |\lambda_N|$ is proportional to the ratios $\lambda_m / \lambda_{m-1}$ and $\lambda_{m+1} / \lambda_m$. One can exploit this knowledge by shifting the matrix, so replacing A by $A - \sigma I$, which shifts all the eigenvalues by σ. If we are able to shift one eigenvalue such that it is sufficiently smaller than all the shifted others, then we will get a good convergence for that eigenvalue in that step. This idea leads to the shifted QR method given in Algorithm 2. Since we shifted an eigenvalue such that it becomes the smallest, that one will occur at the last position. To reiterate on that eigenvalue we pick the last element of the matrix. An alternative is to choose an eigenvalue of the last 2×2 block, which also makes it possible to go for complex eigenvalues. During the process we get a better approximation on that eigenvalue and this enables us to shift it closer to zero, speeding up the convergence. Once the last element in the matrix has converged, the whole last row, except for the last element, will be zero and hence we can proceed on the sub-matrix consisting of the first $N - 1$ rows and columns. To reduce computation time one first performs an orthogonal similarity transformation to get the matrix in upper Hessenberg form; for symmetric matrices the Hessenberg form becomes a tridiagonal matrix. It can be shown that this also leads to a sparse Q and that all iterates will be of Hessenberg form. In the case of a non-symmetric real matrix one can still make use of real computations if one applies a double shift step. In Golub and van Loan (1996) this is worked out in detail for the so-called Francis QR step.

The QR method is for standard eigenvalue problems, but in most cases we have to deal with generalized eigenvalue problems $A\mathbf{x} = \lambda B\mathbf{x}$. The aim is to get this into the generalized Schur normal form which looks like $QRZ\mathbf{x} - \lambda QSZ\mathbf{x} = 0$ where both Q and Z are orthogonal and R and S upper triangular. The eigenvalues

---

Algorithm 2 The QR method with shift for a symmetric matrix

Set $A^{(0)} = A$
while $N > 1$
    for $n = 1, 2, \ldots$ **until convergence of last entry**
        $\sigma_n = A_{N,N}^{(n)}$
        $Q^{(n)} R^{(n)} = A^{(n)} - \sigma_n I$
        $A^{(n+1)} = R^{(n)} Q^{(n)} + \sigma_n I$
    end
    $\lambda_N = A_{N,N}$
    $N = N - 1$
    $A^{(0)} = A^{(n)}(1 : N, 1 : N)$
end

---

follow easily by observing the diagonal of $R - \lambda S$. Note that for non-singular B the generalized eigenvalue problem can be cast into a standard eigenvalue problem with matrix $B^{-1}A$ or the similar matrix AB^{-1}. One has constructed a so-called QZ-algorithm generating a Q and Z from the QR-algorithm applied to AB^{-1}. For details, we refer again to Golub and van Loan (1996)

In summary, with the QR and QZ methods the complete eigensolution of a dense matrix can be computed in a modest (but matrix-dependent) factor times N^3 floating-point operations and using $\mathcal{O}(N^2)$ memory locations. Note that this order of operations is equal to that of direct methods for solving dense linear systems. Both QR and QZ methods are in general too expensive for large, sparse matrices and an overkill if we only want to find a few eigenvalues, for example only the ones closest to the origin or the imaginary axis. In this case one uses iterative sub-space projection techniques. In these methods the QR/QZ method is used to find eigenvalues of the small projected matrices (see Section 3.4) which arise. The iterative sub-space projection techniques attempt to detect partial eigensolution information using much less than $\mathcal{O}(N^3)$ computational work.

6.6 Summary

- Since Newton's method is at the heart of continuation, one needs the Jacobian of the right-hand side of the governing equations. Explicit programming of the Jacobian usually gives the fastest code, but can be difficult to extract from a general-purpose one. Moreover, even if one is programming the Jacobian, then it is wise to check its consistency with the right-hand side by comparing it to a Jacobian which is obtained by numerical differentiation.

- Numerical differentiation comes with severe propagation of round-off errors if the step size is made smaller and smaller. Hence, a trade-off has to be made between approximation error and round-off error. The problem of round-off error can be prevented if the right-hand side is real by using complex arithmetic. An alternative is to use automatic differentiation packages which compute the Jacobian from a given right-hand side automatically.
- In pseudo-arclength continuation one will encounter a bordered matrix, that is, a sparse matrix, being the Jacobian of the right-hand side, extended by one or more full rows and columns. If one has available an LU of the sparse part, then that can be used repeatedly to solve the system with the complete matrix. This approach might fail near a bifurcation point, since at the bifurcation point the Jacobian is singular. The standard approach to accommodate this problem is to use iterative refinement. However, if the matrix is irreducible, then one can also use all but the last row and column of the LU factorization, adding the omitted row and column to the border. Improvement of this is possible using multi-grid and multi-level methods, where one makes a direct solve of the bordered matrix on the coarsest level only.
- There are various ways to compute a periodic orbit: the extended system method, the Galerkin projection method, and a third one is to look for a fixed point of the Poincaré map. The latter two approaches can be transformed into a multiple shooting method, where one, as in the extended method, introduces intermediate points. The advantage of this is that one can do the propagation for each intermediate point in parallel, which can be exploited on parallel computers.
- Solving large, sparse linear systems by direct methods is quite well developed nowadays. Packages are available that can do this job. The key ingredient is the fill-reducing ordering which is needed to keep computation time and memory consumption at bay. The optimal complexity for a 2D and 3D problem is $O(N\sqrt{N})$ and $O(N)$, respectively. Pivoting, needed to deal with instabilities in the factorization, may conflict with the fill-reducing order.
- The computation of all eigenvalues can be done using the QR method and the QZ method for the standard and generalized eigenvalue problems, respectively. The linear convergence of the method is dictated by the ratio of subsequent eigenvalues, ordered

in magnitude. These ratios can be influenced by shifts, and by an appropriate choice the convergence speed can be made super-linear. For the problems treated in this book we are only interested in the eigenvalues passing the imaginary axis, so computing all eigenvalues is usually not needed. Therefore, these methods are not used as such but included in iterative procedures where in many cases we have at the heart of it a (Petrov-) Galerkin projection of the original eigenvalue problem.

6.7 Exercises

Exercise 6.1 *As we have seen in Chapter 2, the Ginzburg–Landau equation can be written as*

$$\frac{\partial a}{\partial \tau} = a + (1 + i\alpha_1)\frac{\partial^2 a}{\partial \chi^2} - (1 + i\alpha_2)a|a|^2.$$

a. Discretize the real one-dimensional version of this equation.

b. Determine the Jacobian matrix directly 'by hand'.

When computing the Jacobian by hand, it is extremely important to check whether its implementation is compatible with the right-hand side function. Check this using the file compat.m; use a variety of ϵ's to observe the convergence effect and the effect of rounding errors.

Exercise 6.2 *In AUTO, a clever method is used to determine Floquet multipliers, determining the stability of the periodic orbit computed.*

a. Write down, schematically, the Jacobian of the pseudo-arclength method for the case $n = 2, N = 3, m = 3$.

b. Next, use Gaussian elimination to remove all local quantities, so that only quantities at the mesh points need to be computed.

c. Show that the discrete Poincaré map arises from a sub-system of the condensed Jacobian, that is, from two particular matrices A_0 and A_1.

d. Show that the Floquet multipliers can be computed as the eigenvalues of $-A_1^{-1}A_0$.

Exercise 6.3 *Suppose one is looking for periodic orbits of a certain system of autonomous equations, but there is no fixed-point branch containing a Hopf bifurcation available. Suggest two methods on how a starting periodic orbit for a periodic orbit continuation can be generated.*

Exercise 6.4 *Perform the continuation of the Ginzburg–Landau equation solution from Example 6.1 in the parameter γ_3 using the file main.m.*

Exercise 6.5 *In many practical problems, a conservation equation for a tracer X has to be solved of the form*

$$\frac{\partial X}{\partial t} + \nabla \cdot (\mathbf{v}X) = K\nabla^2 X$$

where $\mathbf{v}$ is a given divergence free velocity field and K is a diffusion constant. The boundary conditions are at the boundary Γ:

$$\nabla X \cdot \mathbf{n} = F_\Gamma; \ \mathbf{v} \cdot \mathbf{n} = 0$$

where F_Γ is a prescribed function such that

$$\int_\Gamma F_\Gamma \, d\Gamma = 0.$$

a. Show that the steady solution X is determined up to an additive constant.

b. What is the condition to fix the constant in a. ?

c. Discretise the equations for X on a two-dimensional square domain $[0, L] \times [0, L]$ and formulate the linear system of equations, for a given field $\mathbf{v}$, to solve for a steady state.

d. Find an efficient direct method to solve this system, while taking the condition in b. into account.

7

Stationary Iterative Methods

If you can't do it in one big step,
you can try to do it in a sequence of small steps.

In this chapter we will introduce iterative methods to solve linear systems and iterative methods to solve eigenvalue problems. These problems are treated together since there is a strong relationship between them. It is convenient to introduce two classes of methods: stationary and non-stationary methods. In the former, the iteration is the same in every step while it changes from step to step in non-stationary methods. In this chapter, we discuss the stationary methods, and in the next chapter the non-stationary ones.

7.1 Stationary Methods for Linear Systems

We consider the problem $A\mathbf{x} = \mathbf{b}$ and write $A = K - R$, where K is non-singular. This is called a *splitting* of A. Now re-write the system as

$$K\mathbf{x} = R\mathbf{x} + \mathbf{b},$$

and next to

$$\mathbf{x} = C\mathbf{x} + K^{-1}\mathbf{b}. \tag{7.1}$$

By adding superscripts this expression transforms into the general form of a stationary iteration,

$$\mathbf{x}^{(n+1)} = C\mathbf{x}^{(n)} + K^{-1}\mathbf{b}, \text{ for } n = 0, 1, 2, \ldots, \tag{7.2}$$

where $\mathbf{x}^{(0)}$ is given, and

$$C = K^{-1}R \tag{7.3}$$

is the *iteration matrix*.

If $\|\mathbf{x}^{(n)} - \mathbf{x}\| \to 0$ in some vector norm, then the method (7.2) is said to be *convergent*. One can analyze the convergence by subtracting (7.2) from (7.1), giving *See example Ex. 7.1*

$$\mathbf{e}^{(n+1)} = C\mathbf{e}^{(n)}, \tag{7.4}$$

where $\mathbf{e}^{(n)} = \mathbf{x} - \mathbf{x}^{(n)}$ is the *iteration error*. This is a recursion and it is easily shown that

$$\mathbf{e}^{(n)} = C^n\mathbf{e}^{(0)}. \tag{7.5}$$

For the iterative method to be convergent the repetitive application of C to an arbitrary initial error $\mathbf{e}^{(0)}$ should tend to zero. This leads to the following theorem (see also Theorem 7.17 in Burden and Faires (2001)) for the spectral radius $\rho(C) \equiv \max_{i=1}^{N} |\lambda_i(C)|$ where $\lambda_i(C)$, $i = 1, \ldots, N$ denote the eigenvalues of C (see Definition B.2).

Theorem 7.1 *The stationary iterative method is convergent if and only if the spectral radius $\rho(C)$ of the iteration matrix C is less than one.*

Proof We only sketch the main line of the proof; details can be found elsewhere (Varga, 1962; Young, 1971).

If $\rho(C) \geq 1$, then there is an eigenvalue λ and an eigenvector $\mathbf{q}$ with $|\lambda| \geq 1$ and $C\mathbf{q} = \lambda\mathbf{q}$. Hence, $C^n\mathbf{q} = \lambda^n\mathbf{q}$, and $\|C^n\mathbf{q}\| = |\lambda|^n\|\mathbf{q}\|$. So, if $\mathbf{e}^{(0)} = \mathbf{q}$, then the method will not converge.

Now suppose that $\rho(C) < 1$. There exists always a non-singular matrix V such that $V^{-1}CV$ is in *Jordan-normal form J* given by

$$J = \begin{bmatrix} J_1 & & & 0 \\ & J_2 & & \\ & & \ddots & \\ 0 & & & J_m \end{bmatrix}$$

in which every *Jordan block J_i* of order p_i, with $\sum_{i=1}^{m} p_i = N$, is of the form *See example Ex. 7.2*

$$J_i = \begin{bmatrix} \lambda_i & 1 & 0 & \cdots & 0 \\ 0 & \lambda_i & 1 & & \\ \vdots & & \ddots & & \\ 0 & & & \lambda_i & 1 \\ 0 & \cdot & \cdot & \cdot & \lambda_i \end{bmatrix} \tag{7.6}$$

where λ_i is an eigenvalue of C. Instead of C^n we can now consider J^n, more specifically J_i^n. A further study (see Theorem B.26) reveals that for $\lambda_i \neq 0$

$$\|J_i^n\| \approx cn^{p-1}|\lambda_i|^n. \tag{7.7}$$

Hence, the power of the Jordan block will tend to zero for n to infinity, from which it follows that $\|C^n\| \to 0$. Now from

$$\|C^n \mathbf{e}^{(0)}\| \le \|C^n\|\|\mathbf{e}^{(0)}\|$$

the convergence of the iterative method follows for $\lambda_i \neq 0$. For $\lambda_i = 0$ the J_i is idempotent, that is, $J_i^n = 0$ for $n \ge p_i$. ∎

From the asymptotic behaviour of the power of J in (7.7), we deduce that for sufficiently large n the behaviour of $\|J_i^n\|$ is determined by $|\lambda_i|$. Hence, the behaviour of $\|C^n\|$ is also of the form (7.7) with $|\lambda_i|$ replaced by $\rho(C)$ and p the biggest occurring order of Jordan blocks with $|\lambda_i| = \rho(C)$. For large values of p, initially the behaviour can be much different from the asymptotic behaviour, that is, the behaviour for n tending to infinity.

From the preceding we conclude that the best convergence of (7.2) occurs if $\rho(K^{-1}R)$ is as small as possible, but in any case we must have for convergence that

$$\rho(K^{-1}R) < 1. \tag{7.8}$$

7.2 The Classical Iterative Methods

The classical iterative methods emerge from the splitting of A as

$$A = D - L - U = K - R, \tag{7.9}$$

where D is the diagonal of A, L the strict lower triangular part of A, and U the strict upper triangular part. In what follows we will consider them for the five-point stencil following from the discretization of the Poisson equation where a lexicographical ordering is used and both i and j increase, similar to that given in (4.17).

Additional Material

- Classical iterative methods are extensively discussed in Young (1971) and Burden and Faires (2001).

For the *Jacobi method* $K = D$ and $R = L + U$. Hence, the Jacobi iteration matrix (7.3) is given by

$$C = D^{-1}(L + U). \tag{7.10}$$

For the five-point stencil (4.18), the iteration (7.2) can be written as

$$C_P U_{i,j}^{(n+1)} = C_W U_{i-1,j}^{(n)} + C_S U_{i,j-1}^{(n)} + C_E U_{i+1,j}^{(n)} + C_N U_{i,j+1}^{(n)} + f_P. \tag{7.11}$$

(Don't be confused by the double but different use of U in (7.10) and (7.11)). Note that in this case it does not matter for the result in which order we traverse the unknowns $U_{i,j}$; this makes this method very fit for parallelization.

The *Gauss–Seidel method* is an improvement of Jacobi's method; an update of $U_{i,j}$ will be used for updating one or more of its neighbours. In this case, $K = D - L$, hence a lower triangular matrix, and $R = U$. Hence, the iteration matrix is

$$C = (D - L)^{-1} U. \tag{7.12}$$

It can be shown that the convergence of the Gauss–Seidel method and the Jacobi method are related for an important class of matrices (Young, 1971). If for that class Jacobi's method is converging, then the method of Gauss–Seidel is converging twice as fast.

The application of the Gauss–Seidel method to the five-point stencil is given by

$$C_P U_{i,j}^{(n+1)} = C_W U_{i-1,j}^{(n+1)} + C_S U_{i,j-1}^{(n+1)} + C_E U_{i+1,j}^{(n)} + C_N U_{i,j+1}^{(n)} + f_P. \tag{7.13}$$

In contrast to the Jacobi method, in a program we can do with one array to store the $U_{i,j}$'s, because once a value is updated the old value is not used anymore.

For large problems the convergence of Gauss–Seidel's method may still be too slow, and, in some cases one can speed up the method by over-relaxation. This method is coined the *successive over-relaxation method (SOR)*. In general, the idea of relaxation is to multiply the correction one wants to add to the old value by a factor. Let us consider it for the the problem on the five-point stencil first before discussing the matrix form. At a certain point (i, j) the new value proposed by Gauss–Seidel's method is

$$C_P \hat{U}_{i,j} = C_W U_{i-1,j}^{(n+1)} + C_S U_{i,j-1}^{(n+1)} + C_E U_{i+1,j}^{(n)} + C_N U_{i,j+1}^{(n)} + f_P.$$

The correction of $U_{i,j}^{(n)}$ is now $\hat{U}_{i,j} - U_{i,j}^{(n)}$ which we multiply by a factor ω and add to $U_{i,j}^{(n)}$ resulting in

$$U_{i,j}^{(n+1)} = U_{i,j}^{(n)} + \omega(\hat{U}_{i,j} - U_{i,j}^{(n)}) = (1 - \omega)U_{i,j}^{(n)} + \omega \hat{U}_{i,j}.$$

In matrix form it is written as

$$(D - \omega L)\mathbf{x}^{(n+1)} = [(1 - \omega)D + \omega U]\mathbf{x}^{(n)} + \omega \mathbf{b}.$$

Also this stationary iteration results from a splitting of A, which is

$$K = D/\omega - L \quad \text{en} \quad R = (1/\omega - 1)D + U,$$

and the iteration matrix is

$$C_\omega = (D - \omega L)^{-1}[(1 - \omega)D + \omega U]. \tag{7.14}$$

Observe that the method is equal to the Gauss–Seidel method for $\omega = 1$. In many cases, among which is the five-point discretization for the Poisson equation, the convergence is faster than the Gauss–Seidel method for $1 < \omega < 2$, from which the prefix 'over' in 'over-relaxation' originates. Also in some cases where Gauss–Seidel does not converge the SOR method may converge for some $\omega < 1$; in this case, it would be better to speak of under-relaxation.

In all the preceding methods, a point can be generalized to a line, a plane, or a group of unknowns. Doing this we get *block variants*, in which in every step the unknowns associated to a certain block are solved together. For example, we have SLOR where the 'L' stands for line. Block Jacobi is especially interesting for parallel processing, because the new values in every block can be computed independently of those in other blocks.

For the line Jacobi method, (7.11) transforms into

$$C_P U_{i,j}^{(n+1)} = C_W U_{i-1,j}^{(n+1)} + C_S U_{i,j-1}^{(n)} + C_E U_{i+1,j}^{(n+1)} + C_N U_{i,j+1}^{(n)} + f_P. \tag{7.15}$$

So, we need to solve $U_{i,j}^{(n+1)}$, $U_{i-1,j}^{(n+1)}$, and $U_{i+1,j}^{(n+1)}$ simultaneously, which leads to a tridiagonal system for each j. This means that D becomes a block diagonal matrix containing these tridiagonal matrices on its diagonal. Walking from south to north over the domain, we can also exploit the already computed values on the previous line resulting in line Gauss–Seidel;

$$C_P U_{i,j}^{(n+1)} = C_W U_{i-1,j}^{(n+1)} + C_S U_{i,j-1}^{(n+1)} + C_E U_{i+1,j}^{(n+1)} + C_N U_{i,j+1}^{(n)} + f_P. \tag{7.16}$$

In a *domain decomposition* approach, the grid is split up in a number of non-overlapping parts, say Ω_k for $k = 1, \ldots, n_k$. Let the grid point (i, j) be in Ω_k; then the associated block-Jacobi method can be written as

$$C_P U_{i,j}^{(n+1)} = C_W U_{i-1,j}^{(n+\chi_k(i-1,j))} + C_S U_{i,j-1}^{(n+\chi_k(i,j-1))}$$
$$+ C_E U_{i+1,j}^{(n+\chi_k(i+1,j))} + C_N U_{i,j+1}^{(n+\chi_k(i,j+1))} + f_P, \tag{7.17}$$

where $\chi_k(l, m) = 1$ if (l, m) is in Ω_k and zero if not.

As has been mentioned before, solving linear systems is usually the most time consuming part in computing bifurcation diagrams. So, especially here it is important to exploit computer architecture. Hence, it is useful to have a basic understanding of the working of a parallel computer. In essence, there are three important ingredients to deal with: (i) the data-communication time, (ii) the waiting time, and (iii) the data-processing time. For an introduction to parallel computing we refer to the textbook of Hager and Wellein (2011).

In the data-processing time, there are two contributions: the amount of data to be processed and the efficiency of the numerical method. For an iterative solver, one could think of the amount of data as being the grid of unknowns times the number of iterations the algorithm will take. Then, on a single processor, one obviously will use the method that needs the lowest number of iterations. The multi-grid method, described in the next section, is a good example of that. On a parallel computer, the data-processing time can be brought down by letting multiple processors work on the data simultaneously. However, very often data is needed from other processors and this data may not yet be available since the computation of it is still going on, which leads to waiting time. Moreover, sending such data between processors also takes time, leading to data-communication time.

The block Jacobi method is an example where almost no data is needed from other processors. Suppose (i, j) is on the boundary of Ω_k; then from (7.17) we see that some of $\chi_k(l, m)$ will be zero. So, we need to copy the associated values of the unknowns to the processor holding domain Ω_k. Hence, one iteration of block-Jacobi will consist of two steps. In the first one, data is exchanged between neighbouring domains simultaneously, and in the second one all the domains compute their contribution to the update simultaneously. But if we transform this block Jacobi method into a block Gauss–Seidel method, then a processor has to wait for the results of a neighbouring domain computed on another processor in order to start its contribution to the update. So, here a trade-off has to be made. The block Gauss–Seidel method will need fewer iterations to converge than the block Jacobi method, so there is less data processing. However, per iteration it takes longer because of the waiting. If it takes too long, the Jacobi method might still be faster.

The holy grail in parallel solution of linear systems is to find a method with low amounts of data to be processed and low waiting times for other processors.

7.3 The Geometric Multi-grid Method

By using a sequence of coarse to fine grids it is possible to create a method that has optimal complexity, that is, the amount of work is proportional to the number of unknowns. We will introduce the idea for the Poisson equation $-\Delta u = f$ defined on the unit square with Dirichlet boundary conditions. On this square, we define a uniform grid with mesh widths $h = 1/(m + 1)$ and $k = h$ in the x and y direction, respectively. After a standard finite-difference discretization we obtain the equations

$$U_{i,j} = \frac{1}{4} \left\{ U_{i-1,j} + U_{i+1,j} + U_{i,j-1} + U_{i,j+1} + h^2 f_{i,j} \right\} \text{ for } i, j = 1, \dots, m. \quad (7.18)$$

If we apply the Gauss–Seidel method to this equation, then the error $\mathbf{e}^{(n)}$ satisfies $\mathbf{e}^{(n+1)} = C\mathbf{e}^{(n)}$, with C the Gauss–Seidel iteration matrix (7.12). Figure 7.1 shows

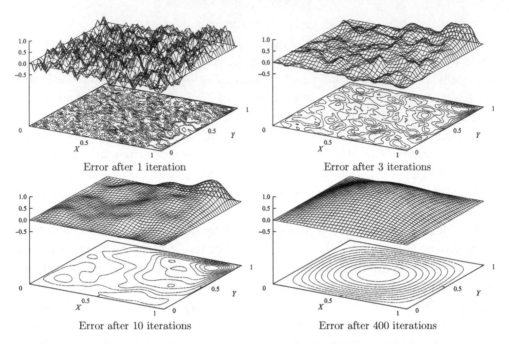

Figure 7.1 Field view of the error after 1, 3, 10, and 400 Gauss–Seidel iterations.

what this error looks like after a specified amount of iterations starting from a random initial error on a 40×40 grid. These plots show that the error is already rather smooth after a few iterations (here after 10 iterations), but after 400 iterations the method is still working hard on reducing a smooth component in the error. It can also be shown analytically that the Gauss–Seidel iteration is good at removing high-frequency components from the error and less good in handling low-frequency components. Here high frequencies are the highest frequencies that can be represented on the grid. Observe that for a smooth error we could do with fewer grid points to represent it. Hence, the crux of the multi-grid method is to get rid of this error on a coarser grid. On a coarser grid the error contains frequencies that are closer to the highest frequency that can be represented on the coarse grid. Hence, on this coarse grid the Gauss–Seidel iteration will do a better job of getting rid of these frequencies. The argument can be repeated leading to a nested series of iterations which is displayed in Algorithm 3. Here the function 'smooth' is a smoothing iteration like the Gauss–Seidel iteration which updates $\mathbf{x}$. The function 'restrict' computes from its vector argument on the current grid the approximation of that on the next coarser grid. Furthermore, the function 'prolongate' is an interpolating function to obtain values in the finer grid points that are not present on the coarser grid, A_c is the approximation of A on a coarser grid, and $\mathbf{r}_c$ is the restriction of $\mathbf{r}$ on a coarser grid.

Algorithm 3 One multi-grid cycle written as a recursive function. Here $\mathbf{r} \equiv \mathbf{b} - A\mathbf{x}$

> **function x** = solve$(A, \mathbf{b})$
>> **if** (coarsest grid is not reached yet) **then**
>>> $\mathbf{x} = 0$
>>> $\mathbf{x} = \text{smooth}(\mathbf{r})$
>>> $\mathbf{r}_c = \text{restrict}(\mathbf{r})$
>>> $\Delta\mathbf{x}_c = \text{solve}(A_c, \mathbf{r}_c)$
>>> $\Delta\mathbf{x} = \text{prolongate}(\Delta\mathbf{x}_c)$
>>> $\mathbf{x} = \mathbf{x} + \Delta\mathbf{x}$
>>> $\mathbf{x} = \text{smooth}(\mathbf{r})$
>> **else**
>>> $\mathbf{x} = \text{solve\_exactly}(A, \mathbf{b})$
>> **end**
>> **return x**
> **end**

In order to demonstrate the typical behaviour of the multi-grid method, we show a comparison between multi-grid and SLOR applied to a discretized Poisson problem of the type (7.18) for an increasing number of grid points. In Table 7.1, the number of iterations needed to bring down the residual to a value indicated in the first column for a 16×16, 32×32, and 64×64 grid, respectively, is listed. In Table 7.2, the amount of time needed to solve the problem is given.

For SLOR in Table 7.1, we see that the required number of iterations becomes twice as big when the number of grid points is doubled in each direction, whereas for the multi-grid method the number of iterations remains constant. Hence, the convergence of the multi-grid method is independent of the grid size. Both behaviours can be proved for the Poisson equation (Young, 1971; Brand, 1992). With respect to the computation times in Table 7.2, we see that those of multi-grid are proportional to the number of grid points, which is due to the fact that the amount of work is just the number of iterations times the work per step. Since the former is nearly constant and the latter depends linearly on the number of unknowns, the product is also linear in the number of unknowns. This behaviour is optimal, since in any iterative method for an update we have to visit all unknowns. So, one cannot do better than a computational complexity proportional to the number of unknowns. It is an attractive property for very large problems. For SLOR we find that, upon refinement with a factor of 2 in both directions, the time increases by a factor of 8 instead of the optimal value of 4. In the columns we see always a relatively fixed increase. This is typical for stationary methods. One such increment (say m) shows the number of iterations needed to gain three digits. From this

Table 7.1 *Number of iterations for multi-grid (MGRD) and successive line over relaxation (SLOR) for three different grid sizes and three different stop tolerances.*

	MGRD			SLOR		
$\|r\|_2$	16×16	32×32	64×64	16×16	32×32	64×64
10^{-3}	6	5	6	10	21	41
10^{-6}	10	9	11	17	40	81
10^{-9}	14	13	17	24	58	121

Table 7.2 *Needed CPU time for multi-grid (MGRD) and Successive Line Over Relaxation (SLOR) for three different grid sizes and three different stop tolerances.*

	MGRD			SLOR		
$\|r\|_2$	16×16	32×32	64×64	16×16	32×32	64×64
10^{-3}	0.3	1.1	4.7	0.3	2.4	18.0
10^{-6}	0.6	1.9	8.6	0.5	4.6	35.5
10^{-9}	0.8	2.8	13.2	0.7	6.7	53.0

number of iterations, one can estimate the spectral radius of the iteration matrix of each method since we have the relation $\rho^m \approx 0.001$. As an example, for MGRD 16×16, we see the number of iterations increase by 4. So $m = 4$ and hence the spectral radius is about 0.17. *See example Ex. 7.3*

7.4 Power Iteration

Next, we turn to the solution of eigenvalue problems $A\mathbf{x} = \lambda\mathbf{x}$. In this section, we will discuss the power iteration

$$\mathbf{v}^{(n)} \equiv A^{n-1}\mathbf{v}^{(1)} \tag{7.19}$$

where $\mathbf{v}^{(1)}$ is usually a random vector. We will show that, for large n, $\mathbf{v}^{(n)}$ will converge to the eigenvector related to the eigenvalue with the largest modulus of A, provided that there is only one such eigenvalue. We will analyse this iteration in some detail because it is also the basis for many other methods to compute eigenvalues. Moreover, as (7.5) is of the same shape as (7.19), it also sheds light on the convergence behaviour of iterative methods for linear systems. In Section 7.6 we will see that this understanding makes it possible to speed up convergence of those methods by extrapolation. Also note that the iteration occurs in hidden form in the stability analysis of time integration methods for linear problems, for example by applying the forward Euler method to (4.99), leading to a discretized form of (4.100), which again will be of the shape (7.19).

Assume that A is simple, that is, not defective, then it has a complete set of eigenvectors:

$$A\mathbf{q}_k = \lambda_k \mathbf{q}_k \ , \quad \|\mathbf{q}_k\|_2 = 1 \ (k = 1, 2, \ldots, N).$$

We further assume that the largest eigenvalue in the absolute sense is single and that

$$|\lambda_1| > |\lambda_2| \geq \cdots \geq |\lambda_N|.$$

Hence, an arbitrary vector $\mathbf{v}^{(1)}$ can be expressed in terms of the eigenvectors as $\mathbf{v}^{(1)} = \sum_i \gamma_i \mathbf{q}_i$. For the following analysis we assume that $\gamma_1 \neq 0$, hence, $\mathbf{v}^{(1)}$ has a non-zero component in the direction of the eigenvector corresponding to λ_1.

Next, we write $\mathbf{v}^{(n)}$ defined by (7.19) in terms of the eigenvectors of A:

$$\mathbf{v}^{(n)} = \sum_{i \geq 1} \gamma_i \lambda_i^{n-1} \mathbf{q}_i = \lambda_1^{n-1} \left\{ \gamma_1 \mathbf{q}_1 + \sum_{i \geq 2} \gamma_i \left(\frac{\lambda_i}{\lambda_1} \right)^{n-1} \mathbf{q}_i \right\}. \tag{7.20}$$

This expression reveals that $\mathbf{v}^{(n)}$ converges linearly to a multiple of $\mathbf{q}_1$ with rate $|\lambda_2/\lambda_1|$.

Once one has an eigenvector, it is easy to compute the eigenvalue, for example via the Rayleigh quotient. Hence, we will have

$$\lim_{n \to \infty} \frac{(\mathbf{v}^{(n)}, A\mathbf{v}^{(n)})}{(\mathbf{v}^{(n)}, \mathbf{v}^{(n)})} = \lambda_1,$$

with inner product $(\mathbf{x}, \mathbf{y}) = \sum_{i=1}^N \bar{x}_i y_i$, where the bar denotes the complex conjugate. The convergence rate of the sequence of *Rayleigh quotients* follows by substituting (7.20) in the Rayleigh quotient, leading to

$$\frac{(\mathbf{v}^{(n)}, A\mathbf{v}^{(n)})}{(\mathbf{v}^{(n)}, \mathbf{v}^{(n)})} = \lambda_1 \frac{|\gamma_1|^2 + \sum_{i,j \geq 1/i=j=1} \bar{\gamma}_i \gamma_j \left(\frac{\bar{\lambda}_i}{\lambda_1} \right)^{n-1} \left(\frac{\lambda_j}{\lambda_1} \right)^n (\mathbf{q}_i, \mathbf{q}_j)}{|\gamma_1|^2 + \sum_{i,j \geq 1/i=j=1} \bar{\gamma}_i \gamma_j \left(\frac{\bar{\lambda}_i}{\lambda_1} \right)^{n-1} \left(\frac{\lambda_j}{\lambda_1} \right)^{n-1} (\mathbf{q}_i, \mathbf{q}_j)}.$$

Algorithm 4 The power method

$\mathbf{v}^{(1)}$ such that $||\mathbf{v}^{(1)}|| = 1$

for $n = 1, 2, \ldots$ **until convergence**

$\quad \mathbf{z}^{(n+1)} = A\mathbf{v}^{(n)}$

$\quad \theta^{(n)} = (\mathbf{v}^{(n)}, \mathbf{z}^{(n+1)})$

$\quad \mathbf{v}^{(n+1)} = \mathbf{z}^{(n+1)}/||\mathbf{z}^{(n+1)}||_2$

end

The slowest convergence of the sum parts in the numerator and denominator occurs if $i = 1$ and $j = 2$ or the other way around and the convergence rate is $|\lambda_2/\lambda_1|$. However, if the matrix A is normal (see Definition B.5), then the eigenvectors are mutually orthogonal and

$$\frac{(\mathbf{v}^{(n)}, A\mathbf{v}^{(n)})}{(\mathbf{v}^{(n)}, \mathbf{v}^{(n)})} = \lambda_1 \frac{|\gamma_1|^2 + \sum_{i \geq 2} |\gamma_i|^2 \left(\frac{|\lambda_i|}{|\lambda_1|}\right)^{2(n-1)} \left(\frac{\lambda_i}{\lambda_1}\right)}{|\gamma_1|^2 + \sum_{i \geq 2} |\gamma_i|^2 \left(\frac{|\lambda_i|}{|\lambda_1|}\right)^{2(n-1)}}. \tag{7.21}$$

Hence, the convergence rate becomes $|\lambda_2/\lambda_1|^2$.

Mathematically this would define the algorithm. However, if $|\lambda_1| < 1$ on a computer eventually $\mathbf{v}^{(n)}$ can become so small that it becomes less than the smallest non-zero vector that can be represented by the floating-point representation and hence it will be rounded to 0; this situation is called underflow. Similarly, if $|\lambda_1| > 1$, we can get overflow. To preclude this, we just can normalize the eigenvector in every step, which leads to the power method given in Algorithm 4. The sequence $\theta^{(i)}$ converges, under the above assumptions, to the dominant (in absolute value) eigenvalue of A. *See example Ex. 7.4* Note that $\theta^{(n)}$ is a by-product of the iteration; the iteration is on the vector. This means that there is no need to compute $\theta^{(n)}$ in each step. Note also that an alternative and cheaper way of approximating $\theta^{(n)}$ is $\theta^{(n)} = \mathbf{z}_i^{(n+1)}/\mathbf{v}_i^{(n)}$, where we take for i that index where $|\mathbf{v}^{(n)}|$ has its maximal element. However, this estimate will not give the twice-as-fast convergence we have seen for the Rayleigh quotient approximation for normal matrices.

For general matrices, including defective ones, one can consider the Jordan normal form J of A related by $A = QJQ^{-1}$, then $\mathbf{v}^{(n)} = A^{n-1}\mathbf{v} = QJ^{n-1}Q^{-1}\mathbf{v}^{(1)}$. Let the largest eigenvalue be denoted by λ_1, in modulus; then we can write $\mathbf{v}^{(n)} = \lambda_1^{n-1}Q(J/\lambda_1)^{n-1}Q^{-1}\mathbf{v}^{(1)}$. The $(n-1)$-th power of all diagonal blocks J_i/λ_1 (see (7.6)) for which $|\lambda_i/\lambda_1| < 1$ will go to zero for $n \to \infty$. If $|\lambda_i| = |\lambda_1|$, then the associated Jordan block will remain and $\mathbf{v}^{(n)}$ will eventually dwell in the space spanned by the generalized eigenvectors of each of these remaining blocks. In fact, these spaces are built by the associated columns in Q for which the power of the scaled Jordan blocks remains. We treat a few special cases. When there is only one single eigenvalue which has the maximum modulus as assumed in the foregoing, then there is only one Jordan block involved of size 1 and hence we have convergence to a single vector. If this eigenvalue would have been complex and the matrix is real, then the complex conjugate also is an eigenvalue and hence there must be another Jordan block related to this conjugate eigenvalue. If both blocks have size 1, then $\mathbf{v}^{(n)}$ will eventually dwell in the space spanned by the two corresponding eigenvectors. This is especially the case if we start off with a real $\mathbf{v}^{(1)}$; then all iterates will be real and $\mathbf{v}^{(n)}$ will eventually dwell in the space spanned by the real and imaginary parts of the eigenvector associated to the complex eigenvalue. If $\mathbf{v}^{(n)}$ dwells in a sub-space for large n, then we can extract information on the sub-space by looking at successive iterates (Wilkinson, 1965, p. 579).

As we have seen, the power method provides a way to compute the eigenvalue with largest modulus of a matrix A. However, for studying the stability of solutions of PDEs we are usually interested in the eigenvalues near the origin. Fortunately, we can also compute the smallest eigenvalue in modulus by the power method by replacing A by its inverse, provided that this eigenvalue λ_N is single. Using the inverse, the power method will converge to the eigenvector associated to $1/\lambda_N$. This eigenvector is the same as that for λ_N, and since $\theta^{(n)}$ in the power method will converge to $1/\lambda_N$ we simply have that $1/\theta^{(n)}$ will converge to λ_N. The resulting method is called *inverse power method*. The method can be extended by a shift by replacing A in the power method by $(A - \sigma I)^{-1}$ with a user given σ; the transformation of A is named *shift-and-invert*. The eigenvalues of this matrix are $1/(\lambda_i - \sigma)$, $i = 1, \ldots, N$. We will call σ the *target* since the eigenvalues λ_i of A closest to σ will become the biggest in modulus of the transformed A.

Suppose that λ_k is the eigenvalue closest to the target; then the speed of convergence will be the ratio of the transformed eigenvalue of the second closest eigenvalue and that of the closest one. So, if $l = \text{argmin}_{i, i \neq k} |\lambda_i - \sigma|$, then the convergence rate is

$$\frac{1/|\lambda_l - \sigma|}{1/|\lambda_k - \sigma|} = \frac{|\lambda_k - \sigma|}{|\lambda_l - \sigma|}. \tag{7.22}$$

Note that a good convergence rate is obtained when σ is close to an eigenvalue. So, one of the applications of the inverse power methods is to compute an eigenvector connected to an already known eigenvalue by simply choosing σ to be this eigenvalue. Note that it is not necessary to compute the inverted matrix explicitly. In the computation of the next vector $\mathbf{z}^{(n+1)} = (A - \sigma I)^{-1}\mathbf{v}^{(n)}$, the vector $\mathbf{z}^{(n+1)}$ can be solved from

$$(A - \sigma I)\mathbf{z}^{(n+1)} = \mathbf{v}^{(n)}.$$

For instance, using a direct method, one creates an LU factorization of $A - \sigma I$ once, and uses that in every step of the power method to solve this system.

Shift-and-invert together with its sister the Cayley transform (see Section 8.6) constitute the most important tools to get eigenvalues near a target. Since we get better and better approximations of the eigenvalue closest to the target during the iteration process with shift-and-invert, we can speed up the convergence to this eigenvalue and the associated eigenvector by updating the target σ. This technique is known as *Rayleigh-quotient iteration*. Its convergence is ultimately cubic for symmetric matrices and quadratic for non-symmetric systems. This alludes already to a Newton-like process, and indeed the eigenvalue problem is a relatively simple non-linear problem and one can apply Newton's method to it. We will get back to this when treating the Jacobi–Davidson method in Section 8.5, which can be seen as a sub-space–accelerated Newton process for the eigenvalue problem.

7.5 Simultaneous and Orthogonal Subspace Iteration

We have seen that with the power method one eigenvalue can be computed. In general, we like to find multiple eigenvalues near a target. One way of getting those is by applying the power iteration to multiple vectors at the same time. Assume the independent vectors $V_m^{(1)} = [\mathbf{v}_1, \mathbf{v}_2, \ldots, \mathbf{v}_m]$; then

$$V_m^{(n)} = A^{n-1} V_m^{(1)}.$$

Because each of the columns of $V_m^{(1)}$ is effectively used as a starting vector for a single-vector power iteration (hence the name *Simultaneous Iteration*), each of them converges towards the same dominant eigenvector, which is unwanted. Analytically, that is, in infinite precision, components of the other eigenvectors always remain in the columns of $V_m^{(n)}$; but in finite precision, where rounding is used, these components become so small that eventually they get lost in the rounding, and this will make the columns of $V_m^{(n)}$ dependent. To avoid this we make them orthonormal after each multiplication with A. This leads to the *orthogonal sub-space iteration* method given in Algorithm 5.

Again as a by-product we can compute the eigenvalues. For this we use the Galerkin approach, that is, test space and search space are the same as discussed in Section 3.4.2, and project the eigenvalue problem on V_m. Similar to Example 3.3, this leads to $V_m^{(n)} \hat{x} = \theta^{(n)} (V_m^{(n)})^T V_m^{(n)} \hat{x}$. Since $V_m^{(n)}$ is orthogonal, the matrix in the right-hand side becomes identity and hence the approximate eigenvalues follow from those of the matrix $B^{(n)} \equiv (V_m^{(n)})^T A V_m^{(n)}$.

The columns of $V_m^{(n)}$ converge to a basis of an invariant sub-space of dimension m, under the assumption that $|\lambda_m| > |\lambda_{m+1}|$. This can be analyzed similarly to the analysis of the power iteration which led to (7.20). In this case, one splits the sum in a part over 1 to m and a remainder over $m + 1$ to N and by the assumption we have that the remainder will go to zero with respect to the first part. In the

---

Algorithm 5 The orthogonal sub-space iteration method

---

   start with orthonormal $V_m^{(1)}$
   for $n = 1, \ldots,$ **until convergence**
    $Z_m^{(n+1)} = A V_m^{(n)}$
    compute all eigenvalues of $B^{(n)} = (V_m^{(n)})^T Z_m^{(n+1)}$
    orthonormalize the columns of $Z_m^{(n+1)}$:
     $V_m^{(n+1)} R_m^{(n+1)} = Z_m^{(n+1)}$
  end

---

symmetric case, the Galerkin projection could have also been called the Ritz approach. Therefore the eigenvalues of $B^{(n)}$ are called *Ritz values* of A with respect to $V_m^{(n)}$ (often indicated by $\theta_i^{(n)}$, $i = 1, \ldots, m$), and the corresponding approximate eigenvectors are called *Ritz vectors. See example Ex. 7.5*

In real computations round-off errors are introduced. Therefore, components of each eigenvector of A are present or will be created during the process in every column of the iterates $V_m^{(n)}$. As a consequence, the matrix $B^{(n)}$ will converge to the *partial Schur form* of A, that is, $AV_m = V_m B^{(n)}$, where $B^{(n)}$ is upper triangular and the eigenvalues are ordered from biggest to smallest in modulus on the diagonal. In fact, if we take $m = N$ and $V_N^{(1)} = A$, then one can show that the process is equivalent to the QR method without shift for full matrices (see Section 6.5), which also leads to the Schur normal form of A.

The computation of the partial Schur form may take a long time. For instance, in the case where $V_m^{(1)}$ consists of eigenvectors ordered in the opposite way, starting with the eigenvector corresponding to the smallest eigenvalue in modulus. Then the process of computing the eigenvectors in the right order can be sped up by computing the Schur form of $B^{(n)}$ such that its eigenvalues occur from largest to smallest in modulus on the diagonal, say $B^{(n)} U = UR$. For the example with the eigenvectors in opposite order this will be settled in one step.

| Additional Material |

- Much of the background linear algebra required to understand this chapter can be found in Appendix B.

7.6 Extrapolation

In general, linear convergence of numerical processes can be sped up by exploiting the knowledge gained from the power iteration. Here we will apply it to the iterative solution of linear systems by the splitting approach (7.2).

We start from (7.4) and note that it is in fact a power iteration, but we do not have the vectors $\mathbf{e}^{(n)}$. However, by subtracting from (7.4), the same equation as (7.4) but with n replaced by $n - 1$, we get the equation

$$\mathbf{x}^{(n+1)} - \mathbf{x}^{(n)} = C(\mathbf{x}^{(n)} - \mathbf{x}^{(n-1)}),$$

which is a power iteration on the difference of subsequent iterates. During the iteration we can get better approximations of the eigenvalue of largest modulus of C using the Rayleigh quotient as in Algorithm 4. So, after sufficient iterations we can approximate (7.4) by

$$\mathbf{x} - \mathbf{x}^{(n+1)} \approx \theta^{(n)}(\mathbf{x} - \mathbf{x}^{(n)}) \tag{7.23}$$

or re-writing

$$\mathbf{x} \approx (\mathbf{x}^{(n+1)} - \theta^{(n)}\mathbf{x}^{(n)})/(1 - \theta^{(n)}) \tag{7.24}$$

or

$$\mathbf{x} \approx \mathbf{x}^{(n+1)} + \frac{\theta^{(n)}}{1 - \theta^{(n)}}(\mathbf{x}^{(n+1)} - \mathbf{x}^{(n)}). \tag{7.25}$$

So, if $\theta^{(n)}$ and $\mathbf{x}^{(n+1)} - \mathbf{x}^{(n)}$ are an accurate eigenpair, then the right-hand side will be a good approximation of the solution. In Example 7.1, this extrapolation is used for a mildly diverging method, that is, $\theta^{(n)} > 1$. Here, after 20 standard iterations we extrapolate and continue the iteration starting with that result. We see that the extrapolation turns the method into a converging one. Note that the extrapolation interval will depend on the convergence rate of the power iteration. Moreover, the extrapolation is effective only if the convergence is sufficiently regular, which means here that there should be a single dominant eigenvalue of the iteration matrix.

Extrapolation is a rather powerful tool, since mildly diverging iterations are all around in bifurcation analysis. For instance, consider time integration near an unstable steady state, where one eigenvalue is positive. Then we can detect this steady state using extrapolation. The iteration consists then of a time integration over a fixed short time interval. Note that (7.25) is the same as the one called Aitken's extrapolation (Quarteroni *et al.*, 2007, Section 6.6.1). This extrapolation is applied to speed up the convergence of fixed point methods for *scalar* non-linear equations in the case that they converge linearly near the fixed point. In that case, when only one eigenvalue is involved, one can extrapolate every three iterations, leading to a quadratically converging method. This convergence behaviour occurs due to the non-linearity; if the scalar equation is linear, one extrapolation will already give the exact solution.

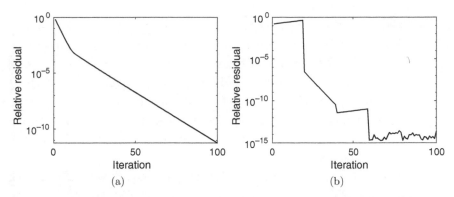

Figure 7.2 Convergence of a stationary method on system with Jacobian at peri-odic solution from Ginzburg–Landau equations with (a) block splitting using SOR with $\omega = 0.5$ and (b) Gauss–Seidel with extrapolation after 20 steps. Parameters $\gamma_1 = 12 + i$, $\gamma_2 = \gamma_3 = 1$, $A(0, t) = 0$, $A(1, t) = 0.1(1 + i)$, $m = 10$, $\Delta t = 1/20$.

Example 7.1 (Ginzburg–Landau) *Following our discussion on applicability of direct methods (see Example 6.3), we have in a similar fashion that the preceding methods will work for the computation of the steady-state and time-dependent computation, using implicit methods, as long as the Jacobian of the matrix is diagonally dominant. If it loses diagonal dominance, then these methods will in most cases fail.*

For example, the system originating for the computation of the periodic orbit treated in Example 6.1, using the extended system method of Section 6.3.1, is not diagonal dominant. The matrix is given by (6.19), and one can apply a block method to it. In this case, in the splitting of A we just omit in K the matrix $-M$ in the upper right corner which is responsible for the periodicity in the system. The Gauss–Seidel method based on this splitting will, however, not converge; one can observe the mild divergence in the first 20 iterations in the plot in part (b) of Fig. 7.2 computed with the MATLAB/Octave package BifAnFF.[1] We can make it converging by converting the block Gauss–Seidel method in a block SOR method using under-relaxation. After some trial and error, one finds that convergence occurs if we choose $\omega = 0.5$, which is shown in part (a) of Fig. 7.2. The right-hand side is chosen randomly. An alternative is to apply extrapolation to the block Gauss–Seidel method. The only parameter to tune is the amount of iterations before we extrapolate which is set to 20 here. Extrapolation turns the diverging block Gauss–Seidel method into a converging one.

7.7 Summary

- Stationary iteration methods for $A\mathbf{x} = \mathbf{b}$ can always be found from a splitting $A = K - R$ leading to $\mathbf{x}^{(n+1)} = R\mathbf{x}^{(n)} + \mathbf{b}$.

- The classical methods are based on the splitting $A = L + D + U$ with L and U the strict lower and upper part of the matrix, respectively, and D the diagonal. Choosing $K = D$ or $K = L + D$ leads to the Jacobi or the Gauss–Seidel method, respectively. For the latter one can also choose $K = D + U$. If the Jacobi method converges, then the Gauss–Seidel method also will converge. And it will converge faster. Loosely speaking, the more K resembles A, the faster the convergence. The Gauss–Seidel method can be accelerated by introducing a parameter leading to the Successive Over-Relaxation method (SOR).

- The classical methods can be generalized to block methods. Block methods are interesting for parallel computation. In that case, the domain is split up in sub-domains and each processor is assigned its own sub-domain. Block–Jacobi only needs to communicate between processors after each iteration and therefore the systems on all sub-domains can be solved simultaneously. This does not hold for the Gauss–Seidel process, where during the iteration there has to be communication between sub-domains and, moreover, the sub-domains have to be solved in a certain order.

- For an iterative method the best one can think of is that the amount of operations is proportional to the number of unknowns. This requires that each iteration should have that property and that the number of iterations to converge should be independent of the number of unknowns. The multi-grid method achieves this.

- The eigenvector of the dominant eigenvalue of a matrix A can be found from power iteration $A^n \mathbf{v}$, where $\mathbf{v}$ is arbitrary. By avoiding over- or under-flow in this process by scaling, one ends up with the power method. The power method can be generalized to iteration on sub-spaces leading to the eigenvectors of the dominant eigenvalues of A. This leads to the orthogonal sub-spaces iteration method. By replacing A by $(A - \sigma I)^{-1}$ one makes the eigenvalues of A close to σ dominant. Applying this in the preceding methods is called inverse iteration.

7.8 Exercises

Exercise 7.1 *Show that* $\mathbf{d}^{(n)} = \mathbf{x}^{(n)} - \mathbf{x}^{(n-1)}$ *also satisfies the relations (7.4) and (7.5).*

Exercise 7.2 *Determine the eigenvalues and eigenvectors of*

$$
\begin{bmatrix} -\frac{1}{2} & 1 \\ -1 & \frac{3}{2} \end{bmatrix}.
$$

Exercise 7.3 *Suppose we have a discretized elliptic equation and you have to advise on the method to choose for solving this equation. Which is the prefered method? Motivate your answer.*

Exercise 7.4

a. *Show that $\theta^{(i)}$ in the power method is the eigenvalue of the Galerkin projection of $A - \lambda I$ on the space spanned by $\mathbf{v}^{(i)}$.*

b. *Suppose we replace the matrix A in the power method by a shifted variant $A - \sigma I$. Assume that A is real and symmetric positive definite. If you already know the eigenvalues of A, how would you choose σ in order to let the power method converge fastest to the eigenvector associated to the largest (here including sign) eigenvalue of A? And how would you choose it for the eigenvector corresponding to the eigenvalue which is smallest?*

Exercise 7.5

a. *Give the Galerkin projection of the eigenvalue problem $A\mathbf{x} = \lambda\mathbf{x}$ on the space $V_m^{(n)}$ generated by simultaneous iteration and identify in that the matrix $\hat{B}^{(n)}$ whose eigenvalues approximate the biggest eigenvalues of A.*

b. *Show that the eigenvalues of $B^{(n)}$ in orthogonal sub-space iteration are the eigenvalues of the Galerkin approximation of the eigenvalue problem $A\mathbf{x} - \lambda\mathbf{x}$ on $V_m^{(n)}$. Show also that the eigenvalues are the same as those of $\hat{B}^{(n)}$, determined in the previous part, if in simultaneous iteration we start off both methods with the same (orthogonal) space $V_m^{(1)}$.*

8

Non-stationary Iterative Methods

Making use of history
may help a lot.

In this chapter we will describe several Krylov sub-space methods for the iterative solution of eigenvalue problems and linear systems of equations. In addition to the Krylov methods we will also present the Jacobi–Davidson method for eigenvalue problems and discuss a number of pre-conditioning techniques. We first discuss the Krylov sub-space, then the methods for eigenvalue problems and finish with methods for solving linear systems.

8.1 Krylov Sub-spaces

In this section, we introduce the Krylov sub-space and mention some of its properties. Next, we consider the construction of an orthogonal basis for it.

8.1.1 Definition and Properties

When we keep all the vectors of the power iteration (see Section 7.4), these vectors will span the *Krylov sub-space*, that is, after $m - 1$ applications of A to a vector $\mathbf{v}^{(1)}$ we have

$$\mathcal{K}^m(A; \mathbf{v}^{(1)}) \equiv \text{span}\{\mathbf{v}^{(1)}, A\mathbf{v}^{(1)}, \ldots, A^{m-1}\mathbf{v}^{(1)}\}.$$

Before discussing specific methods, we discuss a number of properties of this space. We assume that the Krylov sub-space has been generated in $\mathbb{C}^N$.

Properties:

1. In exact arithmetic, the dimension of the space will be at most N because this is the dimension of the whole space. So, for $m \geq N$ the vectors will be

dependent. However, there are special cases where the dimension is (much) lower.

2. Suppose A has simple eigenvalues, then the maximum dimension of the space $\mathcal{K}^m(A; \mathbf{v}^{(1)})$ is k if the starting vector $\mathbf{v}^{(1)}$ has components only in the directions of eigenvectors corresponding to k *different* eigenvalues. Furthermore, if A has only k different eigenvalues and $\mathbf{v}^{(1)}$ arbitrary, $\mathcal{K}^m(A; \mathbf{v}^{(1)})$ will reach its maximum dimension anyway after k steps. An illustrative example here is $A = I$. Then the dimension of the space is 1. *See example Ex. 8.1a*

3. If an eigenvalue of A is not simple, then the maximum dimension of the Krylov sub-space is reached after at most $p - 1$ extra steps where p is the size of the associated Jordan block (see (7.6)). To see the logic of this, we remark that for any ϵ there is a diagonizable matrix $A(\epsilon)$ with $||A(\epsilon) - A|| \leq \epsilon$ creating p distinct eigenvalues from the single eigenvalue of the Jordan block. Hence, it also gives rise to $p - 1$ extra steps to reach the maximum dimension of the Krylov sub-space according to the previous property. *See example Ex. 8.1b*

4. From the definition of $\mathcal{K}^m(A; \mathbf{v}^{(1)})$, it is clear that

$$\mathcal{K}^m(A - \sigma I; \mathbf{v}^{(1)}) = \mathcal{K}^m(A; \mathbf{v}^{(1)}). \tag{8.1}$$

This shows that the Krylov sub-space is the same when A is shifted. The implication is that the rate of convergence of eigenvectors of A in $\mathcal{K}^m(A; \mathbf{v}^{(1)})$, for $m = 1, 2, 3, \ldots$, is invariant under translations of A.

Additional Material

- An introduction to Krylov methods and their applications can be obtained from van der Vorst (2003), Saad (2003), and Liesen and Strakoš (2013).

8.1.2 Basis Construction

The basis $\mathbf{v}^{(1)}, A\mathbf{v}^{(1)}, \ldots, A^{m-1}\mathbf{v}^{(1)}$ for $K^m(A; \mathbf{v}^{(1)})$ has a flaw when rounding is involved. Obviously vectors $A^n\mathbf{v}^{(1)}$ point more and more in the direction of the dominant eigenvector for increasing n (the power iteration), and hence the components of other eigenvectors get lost in the rounding eventually. This will lead to dependency in the basis. As with simultaneous iteration, treated in Section 7.5, the way out is to generate an orthogonal basis for the Krylov sub-space, which was first described by Arnoldi (1951). He used a modified Gram–Schmidt procedure (Golub and van Loan, 1996; Quarteroni *et al.*, 2007) to generate the orthonormal basis simultaneously with creating the basis of the Krylov sub-space; it is given in Algorithm 6.

Algorithm 6 The Arnoldi algorithm with modified Gram–Schmidt to construct the reduced matrix H of A

$\mathbf{v}^{(1)}$ is a starting vector with $||\mathbf{v}^{(1)}|| = 1$

for $j = 1, \ldots, m-1$

$\quad \mathbf{z}^{(j+1)} = A\mathbf{v}^{(j)}$

$\quad$ **for** $i = 1, \ldots, j$

$\quad\quad h_{i,j} = (\mathbf{v}^{(i)})^* \mathbf{z}^{(j+1)}$

$\quad\quad \mathbf{z}^{(j+1)} = \mathbf{z}^{(j+1)} - h_{i,j}\mathbf{v}^{(i)}$

$\quad$ **end**

$\quad h_{j+1,j} = ||\mathbf{z}^{(j+1)}||_2$

$\quad \mathbf{v}^{(j+1)} = \mathbf{z}^{(j+1)}/h_{j+1,j}$

end

Observe that in the algorithm (Algorithm 6) the coefficients $h_{i,j}$ of a matrix H are being computed. These allow one to write the relation constructed in the algorithm in a compact form by

$$AV_{m-1} = V_m H_{m,m-1}, \qquad (8.2)$$

where $V_j \equiv [\mathbf{v}^{(1)}, \mathbf{v}^{(2)}, \ldots, \mathbf{v}^{(j)}]$. From the algorithm, we also see that the m by $m-1$ matrix $H_{m,m-1}$ is upper Hessenberg, in fact a pictorial view of (8.2) is as follows:

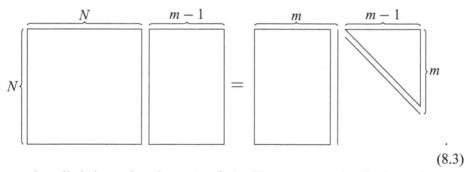

$$(8.3)$$

$H_{m,m-1}$ is called the *reduced matrix* of A with respect to the Krylov sub-space $\mathcal{K}^m(A; \mathbf{v}^{(1)})$. In fact, we have that $V_m^T A V_{m-1} = H_{m,m-1}$ and hence the Arnoldi algorithm is a variant of a Galerkin projection, where the test space has a bigger dimension than the search space. Note that the orthogonalization becomes increasingly expensive for increasing dimension of the sub-space, since the inner loop, running over i, increases in length each next iteration.

In the presence of the round-off errors, the Arnoldi algorithm is vulnerable to loss of orthogonality in the basis V_m. For example, if $h_{j+1,j}$ is nearly zero, then $\mathbf{z}^{(j+1)}$ will be largely contaminated by round-off errors due to the Gram–Schmidt process. These errors will be multiplied to find $\mathbf{v}^{(j+1)}$, and hence this vector will not be perpendicular to the space spanned by $\mathbf{v}^{(1)}, \ldots, \mathbf{v}^{(j)}$. The outcome is that the matrix $H_{m,m-1}$ is not very accurate, which has consequences for the results we like to deduce from it.

8.2 Arnoldi's Method

As we have seen in Section 8.1.2, the construction of a basis with Arnoldi's algorithm for the Krylov sub-space $\mathcal{K}^m(A; \mathbf{v}^{(1)})$ leads to an upper Hessenberg matrix that describes the relation between the basis vectors. If we increase the indices in (8.2) by one, we obtain

$$AV_m = V_{m+1}H_{m+1,m}, \tag{8.4}$$

which we do here for convenience. If we use as test space the space generated by the columns of V_m, then because of orthogonality we have that

$$V_m^T A V_m = V_m^T [V_m, \mathbf{v}^{(m+1)}] H_{m+1,m} = [I, \mathbf{0}] H_{m+1,m} = H_{m,m}. \tag{8.5}$$

Hence, $H_{m,m}$ is the Galerkin projection of A onto $K^m(A; \mathbf{v}^{(1)})$. Assume that $(\theta, \hat{\mathbf{y}})$ is an eigenpair of $H_{m,m}$; then

$$H_{m,m}\hat{\mathbf{y}} = \theta \hat{\mathbf{y}}.$$

Expanding $H_{m,m}$ gives respectively

$$V_m^T A V_m \hat{\mathbf{y}} - \theta V_m^T V_m \hat{\mathbf{y}} = 0,$$
$$V_m^T (A V_m \hat{\mathbf{y}} - \theta V_m \hat{\mathbf{y}}) = 0, \text{ and}$$
$$V_m^T (A\mathbf{y} - \theta \mathbf{y}) = 0,$$

where $\mathbf{y} = V_m \hat{\mathbf{y}}$. So, the eigenvalues θ of the matrix $H_{m,m}$ are the eigenvalues of the (Galerkin) projected eigenvalue problem (cf. Section 3.4). Also here, the value θ is called a *Ritz value* of A with respect to the Krylov sub-space $K^m(A; \mathbf{v}^{(1)})$, and $\mathbf{y}$ is the corresponding *Ritz vector. See example Ex. 8.2*

As mentioned earlier, the Arnoldi algorithm in Algorithm 6 needs to store the whole basis which is increased by one vector in each step. To restrict both memory consumption and computational work, re-starts are necessary. Several ways to do this have been proposed in the past, but currently the most widely used one is an implicit re-start process devised by Sorensen (1992). We will not go into the details of this method here.

For symmetric matrices, we will show that the convergence to the two extreme eigenvalues occurs from the inside and is monotonous. Recall that for a real symmetric matrix it holds that

$$\max_{\mathbf{x}} \frac{(\mathbf{x}, A\mathbf{x})}{(\mathbf{x}, \mathbf{x})} = \max_{\lambda \in \sigma(A)} \lambda$$

and likewise for the minimum. This is just a special case of the Courant–Fisher min-max theorem (Horn and Johnson, 1985). Now we have that

$$\max_{\theta \in \sigma(H)} \theta = \max_{\hat{\mathbf{x}}} \frac{(\hat{\mathbf{x}}, H\hat{\mathbf{x}})}{(\hat{\mathbf{x}}, \hat{\mathbf{x}})} = \max_{\hat{\mathbf{x}}} \frac{(\hat{\mathbf{x}}, V^T A V \hat{\mathbf{x}})}{(\hat{\mathbf{x}}, V^T V \hat{\mathbf{x}})}$$

$$= \max_{\mathbf{x} \in V} \frac{(\mathbf{x}, A\mathbf{x})}{(\mathbf{x}, \mathbf{x})} \leq \max_{\mathbf{x}} \frac{(\mathbf{x}, A\mathbf{x})}{(\mathbf{x}, \mathbf{x})} = \max_{\lambda \in \sigma(A)} \lambda.$$

This shows that the maximum Ritz value will converge monotonously from below to the maximum eigenvalue while increasing the dimension of the Krylov sub-space. Since the Krylov sub-space is just an extension of the power method, we know that the eigenvector related to the maximum eigenvalue is one of the first to become strongly present in the sub-space. Or, more precisely, the largest Ritz value of A will converge faster to the largest eigenvalue of A than the Rayleigh quotient does in the power method. But this is not all; due to the invariance of the Krylov sub-space for shifting of A (Section 8.1.1, Property 4) the eigenvalues also which can be made extreme by shifting can be found faster than applying the power method to the corresponding shifted matrix. So, in the real symmetric case also the minimum Ritz value will converge (among the first) to the minimum eigenvalue. The preceding can be extended to normal matrices, and to some extent it also holds for non-normal matrices, depending on the so-called departure of normality (Golub and van Loan, 1996).

One can show that the largest eigenvalues converge faster than what can be obtained by an optimal shift σ in the power method. Note that the vector $(A - \sigma I)^{m-1} \mathbf{v}^{(1)}$ is in the Krylov sub-space and will converge, according to our knowledge of the power iteration in Section 7.4, to the eigenvector corresponding to the largest eigenvalue in modulus of $A - \sigma I$. One can optimize the shift σ and find that for such an eigenvalue the associated eigenvector converges at least as fast as the optimal shift one can perform in the power method. For example, consider a symmetric matrix with eigenvalues $\lambda_1 \leq \lambda_2 \leq \cdots \leq \lambda_N$ (see Fig. 8.1); then the eigenvalues of the shifted matrix $A - \sigma I$ are $\lambda_i - \sigma$, for $i = 1, \ldots, N$. By choosing $\sigma = (\lambda_2 + \lambda_N)/2$, we force that $\lambda_2 - \sigma$ and $\lambda_N - \sigma$ have the same magnitude and

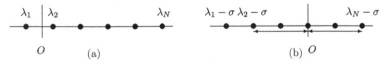

Figure 8.1 Original (a) and shifted eigenvalues (b) such that the average of λ_2 and λ_N is shifted to the origin, that is, $\sigma = (\lambda_2 + \lambda_N)/2$.

become the next biggest eigenvalues. According to the theory in Section 7.4, the convergence rate becomes

$$\frac{|\lambda_2 - \sigma|}{|\lambda_1 - \sigma|}.$$

This shows that if λ_2 is close to σ, that is, the eigenvalues $\lambda_2, \ldots, \lambda_N$ are close, and λ_1 is not, then there is a fast convergence to the eigenvector associated to λ_1.

Note that we can do the same to analyze the convergence of the eigenvector corresponding to λ_N in the Krylov sub-space. Then the optimal shift is $\sigma = (\lambda_1 + \lambda_{N-1})/2$. In this symmetric case there are only two eigenvalues that can be made largest in modulus. In non-symmetric cases there may be more, for example if all eigenvalues are on a circle in the complex plane.

The shift invariance of the Krylov sub-space definition implies that, instead of speaking of eigenvalues largest in modulus, one uses the notions *exterior* and *interior* eigenvalues, the former being the ones that can be made the largest in modulus by a suitable shift while the latter cannot. In general, the eigenvectors corresponding to the exterior eigenvalues converge first in the Krylov sub-space, but this will also depend on the given initial vector $\mathbf{v}_1$; see also Property 2 in Section 8.1.1. *See example Ex. 8.3.a*

Recall that the preparatory step for the QR-method to compute all eigenvalues of full matrices (see Section 6.5) is to bring the matrix to Hessenberg form. The Arnoldi approach performs this for sparse matrices. Note that in both cases it would be desirable to make the sub-diagonal entries as small as possible, leading to good approximations of the eigenvalues on the diagonal.

As explained at the end of the previous section, the loss of orthogonality in the basis of the Krylov sub-space leads to an inaccurate matrix $H_{m-1,m-1}$ and hence to inaccurate eigenvalues. In the extreme case, one can even find an approximate eigenvalue which was found already before. Remedies have been formulated to overcome this problem (Parlett, 1980, 1990).

Additional Material
• Much of the background linear algebra required to understand this chapter can be found in Appendix B.

8.3 The Lanczos Method

For symmetric real matrices the Arnoldi method can be simplified, since, if A is symmetric and real, then so is

$$H_{m-1,m-1} = V_{m-1}^T A V_{m-1}.$$

Algorithm 7 The Lanczos algorithm for the construction of the reduced matrix

---

$\mathbf{v}^{(1)}$ is a starting vector with $||\mathbf{v}^{(1)}|| = 1$
Set $\mathbf{v}_0 = 0$ and $\beta_0 = 0$,
for $j = 1, 2, \ldots, m-1$
$\quad \mathbf{z}^{(j+1)} = A\mathbf{v}^{(j)} - \beta_{j-1}\mathbf{v}^{(j-1)}$
$\quad \alpha_j = (\mathbf{v}^{(j)})^T \mathbf{z}^{(j+1)}$
$\quad \mathbf{z}^{(j+1)} = \mathbf{z}^{(j+1)} - \alpha_j \mathbf{v}^{(j)}$
$\quad \beta_j = ||\mathbf{z}^{(j+1)}||_2$
$\quad \mathbf{v}^{(j+1)} = \mathbf{z}^{(j+1)}/\beta_j$
end

Hence, since $H_{m-1,m-1}$ and $H_{m-1,m-1}^T$ are equal, they both should be upper Hessenberg and therefore tridiagonal. Herewith, (8.2) turns into

$$AV_{m-1} = V_m T_{m,m-1}, \qquad (8.6)$$

where the matrix $T_{m,m-1}$ is an m by $m-1$ tridiagonal matrix with its leading $(m-1) \times (m-1)$ part symmetric. The jth column of this expression is

$$A\mathbf{v}^{(j)} = [\mathbf{v}^{(j-1)}, \mathbf{v}^{(j)}, \mathbf{v}^{(j+1)}] \begin{bmatrix} \beta_{j-1} \\ \alpha_j \\ \beta_j \end{bmatrix}$$

which can be re-written in the three-term recurrence

$$\beta_j \mathbf{v}^{(j+1)} = A\mathbf{v}^{(j)} - \beta_{j-1}\mathbf{v}^{(j-1)} - \alpha_j \mathbf{v}^{(j)}$$

where α_j assures that $\mathbf{v}^{(j+1)}$ is orthogonal to $\mathbf{v}^{(j)}$ and β_j is a normalization. The resulting Lanczos algorithm is given in Algorithm 7.

The miraculous consequence of this method is that in contrast to the Arnoldi algorithm where each new vector has to be orthogonalized with respect to all the previous vectors, here it needs to be made orthogonal to the last vector since it is apparently already orthogonal to all the other vectors in the space. So, contrary to the Arnoldi algorithm we do not need to store the whole hitherto computed space but only the last two vectors. Hence, the memory consumption does not increase during the process. Also the number of computations remains the same in each step of the process and the computation time does also not increase per step.

The computation of approximate eigenvalues here is similar to that of the Arnoldi method discussed in the previous section, that is, we can find Ritz values by computing the eigenvalues of $T_{m,m}$. However, we cannot compute the Ritz vector immediately since we have dropped all but the last two vectors in the space.

However, by inverse power iteration using shift-and-invert as treated at the end of Section 7.4, one quickly finds the associated eigenvector.

8.4 The Two-Sided Lanczos Method

A generalization of the Lanczos method for non-symmetric matrices has been developed in order to avoid the increasing storage demand and computational work in the orthogonalization process needed in Arnoldi's method. This generalization is known as the *two-sided Lanczos method*, or Bi-Lanczos method. This method is in fact a Petrov–Galerkin approach (see Section 3.4.2) to determine approximate eigenpairs. Instead of only one set of basis vectors contained in V_m there is another set contained in W_m, where V_m and W_m are bi-orthogonal, that is, $W_m^T V_m$ is a diagonal matrix. In this section, we will restrict ourselves to the real case. Following Wilkinson (1965, Chapter 6.36), both spaces are generated by Arnoldi-like recursions in order to create bases for the respective Krylov sub-spaces $K^m(A; \mathbf{v}^{(1)})$ and $K^m(A^T; \mathbf{w}^{(1)})$:

$$h_{j+1,j}\mathbf{v}^{(j+1)} = A\mathbf{v}^{(j)} - \sum_{i=1}^{j} h_{i,j}\mathbf{v}^{(i)},$$

$$g_{j+1,j}\mathbf{w}^{(j+1)} = A^T\mathbf{w}^{(j)} - \sum_{i=1}^{j} g_{i,j}\mathbf{w}^{(i)},$$

and require that $\mathbf{v}^{(j+1)}$ is orthogonal to all previous $\mathbf{w}^{(i)}$ and that $\mathbf{w}^{(j+1)}$ is orthogonal to all previous $\mathbf{v}^{(i)}$. Clearly, this defines, apart from the constants $h_{j+1,j}$ and $g_{j+1,j}$, the vectors $\mathbf{v}^{(j+1)}$ and $\mathbf{w}^{(j+1)}$, once the previous vectors are given. In matrix form the recursions can be written as

$$AV_j = V_{j+1}H_{j+1,j} \quad \text{and} \quad A^T W_j = W_{j+1}G_{j+1,j}. \tag{8.7}$$

Now the following holds.

Theorem 8.1 *The matrices $G_{j,j}$ and $H_{j,j}$ are tridiagonal and related through a non-singular diagonal matrix D as $D_j H_{j,j} = G_{j,j}^T D_j$. Furthermore, D can be chosen such that $G_{j,j} = H_{j,j}$ and that $D_j H_{j,j}$ symmetric.*

Proof First, we show that indeed V_j and W_j are bi-orthogonal. Since each new $\mathbf{v}^{(j+1)}$ is only orthogonal with respect to the $\mathbf{w}^{(i)}$, for $i < j$, and likewise for $\mathbf{w}^{(j+1)}$ with respect to the $\mathbf{v}^{(i)}$, it follows that

$$W_j^T V_j = L_{j,j} \quad \text{and} \quad V_j^T W_j = K_{j,j},$$

where $L_{j,j}$ and $K_{j,j}$ are lower triangular. Clearly, since the left-hand sides are each other's transpose,

$$K_{j,j}^T = L_{j,j},$$

so that both matrices must be diagonal and hence $K_{j,j} = L_{j,j}$ which we will denote by D_j. This shows the bi-orthogonality of V_j and W_j.

Next, by pre-multiplying the two equations in (8.7) with W_j^T and V_j^T, respectively, we find

$$W_j^T A V_j = D_j H_{j,j}$$

and

$$V_j^T A^T W_j = D_j G_{j,j}.$$

Hence, since the left-hand sides are each other's transpose, $D_j H_{j,j} = G_{j,j}^T D_j$. This shows that $H_{j,j}$ and $G_{j,j}$ must be tridiagonal and we write H as T from now on. So, this means that only the last few columns of V and W need to be retained to construct T.

There is still a freedom here which can be exploited: the entries of D. One can choose it such that $D_j T_{j,j}$ is symmetric, so we require that $d_i t_{i,i+1} = d_{i+1} t_{i+1,i}$. Hence, setting $d_1 = 1$, the diagonal entries follow from $d_{i+1} = d_i t_{i,i+1,i}/t_{i+1,i}$ for $i = 1, \ldots, j - 1$, provided $t_{i+1,i}$ is not zero. But if it is zero, then also $t_{i,i+1} = 0$ due to the observed tridiagonal structure of $T_{j,j}$.

This means that also $G_{j,j}^T D_j$ is symmetric; hence $D_j H_{j,j} = G_{j,j}^T D_j = D_j G_{j,j}$ and thus $G = H = T$.   ∎

So, in the non-symmetric Lanczos method (Lanczos, 1950), we generate dual bases from a three-term recurrence relation with A,

$$\gamma_j \mathbf{v}^{(j+1)} = A \mathbf{v}^{(j)} - \alpha_j \mathbf{v}^{(j)} - \beta_{j-1} \mathbf{v}^{(j-1)},$$

and from one with A^T,

$$\gamma_j \mathbf{w}^{(j+1)} = A^T \mathbf{w}^{(j)} - \alpha_j \mathbf{w}^{(j)} - \beta_{j-1} \mathbf{w}^{(j-1)}.$$

So, the jth column and row of the tridiagonal matrix T is $[\beta_{j-1}, \alpha_j, \gamma_{j+1}]$ where β_{j-1} has been computed in the previous step, α_j will be determined by the requirement that $\mathbf{v}^{(j+1)}$ should be perpendicular to $\mathbf{w}^{(j)}$, γ_{j+1} such that $\|\mathbf{v}^{(j+1)}\|_2 = 1$. Next, we compute the diagonal element of D, that is, $\delta_{j+1} = (\mathbf{w}^{(j)})^T \mathbf{v}^{(j)}$. Finally, we require that DT becomes symmetric, hence from $\delta_{j+1} \gamma_{j+1} = \delta_j \beta_j$ which gives us β_j for the next step. In Algorithm 8, these operatorions are expressed in an algorithm. Note that the algorithm may fail if $\delta_{j+1} = 0$, but then also β_j will be zero, as shown in the proof of Theorem 8.1. It indicates that we have found an invariant sub-space, that is, a space which is mapped onto itself by A. One could stop the iteration then. However, in finite precision round-off errors may cause that δ_{j+1} is nearly zero in such a case and measures have to be taken to avoid failure of the method; see, for example, Freund and Nachtigal (1990).

Algorithm 8 The two-sided Lanczos algorithm

Select a normalized pair $\mathbf{v}^{(1)}$, $\mathbf{w}^{(1)}$ (for instance, $\mathbf{w}^{(1)} = \mathbf{v}^{(1)}$)
such that $(\mathbf{w}^{(1)})^T \mathbf{v}^{(1)} = \delta_1 \neq 0$
$\beta_0 = 0$, $\mathbf{w}_0 = \mathbf{v}_0 = 0$
for $j = 1, 2, \ldots$.
$\qquad \mathbf{z} = A\mathbf{v}^{(j)} - \beta_{j-1}\mathbf{v}^{(j-1)}$
$\qquad \alpha_j = (\mathbf{w}^{(j)})^T \mathbf{z}/\delta_j$
$\qquad \mathbf{z} = \mathbf{z} - \alpha_j \mathbf{v}^{(j)}$
$\qquad \gamma_{j+1} = \|\mathbf{z}\|_2$
$\qquad \mathbf{v}_{j+1} = \mathbf{z}/\gamma_{j+1}$
$\qquad \mathbf{w}_{j+1} = (A^T \mathbf{w}^{(j)} - \beta_{j-1}\mathbf{w}^{(j-1)} - \alpha_j \mathbf{w}^{(j)})/\gamma_{j+1}$
$\qquad \delta_{j+1} = (\mathbf{w}^{(j+1)})^T \mathbf{v}^{(j+1)}$
$\qquad \beta_j = \gamma_{j+1}\delta_{j+1}/\delta_j$
end;

For the computation of approximate eigenpairs, we may proceed in a similar way as in the standard Lanczos case. We have that

$$AV_j = V_{j+1} T_{j+1,j},\qquad(8.8)$$

but now we will use the matrix $W_j = [\mathbf{w}^{(1)}, \mathbf{w}_2, \ldots, \mathbf{w}^{(j)}]$ as a test space. Hence, we will have a Petrov–Galerkin approach for the eigenvaue problem. Let $V_j\hat{\mathbf{y}}$ be the approximate eigenvector in the space spanned by the columns of V_j, with corresponding eigenvalue approximation θ; then we impose that the residual is perpendicular to W_j, that is,

$$W_j^T(AV_j\hat{\mathbf{y}} - \theta V_j\hat{\mathbf{y}}) = 0,$$

or

$$W_j^T AV_j\hat{\mathbf{y}} - \theta W_j^T V_j\hat{\mathbf{y}} = 0.$$

This shows that $\hat{\mathbf{y}}$ and θ form an eigenpair of the generalized symmetric eigenproblem

$$D_j T_{j,j}\hat{\mathbf{y}} = \theta D_j\hat{\mathbf{y}},\qquad(8.9)$$

or of the standard non-symmetric eigenproblem

$$T_{j,j}\hat{\mathbf{y}} = \theta\hat{\mathbf{y}}.$$

The eigenvalue θ of this eigenvalue problem is called the *Petrov value* and the approximate right eigenvector $\mathbf{y} = V_j\hat{\mathbf{y}}$ the *right Petrov vector*.

An interesting aspect of the two-sided Lanczos approach is that it also admits possibilities for the approximation of left eigenvectors. In this real case we just can take the transpose of (8.9) and use the symmetry of $D_j T_{j,j}$ to obtain

$$\hat{\mathbf{y}}^T D_j T_{j,j} = \theta \hat{\mathbf{y}}^T D_j.$$

Working our way up, we obtain

$$(\hat{\mathbf{y}}^T W_j^T A - \theta \hat{\mathbf{y}}^T W_j^T) V_j = 0.$$

Hence, we have a left approximate eigenvector $\mathbf{z} = W\hat{\mathbf{y}}$ which is called the *left Petrov vector*.

8.5 The Jacobi–Davidson Method

In this section, we discuss the Jacobi–Davidson method as proposed by Fokkema and colleagues (1998). This eigensolution method is *not* a Krylov-sub-space method but it uses a sub-space to accelerate the convergence. The exposition in this section starts from the familiar Newton method treated in Section 6.2.

Consider the ordinary eigenvalue problem $(A - \lambda I)\mathbf{x} = 0$ and suppose we have already computed the partial Schur form

$$AQ = QR,$$

where R is an upper-triangular matrix with the already known eigenvalues of A on its diagonal. Furthermore $Q^*Q = I$. Now we want to extend Q by a new vector $\mathbf{q}$ such that we get again a partial Schur form. Hence, we have the equation

$$A[Q\ \mathbf{q}] = [Q\ \mathbf{q}] \begin{bmatrix} R & \mathbf{s} \\ 0 & \lambda \end{bmatrix},$$

where additionally $\mathbf{q}$ should be perpendicular to the current Q and have length 1. This leads to the following set of equations for $\mathbf{q}$, $\mathbf{s}$, and λ:

$$(A - \lambda I)\mathbf{q} - Q\mathbf{s} = 0,$$
$$-(\mathbf{q}^*\mathbf{q})/2 + 1/2 = 0,$$
$$-Q^*\mathbf{q} = 0.$$

When using Newton's method to solve these equations, we define $\mathbf{u} \equiv (\mathbf{q}, \lambda, \mathbf{s})$ and assume we have a good initial guess $\mathbf{u}_0 = (\mathbf{q}_0, \lambda_0, \mathbf{s}_0)$ for $\mathbf{u}$. Then in the ith step of Newton's method, the following correction equation has to be solved:

$$\begin{bmatrix} A - \lambda^{(i)}I & -\mathbf{q}^{(i)} & -Q \\ -(\mathbf{q}^{(i)})^* & 0 & 0 \\ -Q^* & 0 & 0 \end{bmatrix} \begin{bmatrix} \Delta\mathbf{q}^{(i)} \\ \Delta\lambda^{(i)} \\ \Delta\mathbf{s}^{(i)} \end{bmatrix} = \begin{bmatrix} Q\mathbf{s}^{(i)} - (A - \lambda^{(i)}I)\mathbf{q}^{(i)} \\ ((\mathbf{q}^{(i)})^*\mathbf{q}^{(i)})/2 - 1/2 \\ Q^*\mathbf{q}^{(i)} \end{bmatrix}. \quad (8.10)$$

Algorithm 9 The Jacobi–Davidson method for the detection of the smallest eigenvalue in modulus of A

Initialize $\mathbf{t}$, and void spaces Q and V.

while dim(Q)$< k$

    repeat

        Orthonormalize $\mathbf{t}$ with respect to current Q and V and extend V with it.

        Construct the reduced matrix on the space spanned by V: $M = V^*AV$.

        Compute the smallest eigenvalue of M, say θ, and the corresponding eigenvector s.

        Compute the eigenpair residual: $\mathbf{r} = A\mathbf{q} - \theta\mathbf{q}$, where $\mathbf{q} = V\mathbf{s}$.

    until $\|\mathbf{r}\| \leq \epsilon$

    Extend Q by $\mathbf{q}$.

    Make V also orthogonal to $\mathbf{q}$

    Solve the correction equation for a new $\mathbf{t}$

endwhile

Next, one updates

$$\mathbf{u}^{(i+1)} = \mathbf{u}^{(i)} + \Delta\mathbf{u}^{(i)}. \tag{8.11}$$

In principle this algorithm will converge to one of the eigenpairs of A. In the following we will try to speed it up by using sub-space acceleration.

Exploiting the properties of the problem, we adapt the Newton correction equation a bit. First, we assume that $\mathbf{q}^{(i)}$ has already length one and is perpendicular to Q. Hence, the last two entries of the right-hand side of (8.10) become zero. Of course, we have to make sure that $\mathbf{q}^{(i+1)}$ also has these properties, which can be done by applying the Gram–Schmidt process after (8.11) has been performed. Moreover, observe that we could take $\mathbf{s}^{(i)} = 0$ by computing immediately $\mathbf{s}^{(i+1)}$ instead of $\Delta\mathbf{s}^{(i)}$.

Instead of using $\Delta\mathbf{q}^{(i)}$ in the update of $\mathbf{u}^{(i)}$ one can also add it to the sub-space spanned by the hitherto found corrections $\Delta\mathbf{q}^{(j)}$, $j = 1,\ldots,i - 1$, and compute the Ritz values of A with respect to this sub-space. For example, to compute the k smallest eigenvalues in modulus of A one can apply the algorithm in Algorithm 9. Here the columns of V span the space of the corrections, that is, $V = \text{span}\{\Delta q^{(1)}, \ldots, \Delta q^{(i-1)}\}$.

In practice a re-start strategy also is added in order to limit the dimension of the sub-space V. Moreover, variants for the reduced eigenvalue problem are also considered in order to avoid difficulties in computing interior eigenvalues. The correction equation (8.10) can also be solved approximately, for example, by applying an iterative method to solve the system. Only a few steps may suffice to get a useful correction to extend the space V. In the next section we will consider the generalized eigenvalue problem $A\mathbf{x} = \lambda B\mathbf{x}$, which commonly occurs in linear stability analyses of solutions in CFD problems. The Jacobi–Davidson method can straightforwardly be generalized to that case and is called JDQZ, since in that case we will work towards a partial generalized Schur form (see Section 6.5).

8.6 Generalized Eigenproblems

In this section several techniques are introduced to find a few eigenvalues near a target for the generalized eigenvalue problem

$$A\mathbf{x} = \lambda B\mathbf{x} \tag{8.12}$$

where A and B are in $\mathbb{R}^{n \times n}$. For the dense generalized eigenvalue problem the QZ method has been devised, a generalization of the QR method; see Section 6.5. This is still the most powerful method to find all the eigenvalues, but it is limited to moderate-size problems because of the needed computational work and the required memory. *See example Ex. 8.4*

Generalized problems can be solved by the orthogonal sub-space iteration, and the Arnoldi, Lanczos, and two-sided Lanczos methods by transforming the problem into an appropriate standard eigenvalue problem such that the desired eigenvalues are transformed to the largest in modulus. These methods only require the user to define the matrix-vector multiplication using the transformed matrix. In general, the matrix-vector multiplication will require in each iteration step the solution of a linear system with A or B, or a combination of A and B. *See example Ex. 8.5*

In the following we present several approaches for the transformation of (8.12) to a standard eigenvalue problem.

1. **Shift-and-Invert**: For the generalized eigenvalue problem one can also use the shift-and-invert technique introduced at the end of Section 7.4 to find eigenvalues near a target. To derive it we first transform (8.12) to a standard eigenvalue problem by pre-multiplying it by the inverse of B, assuming B is non-singular. This leads to

$$(B^{-1}A)\mathbf{x} = \lambda\mathbf{x}. \tag{8.13}$$

Now we apply shift-and-inverse to this eigenvalue problem and find the eigenvalue problem

$$(B^{-1}A - \sigma I)^{-1}\mathbf{x} = \mu\mathbf{x}, \tag{8.14}$$

where μ and λ are related through

$$\frac{1}{\lambda - \sigma} = \mu. \tag{8.15}$$

We can re-write (8.14) as

$$(A - \sigma B)^{-1} B\mathbf{x} = \mu \mathbf{x} \tag{8.16}$$

where the inverse of B is gone. So apparently this eigenvalue problem is also defined for singular B, and indeed one can also derive it using a calculation without employing the inverse of B.

Using a Krylov sub-space method, one needs to evaluate $\mathbf{p} = (A - \sigma B)^{-1} B\mathbf{q}$ which we write as

$$(A - \sigma B)\mathbf{p} = B\mathbf{q}.$$

So, one first multiplies $\mathbf{q}$ by B and next solves $\mathbf{p}$ from a linear system with system matrix $A - \sigma B$. For the latter we just make once an LU factorization of $(A - \sigma B)$ before the process and use that in each step. An alternative is to use an iterative method to solve the linear system, but this should be solved to sufficient accuracy in order not to limit the attainable accuracy of the eigenvalues.

Note that when σ is an eigenvalue, $(A - \sigma B)$ is singular, and hence every time we apply the matrix $(A - \sigma B)^{-1} B$ to some vector the result will have a strong component in the direction of the assocociated eigenvector. This leads to loss of orthogonality in the basis of the Krylov sub-space and may produce inaccurate results. A special case is where we know beforehand that A is singular, so there is an eigenvalue zero, and that we also use a target $\sigma = 0$. In such a case it is better to shift the target σ away from the eigenvalue 0. *See example Ex. 8.6*

2. **Cayley transform**: The Cayley transform adds an extra freedom to target eigenvalues in a certain region in the complex plane. In this case, one considers the transformed eigenvalue problem

$$(A - \sigma B)^{-1}(A - \mu B)\mathbf{x} = \gamma \mathbf{x}, \tag{8.17}$$

where the eigenvalue $\gamma \in C$ is related to the original eigenvalue λ through

$$\gamma = \frac{\lambda - \mu}{\lambda - \sigma}. \tag{8.18}$$

In Fig. 8.2 a contour plot of γ over the complex plane is shown for $\sigma = -\mu = 100$. For an eigenvalue λ located at any spot in this plane, the colour contour indicates the magnitude of the corresponding eigenvalue γ after undergoing the Cayley transformation. We observe that γ gets small for λ close to μ and large

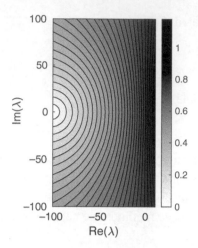

Figure 8.2 Contour plot over the complex plane of γ as a function of λ for $\sigma = -\mu = 100$.

for λ close to σ. So compared to shift-and-invert we can now also hinder the convergence to eigenvectors in the Krylov sub-space associated to eigenvalues close to μ.

Applying the matrix in (8.17) to a vector is similar to how it is done with Shift-and-Invert.

3. **General transform**: For a general transformation, we consider the rational function $q_{m,n}(x) = p_m(x)/p_n(x)$ where $p_m(x)$ is a polynomial of degree m. Then the standard eigenvalue problem $Ax = \lambda x$ can be transformed to $q_{m,n}(A)x = \mu x$ where $\mu = q_{m,n}(\lambda)$. The latter relation does not need to be unique for all μ, but if we want to re-compute λ from it, it should be unique for the values of μ that come out of the Arnoldi or Lanczos process. (An alternative is to use the transformation only to get a desired eigenspace. By a Galerkin projection of A on that space we get the associated eigenvalues.) To get to the generalized eigenvalue problem, we just replace A by $B^{-1}A$, assuming B to be non-singular for the moment. Let $p_m(0) = p_n(0) = 1$; then one can write $p_m(x) = \Pi_{i=1}^{m}(I - \alpha_i x)$ with $\alpha_i \neq 0$ and $p_n(x) = \Pi_{i=1}^{n}(I - \beta_i x)$ with $\beta_i \neq 0$. Furthermore, let $n \geq m$; then one finds

$$
\begin{aligned}
q_{m,n}(B^{-1}A) &= \left(\Pi_{i=1}^{m}(I - \beta_i B^{-1}A)^{-1}(I - \alpha_i B^{-1}A)\right)\Pi_{i=m+1}^{n}(I - \beta_i B^{-1}A)^{-1}\\
&= \left(\Pi_{i=1}^{m}(B - \beta_i A)^{-1}(B - \alpha_i A)\right)\Pi_{i=m+1}^{n}(B - \beta_i A)^{-1}B. \qquad (8.19)
\end{aligned}
$$

Note that the last expression does not explicitly need that B is non-singular and, moreover, the shift-and-invert and Cayley transform are just special cases for $(m, n) = (0, 1)$ and $(m, n) = (1, 1)$, respectively. *See example Ex. 8.7*

As alluded to earlier, in many cases one has to solve a large, sparse, linear system to sufficient accuracy to allow for sufficiently accurate eigenvalues. Therefore, inexact variants of Caley transforms have been proposed (Lehoucq and Meerbergen, 1998; Meerbergen, 1996; Lippert and Edelman, 2000). Note that highly accurate solutions are not needed in the Jacobi–Davidson method; see Section 8.5, where one only wants extensions of the search space that contribute to the eigenvector one is looking for.

Complex eigenvectors are determined up to a scaling and rotation. For uniqueness, one could rotate them such that the real part is maximized. So, let $q = q_R + iq_I$, then we multiply by $\exp(i\phi)$ for a ϕ such that, for instance, the entry k for which $|q_k| = ||q||_\infty$ becomes real and positive (so imaginary part zero). We have observed in experiments that the imaginary part of such an eigenvector can be close to zero. Note also that for real matrices, a zero imaginary part of an eigenvector means that we have a real eigenvalue.

8.7 Linear Systems

In this section we will shortly introduce how the Arnoldi, Lanczos, and two-sided Lanczos algorithms lead to well-known Krylov sub-space methods for linear systems. Before doing so, we discuss two special properties which hold for all methods presented here.

In the special case where we want to solve $A\mathbf{x} = \mathbf{b}$ and we set $\mathbf{v}^{(1)} = \mathbf{b}$, Property 2 from Section 8.1.1 means that the solution $\mathbf{x}$ will be in the Krylov sub-space $\mathcal{K}^k(A; \mathbf{v}^{(1)})$. Consequently, the solution $\hat{\mathbf{x}}$ of the projected linear problem

$$W^T A V \hat{\mathbf{x}} = W^T \mathbf{b}, \tag{8.20}$$

where the columns of V form a basis of $\mathcal{K}^m(A; \mathbf{v}^{(1)})$ and those of W of a space of the same dimension as V and such that $W^T A V$ is non-singular, will give us the solution $\mathbf{x} = V\hat{\mathbf{x}}$ of $A\mathbf{x} = \mathbf{b}$. Note that W can be arbitrary. In what follows we will see some choices one can make that lead to specific methods.

In Section 8.2 we discussed the convergence of eigenvectors in the space $\mathcal{K}^m(A; \mathbf{v}^{(1)})$ and remarked that the initial vector $\mathbf{v}^{(1)}$ will also play a role. If the component of an eigenvector is hardly present in $\mathbf{v}^{(1)}$, it may take a relatively large number of iterations to get this eigenvector sufficiently accurate in the Krylov sub-space. One special case in this regard is when there is an eigenvalue close to zero well separated from nearby positive eigenvalues. Furthermore, assume that A is a symmetric positive-definite matrix (SPD) (see Definition B.9) and that the component of the associated eigenvector has about the same magnitude in the solution of the problem $A\mathbf{x} = \mathbf{b}$ as those of its nearby eigenvalues, where the component of a vector $\mathbf{y}$ in a vector $\mathbf{x}$ is defined by $|(\mathbf{y}, \mathbf{x})|/||\mathbf{x}||$. Then in $\mathbf{b}$ this eigenvector is poorly represented as compared to those of the nearby eigenvalues. So, if the eigenvalues are widely spread, it may take a long while before the exterior eigenvalue close to 0 and its associated eigenvector converge in the Krylov sub-space. *See example Ex. 8.3.b*

Small eigenvalues are commonly introduced in problems where physically the solution is determined up to a constant, for example a Poisson equation with Neumann boundary conditions and the incompressible Navier–Stokes equations in a closed domain where the pressure is determined up to a constant. Often the solution is fixed at one point in such a case. For the Poisson equation in 3D on a regular grid with mesh-size h one finds (by, e.g., the Rayleigh quotient) that there will be an eigenvalue of $O(h)$, while the nearby eigenvalues are $O(1)$, hence slow convergence may be expected. The solution is to just iterate in the space orthogonal to that eigenvector which precludes the problem. This is an example of deflation (Gaul *et al.*, 2013; Ebadi *et al.*, 2016).

8.7.1 The Conjugate Gradient Method

We now turn to solving linear systems using Krylov sub-spaces and start with the conjugate gradient method. First, we make some general remarks on minimization and projection and then show how the conjugate gradient (CG) method can be derived using that. We will make use of the concepts introduced in Section 3.4.

Suppose we have found in some way an orthonormal basis for a linear sub-space of dimension m, denoted by the columns of the matrix V, for which it holds that the reduction of the SPD matrix A of order n can be written as a tridiagonal matrix: $T_m = V^T A V$. Now take some $\mathbf{x} \in \mathbb{R}^N$ and define for $\mathbf{y} \in \mathbb{R}^N$ the function

$$f(\mathbf{y}) = (\mathbf{x} - \mathbf{y}, A(\mathbf{x} - \mathbf{y})). \tag{8.21}$$

We want to find the minimum of this function (which is clearly $\mathbf{y} = \mathbf{x}$). Suppose that we can only take $\mathbf{y}$ from a sub-space $\mathcal{V}$ which is spanned by the columns of V. This leads to the Ritz method:

$$\mathbf{y}^* = \text{argmin}_{\mathbf{y} \in \mathcal{V}} (\mathbf{x} - \mathbf{y}, A(\mathbf{x} - \mathbf{y})). \tag{8.22}$$

This is the same procedure as finding $\mathbf{y}^* \in \mathcal{V}$ for which $((\mathbf{x} - \mathbf{y}^*), A\mathbf{y}) = 0$ for all $\mathbf{y} \in \mathcal{V}$. Here $\mathbf{y}^*$ is the projection of $\mathbf{x}$ on $\mathcal{V}$ under the inner product $(\mathbf{x}, A\mathbf{y})$.

We can write any $\mathbf{y}$ in $\mathcal{V}$ as $\mathbf{y} = V\hat{\mathbf{y}}$. With that, the condition to find the minimum turns into $(V^T A\mathbf{x} - V^T AV\hat{\mathbf{y}}^*, \hat{\mathbf{y}}) = 0$ for all $\hat{\mathbf{y}} \in \mathbb{R}^m$, or equivalently $(V^T A\mathbf{x} - T_m\hat{\mathbf{y}}^*, \hat{\mathbf{y}}) = 0$ for all $\hat{\mathbf{y}} \in \mathbb{R}^m$. Hence, we find $\hat{\mathbf{y}}^*$ from a linear system with a tridiagonal matrix, that is,

$$T_m\hat{\mathbf{y}}^* = V^T A\mathbf{x}. \tag{8.23}$$

The above basis V can be made in a variety of ways, for example using the Lanczos algorithm. *See example Ex. 8.8*

Suppose now that we want to solve $A\mathbf{x} = \mathbf{b}$, with A an SPD matrix. Then (8.21) turns into $f(\mathbf{y}) = (\mathbf{x}, \mathbf{b}) + (\mathbf{y}, A\mathbf{y}) - 2(\mathbf{y}, \mathbf{b})$; and since $\mathbf{x}$ and $\mathbf{b}$ are constants, the minimum of $f(\mathbf{y})$ will be the same as that of

$$g(\mathbf{y}) = (\mathbf{y}, A\mathbf{y}) - 2(\mathbf{y}, \mathbf{b}). \tag{8.24}$$

The minimization of $g(\mathbf{y})$ where $\mathbf{y} \in \mathcal{V}$ leads to the Galerkin approximation $T_m \hat{\mathbf{y}}^* = V^T \mathbf{b}$, which could also have been obtained from (8.23).

Now the following holds.

Lemma 8.1 *The matrix T_m admits a stable LU factorization without pivoting. Moreover, if V is extended with one vector, then T_m and each of its L and U factors can be extended by one row and one column. Finally, an iterative method exists that needs only to store the current solution $\mathbf{x}_m$, the current search vector, and the current residual.*

Proof Clearly, T_m is also SPD, and for this kind of matrices it is known that one can make a stable, with respect to round-off errors, LU factorization without pivoting (Quarteroni *et al.*, 2007, section 3.4.2). Now, suppose $T_m = LU$; then the system to be solved turns into $LU\hat{\mathbf{y}}^* = V^T\mathbf{b}$. When we define $\mathbf{x}_m = V\hat{\mathbf{y}}^*$ we can also write this equation as

$$\mathbf{x}_m = VU^{-1}L^{-1}V^T\mathbf{b} = \hat{V}\hat{\mathbf{b}}$$

where $\hat{V} = VU^{-1}$ and $\hat{\mathbf{b}} = L^{-1}V^T\mathbf{b}$. Note that $\hat{V}$ follows from the system $\hat{V}U = V$ and $\hat{\mathbf{b}}$ from $L\hat{\mathbf{b}} = V^T\mathbf{b}$. Hence, the columns of $\hat{V}$ are solved from left to right and the entries of $\hat{\mathbf{b}}$ are solved from top to bottom. Now, note that, if V is extended by one column, then as a result T_m and the LU-factorization get one extra row and column. Moreover, $\hat{V}$ gets a new column and $\hat{\mathbf{b}}$ a new entry at the bottom. Only the product of the new column of $\hat{V}$ and the new last entry of $\hat{\mathbf{b}}$ needs to be added to $\mathbf{x}_m$ in order to get $\mathbf{x}_{m+1}$. So, we need to store only $\mathbf{x}_m$ to be able to compute the new approximate solution $\mathbf{x}_{m+1}$. ∎

The CG algorithm is shown in Algorithm 10. Observe that α and β depend only on $\mathbf{z}$ and $\mathbf{r}$, and, moreover, that $\mathbf{z}^{(m)}$ with $m = 0, 1, 2, \ldots$ build the search space. Note also that the residual $\mathbf{r}$ is not directly computed from $\mathbf{b} - A\mathbf{x}$ though the relations (i) and (ii) are consistent. The method belongs to the class of gradient methods (Quarteroni *et al.*, 2007, Sections 7.2.2 and 7.2.4). These methods try to find the minimum of a function, here $g(\mathbf{y})$ in (8.24), by using the gradient of $g(\mathbf{y})$, which is up to a factor the residual $\mathbf{r} = \mathbf{b} - A\mathbf{y}$. The standard gradient method uses this direction to find a local minimum in that direction, and the CG method makes a combination of the previous search direction such that the new gradient

---

Algorithm 10 The CG algorithm

Choose $\mathbf{x}^{(0)}$, compute $\mathbf{r}^{(0)} = \mathbf{b} - A\mathbf{x}^{(0)}$ and set $\mathbf{z}^{(0)} = \mathbf{r}^{(0)}$

for $m = 0, 1, 2, \ldots$

$\quad \alpha_m = \frac{(\mathbf{r}^{(m)}, \mathbf{r}^{(m)})}{(\mathbf{z}^{(m)}, A\mathbf{z}^{(m)})}$

$\quad \mathbf{x}^{(m+1)} = \mathbf{x}^{(m)} + \alpha_m \mathbf{z}^{(m)}$

$\quad \mathbf{r}^{(m+1)} = \mathbf{r}^{(m)} - \alpha_m A\mathbf{z}^{(m)}$

$\quad \beta_m = \frac{(\mathbf{r}^{(m+1)}, \mathbf{r}^{(m+1)})}{(\mathbf{r}^{(m)}, \mathbf{r}^{(m)})}$

$\quad \mathbf{z}^{(m+1)} = \mathbf{r}^{(m+1)} + \beta_m \mathbf{z}^{(m)}$

endfor

---

will become perpendicular to the old gradient and even to all previous gradients, hence the name conjugate gradient method.

If the columns of V span the Krylov sub-space associated to A with initial vector $\mathbf{b}$, then this Ritz approach can be built into the conjugate gradient method (Hestenes and Stiefel, 1954) along the lines given in the proof of Lemma 8.1.

For the CG method it is shown (Golub and van Loan, 1996) that the following error estimate holds:

$$\|\mathbf{x} - \mathbf{x}_m\|_A \leq 2 \left(\frac{\sqrt{\kappa} - 1}{\sqrt{\kappa} + 1} \right)^m \|\mathbf{x}\|_A \tag{8.25}$$

where $\|\mathbf{x}\|_A = \sqrt{(\mathbf{x}, A\mathbf{x})}$ and it is assumed that $\mathbf{x}^{(0)} = 0$. Here, κ is the condition number in the 2-norm (see Definition B.3), that is,

$$\kappa = \|A\|_2 \|A^{-1}\|_2, \tag{8.26}$$

which is, due to the symmetry and positive definiteness, equal to the ratio of the largest and smallest eigenvalues of A. In applications considered in this textbook the condition number of the Jacobian matrices will be rather big, leading to slow convergence. Therefore, one would like to modify the system such that a new system matrix A will occur with a smaller condition number. This can be brought about by pre-conditioning which will be discussed in Section 8.8. *See example Ex. 8.9*

Next to the condition number κ, the relative position of the eigenvalues of A also is important. This derives immediately from the convergence of eigenvectors in the Krylov sub-space discussed in Section 8.2. The exterior eigenvalues converge first, and the speed of that convergence is at least that of the ratio of the second largest divided by a largest eigenvalue after an optimal shift. If the extreme eigenvalues are converged to some extent, the observed convergence is that based on the remaining eigenvalues. Repeating this argument, we see that the CG method will converge faster and faster during the iteration. So, in contrast with the stationary methods which converge linearly, the CG method converges super-linearly.

Additional Material

- An extensive treatment on the conjugate-gradient method can be found in van der Sluis and van der Vorst (1986).

If A is not symmetric but still positive definite, so $\mathbf{x}^T A \mathbf{x} > 0$ for all $\mathbf{x} \neq 0$, then there is no minimization process attached to the problem $A\mathbf{x} = \mathbf{b}$; but yet we can make the same Galerkin projection, leading to the problem of finding $\mathbf{y}^*$ in $\mathcal{V}$ such that $(\mathbf{b} - A\mathbf{y}^*, \mathbf{y}) = 0$ for all $\mathbf{y} \in \mathcal{V}$. We require now that V, as before containing an orthogonal basis for $\mathcal{V}$, is such that $H = V^T A V$ is a Hessenberg matrix which can be constructed by Arnoldi's algorithm. This could also be worked out in an iterative method, but here we have to store the whole Krylov space (which follows analogously from the proof of Lemma 8.1). This approach leads to the Full Orthogonalization Method (FOM). However, this method is not commonly used because there are alternatives, for example the GMRES method described later, which also work for general non-singular matrices.

8.7.2 BiCG and BiCGstab

Non-symmetric matrices can be dealt with using methods based on the two-sided Lanczos algorithm of Algorithm 8. Suppose we have thus constructed bases V and W such that $W^T A V$ is tridiagonal. We now want to find $\mathbf{y}^* \in \mathcal{V}$ (search space) such that $(\mathbf{b} - A\mathbf{y}^*, \mathbf{z}) = 0$ for all $\mathbf{z} \in \mathcal{W}$ (test space), leading to the solution of the reduced system $T\hat{\mathbf{y}}^* = W^T \mathbf{b}$. This can be put into an iterative method in very much the same way as the CG method, which leads to the BiCG method if we choose $\mathbf{b}$ as starting vector. The convergence of this method can be quite irregular and therefore the BiCGstab method was developed by van der Vorst (1992). Still, however, the method may break down because the matrix T from the Bi-Lanczos process becomes singular. Therefore, a generalization was made called BiCGstab(l) in which the last l vectors of V and W are used. This leads to a more robust method (Sleijpen and Fokkema, 1993).

8.7.3 GMRES

Suppose we have again the sub-space $\mathcal{V}$ of $\mathbb{R}^N$ with the columns of V as basis. We will now compute the least-squares solution of $A\mathbf{x} = \mathbf{b}$ on $\mathcal{V}$, that is, we are looking for the minimum $\hat{\mathbf{y}}$ of $||\mathbf{b} - AV\hat{\mathbf{y}}||_2$. The standard way to solve such a system is by making a QR factorization of AV, where Q is the orthogonalized form of AV and R is an upper triangular matrix. Herewith,

$$||\mathbf{b} - AV\hat{\mathbf{y}}||_2^2 = ||\mathbf{b} - QR\hat{\mathbf{y}}||_2^2 = ||Q^T\mathbf{b} - R\hat{\mathbf{y}}||_2^2 + ||Q^{\perp^T}\mathbf{b}||_2^2, \qquad (8.27)$$

where $Q^\perp$ is the orthogonal complement of Q, namely a matrix such that $[Q, Q^\perp]$ is a square orthogonal matrix. The first term in (8.27) after the last equality sign can be made zero by solving $R\hat{\mathbf{y}} = Q^T\mathbf{b}$. The remaining term gives the square of the norm of the residual, hence a measure for the accuracy of the found approximation $V\hat{\mathbf{y}}$. We don't need to compute the complement $Q^\perp$ since $Q^{\perp^T}\mathbf{b} = (I - QQ^T)\mathbf{b} = \mathbf{b} - Q(Q^T\mathbf{b})$.

If the basis of $\mathcal{V}$ is generated by the Arnoldi algorithm, then this leads to the Generalized Minimal Residual (GMRES) method developed by Saad and Schultz (1986). The preceding can be optimized a bit in that case. Suppose V is generated by the Arnoldi algorithm with initial vector $\mathbf{b}$; hence it holds that $AV_k = V_{k+1}\tilde{H}$ where $\tilde{H}$ is a $(k + 1) \times k$ Hessenberg matrix and that $\mathbf{b} = ||\mathbf{b}||_2 V_{k+1}\mathbf{e}_1$. Using this, we can write

$$||\mathbf{b} - AV_k\hat{\mathbf{y}}||_2 = ||\,||\mathbf{b}||_2 V_{k+1}\mathbf{e}_1 - V_{k+1}\tilde{H}\hat{\mathbf{y}}||_2. \qquad (8.28)$$

Now we again use that the norm is invariant under orthogonal transformations. Here, we take $[V_{k+1}, V_{k+1}^\perp]$ and find

$$||\mathbf{b} - AV_k\hat{\mathbf{y}}||_2 = ||\,||\mathbf{b}||_2\mathbf{e}_1 - \tilde{H}\hat{\mathbf{y}}||_2. \qquad (8.29)$$

Next, we transform $\tilde{H}$ to upper triangular form using a QR factorization. Hence, if $\tilde{H} = QR$, where we take Q the same size as $\tilde{H}$ and R is upper triangular (note that also Q becomes a Hessenberg matrix), and $\mathbf{q}$ is the vector in the orthogonal complement of Q (the complement has dimension 1), then $||\,||\mathbf{b}||_2\mathbf{e}_1 - \tilde{H}\hat{\mathbf{y}}||_2^2 = ||\,||\mathbf{b}||_2 Q^T\mathbf{e}_1 - R\hat{\mathbf{y}}||_2^2 + ||\mathbf{b}||_2^2||\mathbf{q}^T\mathbf{e}_1||_2^2$. So, finally, we have

$$||\mathbf{b} - AV_k\hat{\mathbf{y}}||_2^2 = ||\,||\mathbf{b}||_2 Q^T\mathbf{e}_1 - R\hat{\mathbf{y}}||_2^2 + ||\mathbf{b}||_2^2||\mathbf{q}^T\mathbf{e}_1||_2^2. \qquad (8.30)$$

Note that the actual computation of $\hat{\mathbf{y}}^*$ can be postponed until $||\mathbf{q}^T\mathbf{e}_1||_2 = ||(I - QQ^T)\mathbf{e}_1||_2$ is small enough.

The GMRES method can be applied to any non-singular matrix. The disadvantage is that the whole Krylov space needs to be retained. One can overcome this by performing a re-start after a fixed number of iterations. An alternative would be to solve the normal equations $A^TAx = A^T\mathbf{b}$ instead of $Ax = \mathbf{b}$. By construction the matrix A^TA is SPD, and we can apply the CG method. However, the condition number of this matrix is usually much larger than that of A. In the worst case, it can even be the square of that of A. This slows down the convergence drastically and therefore this approach is not used in practice. *See example Ex. 8.10*

8.8 Preconditioning

In the previous sections we have seen that the convergence of Krylov sub-space methods depends largely on the condition number of the matrix or more specifically on the distribution of the eigenvalues and, in non-symmetric cases, also on the departure of normality. These properties can be influenced by pre-conditioning. The idea is as follows. If we have a system $A\mathbf{x} = \mathbf{b}$, then we like to find a matrix K for which $K^{-1}A$ has a better eigenvalue distribution and for which the system $K\mathbf{x} = \mathbf{b}$ is much easier to solve than the original system. We perform a *left pre-conditioning* by pre-multiplying the system by the inverse of K, leading to $K^{-1}A\mathbf{x} = K^{-1}\mathbf{b}$. A *right pre-conditioning* is given by $AK^{-1}\mathbf{y} = \mathbf{b}$ where afterwards one has to solve $K\mathbf{x} = \mathbf{y}$. One can also do both left and right pre-conditioning at the same time. This is useful for keeping a matrix symmetric. For instance, if A is SPD, then one can make an incomplete Cholesky factorization of A, defining $K = LL^T$. The equivalent system becomes $(L^{-1}A(L^T)^{-1})\mathbf{y} = L^{-1}\mathbf{b}$ and afterwards $\mathbf{x}$ from $L^T\mathbf{x} = \mathbf{y}$. In the following, we concisely describe several important pre-conditioning approaches.

Additional Material

- Overviews of pre-conditioners are given by Saad (2003) and Chen (2005).

8.8.1 Incomplete LU Factorizations

If one makes an LU factorization of a sparse matrix, then the factors L and U contain usually a large amount of small entries with respect to the biggest element in modulus occurring in it. The intuition is that small entries will not contribute much when solving with L and U; hence, the idea is to get rid of these entries. In general, one can write

$$A = LU + R \tag{8.31}$$

where R is the rest or residual matrix. The standard approach is to just drop the small entries, which is usually called Incomplete LU (ILU) factorization. For M-matrices (see Definition B.15) it was shown by Meijerink and van der Vorst (1977) that this always brings the condition number down. A refinement of this approach is to require that the factorization is exact for certain test or probe vectors. So, if $\mathbf{v}$ is a test vector, then $A\mathbf{v} = LU\mathbf{v}$ or $R\mathbf{v} = 0$. In order to achieve this property, we have to compensate the dropped entry on the remaining entries. If we use just the constant vector as test vector, then one compensates the dropped entries on the diagonal leading to the Modified Incomplete LU (MILU) factorization. Of course, for SPD matrices we can make in a similar way incomplete Cholesky factorizations, and by applying them appropriately one can retain the SPD property in the pre-conditioned matrix, that is, $(L^{-1}A(L^T)^{-1})$ is SPD by construction. We mention a few well-known methods in this class.

The ILU factorization where no fill is allowed outside the pattern of the original matrix is called ILU(0). So, for sparse matrices, this approach leads to a very cheap factorization. The downside of this incomplete factorization is that the speedup is already not spectacular for an M-matrix. Changing it to a modified ILU by requiring that the factorization is exact for the constant vector gives a significant improvement for M-matrices, but may worsen the situation for others. In ILU factorizations one can define the level of fill. The level of fill of ILU(0) is 0, because there is no fill here. For (M)ILU(1) we allow fill at the sparsity pattern of A^2 and in general (M)ILU(k) we allow fill at the sparsity pattern of A^{k+1}.

Another method, ILUT, contains a drop parameter ϵ called the threshold (explaining the T in the name). Here an entry is not added to the L- and U-factor if it is less than ϵ times the norm of the current row in the original matrix. This parameter adds a certain robustness since for ϵ is zero we have simply an exact factorization and then a Krylov method will converge in one step. So, one may expect that by decreasing ϵ the convergence will improve.

ILUTP is ILUT including pivoting (explaining the P in the name). As with direct methods, one likes to preclude pivoting, because it may destroy the sparsity pattern of the original matrix in the LU-factorization. So here a value, say γ, is given which allows partial pivoting (Quarteroni *et al.*, 2007, Section 3.5) if γ times the biggest entry in modulus below the diagonal is bigger than the modulus of the diagonal element. For $\gamma = 1$ we have standard partial pivoting, and for smaller γ we are tempering the pivoting. Usually a value of $\gamma = 0.1$ is chosen.

In MRILU the freedom of ordering also is exploited (hence the name Matrix Renumbering ILU). The aim is to order the matrix in a fill-reducing way (see also Section 6.4) during the process such that the amount of dropped entries is as low as possible. Such factorizations can show near grid-independent convergence for M-matrices.

| Additional Material |

- For more details on the MRILU method, see Botta and Wubs (1999). Another method in this class is ILUpack, which is extensively described in Bollhöfer (2001).

8.8.2 Sparse Approximate Inverse

Another idea is to *not* approximate the matrix, which is done in ILU factorizations, but to approximate the inverse immediately. In general, the inverse of an irreducible matrix (see Definition B.7), is full. The challenge is to get a sparse good approximation of the inverse. It appears that one can quite easily construct approximate inverses using the Frobenius norm of a matrix (which is just the square root of the sum of all squares of the entries of the matrix). It holds, for every matrix M, that

$$||I - AM||_F^2 = \sum_{j=1}^{N} ||\mathbf{e}_j - A\mathbf{m}_j||_2^2 \qquad (8.32)$$

where $\mathbf{e}_j$ is the elementary basis vector with a one at position j and zeros elsewhere and $\mathbf{m}_j$ is the jth column of the matrix M. So if we want to find an M such that the Frobenius norm is less than a certain tolerance and, moreover, M is as sparse as possible, we can try to minimize the norms for the respective columns of M. One way to do this is by solving such a system by a few steps of the GMRES method, since this will entail the repeated multiplication of a sparse matrix to a sparse vector, leading to a sparse result vector if the number of steps is limited (Grote and Huckle, 1997).

8.8.3 Algebraic Multi-grid

This is a big container of multi-level methods (Xu and Zikatanov, 2017) which build a multi-grid-like method using only the matrix, hence without information about the type of problem or the geometry. From a user standpoint this is very attractive, but these methods are not easy to construct, let alone that the method will show grid-independent convergence, as we have seen in Section 7.3 for geometric multi-grids. The earlier mentioned method MRILU and ILUpack could be thought of as being part of this class, as well as a method developed by Notay (2012).

8.8.4 Pre-conditioners from Stationary Methods

All the stationary methods mentioned in the previous chapter can also be used as pre-conditioner. The recursion (7.2) can also be written as

$$\mathbf{x}^{(m+1)} = (I - K^{-1}A)\mathbf{x}^{(m)} + K^{-1}\mathbf{b} = \mathbf{x}^{(m)} + \mathbf{r}^{(m)} \qquad (8.33)$$

where $\mathbf{r}^{(m)} = K^{-1}(\mathbf{b} - A\mathbf{x}^{(m)})$. Hence, an alternative formulation of the iteration, given $\mathbf{x}^{(0)}$, is

$$\mathbf{r}^{(m)} = K^{-1}(\mathbf{b} - A\mathbf{x}^{(m)})$$
$$\mathbf{x}^{(m+1)} = \mathbf{x}^{(m)} + \mathbf{r}^{(m)}. \qquad (8.34)$$

It is clear that, when $K = A$, $\mathbf{x}^{(1)}$ will be the sought solution, hence the K from (7.1) is the pre-conditioner here. In fact, we can show that we iterate in the space

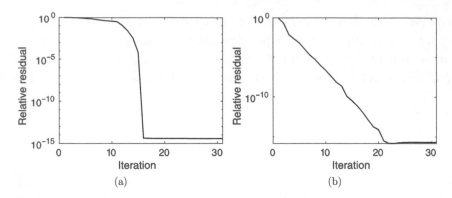

Figure 8.3 Convergence of GMRES with GS splitting pre-conditioner on Ginzburg–Landau problem: (a) for unstable steady solution at $\gamma_1 = 13.6 + i$ and (b) stable orbital solution at $\gamma_1 = 12 + i$ and $\Delta t = 1/20$. Other parameters: $\gamma_2 = \gamma_3 = 1$, $A(0, t) = 0$, $A(1, t) = 0.1(1 + i)$, $m = 10$.

$K^m(K^{-1}A, \mathbf{r}^{(0)})$ as long as m is less or equal to the order of A, and after that (we have reached the full space then) it will just continue to iterate in that space. In the case of Gauss–Seidel and SOR, the associated pre-conditioners will give a non-symmetric matrix even if the original matrix is symmetric. This precludes the use of the CG method. However, there exists a symmetrized variant of SOR (and hence of GS) called the SSOR where both the upper and lower parts of the matrix are used one after another such that the pre-conditioner becomes symmetric.

Example 8.1 (Ginzburg–Landau) *Figure 8.3a shows the convergence of the pre-conditioned residual using GMRES for the solution of a system with the Jacobian matrix at the unstable steady solution for $\gamma_1 = 13.6 + i$ with a random right-hand side computed with the MATLAB/Octave package BifAnFF.[1] The discretization of this case can be found in Example 4.5 and the remaining parameters are just as in Example 6.1. We employ the GS splitting as a pre-conditioner (see Example 7.1). Since the system is of size 20, we may expect an exact solution at last at iteration 20. We see, however, that it converges at step 16. For the solution of the orbital system (b), obtained from Example 6.1, we observe a rapid convergence to machine precision. Both experiments show the enormous advantage of Krylov sub-space methods with respect to the stationary methods, where in the latter we had to tweak the iteration method to get convergence at all. The GMRES method will converge by definition, although there are rare cases where it may stall due to round-off error propagation.*

8.8.5 Domain Decomposition

In domain decomposition methods one splits the computational domain into sub-domains. These sub-domains may be of overlapping type or non-overlapping type.

To these sub-domain splittings, one can apply block forms of the classical iterative solvers (see Section 7.2). In general, the convergence of these is too slow, and
one speeds it up by a so-called coarse-grid correction. This is usually a problem
generated by aggregating the unknowns of each sub-domain to one unknown, leading to an approximation of the original problem on a coarse grid. More advanced
methods generate basis functions by solving suitable local problems on each sub-
domain used to create a coarse-grid approximation for the whole computational
domain. An example in this category is given by FROSch (Heinlein *et al.*, 2020).

8.9   Pre-conditioners in CFD

In implicit approaches in CFD we typically have to deal with matrices of the shape

$$\begin{bmatrix} A & B \\ B^T & 0 \end{bmatrix} \tag{8.35}$$

where B is the discretization of the gradient operator and A comprises that
of the convection and diffusion. Even if A is a positive definite block matrix,
(8.35) is indefinite; it has both positve and negative eigenvalues. Therefore, it is
often called a saddle-point matrix. In general, applying straightforwardly an ILU-
pre-conditioner will not lead to a robust pre-conditioner. It may work for one mesh
size while it fails at another. This is undesirable. Therefore, special methods have
been/are being developed for these matrices. In this section we first introduce the
idea of the Vanka pre-conditioner which is a generalization of Gauss–Seidel, then
we consider block approaches, and finally we describe an integrated approach.

Additional Material
• A review of methods for saddle-point problems is given by Benzi and colleagues (2005).
• An assessment of some methods is made by Ahmed and colleagues (2018).

8.9.1   Vanka Pre-conditioners and Multi-grid

One can see (8.35) as arising from a problem $A\mathbf{x} = \mathbf{b}$ where a constraint is added,
$B^T\mathbf{x} = 0$, as in (4.85). In an incompressible fluid this is the requirement that the
divergence of the velocity should be zero; see (4.63). The idea of the Vanka pre-
conditioner (Vanka, 1986) is to locally satisfy this constraint. In the finite-volume
setting described in Section 4.2.3, the Vanka preconditioner runs over all the control
volumes and adapts the velocities on the boundaries such that divergence freedom

is still satisfied in the current control volume, while it may disturb that in sur-rounding volumes. This pre-conditioner can be used as a standalone but also as a smoother in multi-grid methods. The multi-grid approach on the block system is essentially the same as what is described in Algorithm 3 (John and Tobiska, 2000). An AMG pre-conditioner is discussed in Webster (2018) and an analysis of smoothers in Drzisga and colleagues (2018).

8.9.2 Block Pre-conditioners

The essence and strength of block pre-conditioners (Wathen and Silvester, 1993; Silvester and Wathen, 1994) is that the pre-conditioning is brought back to the pre-conditioning of a number of blocks, where for each block one can use a standard approach, for example an optimized (algebraic) multi-grid method like ML (Gee *et al.*, 2006).

In fact, the block pre-conditioner arises from the block-LU factorization of the block matrix (8.35):

$$
\begin{bmatrix} A & B \\ B^T & 0 \end{bmatrix} = \begin{bmatrix} I & 0 \\ B^T A^{-1} & I \end{bmatrix} \begin{bmatrix} A & 0 \\ 0 & -B^T A^{-1} B \end{bmatrix} \begin{bmatrix} I & A^{-1}B \\ 0 & I \end{bmatrix}. \tag{8.36}
$$

Note that, for a symmetric positive definite A, (8.36) indicates that the original matrix is congruent to the block diagonal matrix in the middle, where the second block is negative semi-definite by definition. Hence, this decomposition shows clearly the indefiniteness of the block system. One can use this block diagonal matrix as a pre-conditioner. In order to study the quality of that pre-conditioner, one would like to locate the eigenvalues of the pre-conditioned system or equivalently, locate the eigenvalues of the generalized eigenvalue problem

$$
\begin{bmatrix} A & B \\ B^T & 0 \end{bmatrix} \mathbf{x} = \lambda \begin{bmatrix} A & 0 \\ 0 & -B^T A^{-1} B \end{bmatrix} \mathbf{x}. \tag{8.37}
$$

We re-write the eigenvalue problem as

$$
\begin{bmatrix} (1-\lambda)A & B \\ B^T & \lambda B^T A^{-1} B \end{bmatrix} \mathbf{x} = 0. \tag{8.38}
$$

It is immediately clear that $\lambda = 1$ is an eigenvalue of this matrix with multiplicity equal to the order of the matrix A. To find eigenvalues different from one, assume $\lambda \neq 1$. This allows one to make an LU-factorization of the matrix, that is,

$$
\begin{bmatrix} I & 0 \\ B^T A^{-1}/(1-\lambda) & I \end{bmatrix} \begin{bmatrix} (1-\lambda)A & B \\ 0 & (\lambda - 1/(1-\lambda))B^T A^{-1} B \end{bmatrix} \mathbf{x} = 0 \tag{8.39}
$$

where we can skip the first non-singular matrix. So the eigenvalue different from one follows from the quadratic problem $\lambda^2 - \lambda + 1 = 0$, which has the solution

$1/2 \pm i\sqrt{3}/2$. So, the pre-conditioner has three different eigenvalues. Hence, a Krylov sub-space method should solve the problem in exactly three iterations. Observe that this also holds for non-symmetric A.

However, the preceding assumes that we will solve the systems with A and $B^T A^{-1} B$ exactly, but this will take too much computer time. We need to use iterative methods for these systems too. So we introduce *inner iterations*. For a system with A one could use one of the methods discussed in the previous section. Note that when we have a time-dependent problem, the diagonal entries of A will increase in magnitude, resulting in a system with a lower condition number not needing a very strong pre-conditioner, but in a steady case one may need a multi-level approach. For the matrix $B^T A^{-1} B$, the situation is more subtle. In the first place, we cannot compute this matrix explicitly because the inverse of A is full. However, in a time-dependent problem one could approximate it by only its diagonal part or by the diagonal matrix arising from putting the absolute sum of each row on its diagonal. In that case, we can evaluate it, and since B^T represents a divergence and B a gradient we obtain a matrix for a Poisson equation. To the resulting problem, one can apply a multi-grid method. In the case of a steady Stokes problem, one can show that the matrix is equivalent to the identity matrix, since, loosely speaking, both $B^T B$ and A represent Laplace operators. In that case, we do not need a pre-conditioner for $B^T A^{-1} B$. The application of a $B^T A^{-1} B$ matrix to a vector can be done one after another (see the solution process for (8.16)). Then we have to solve a system with the matrix A, for which we can use again a multi-grid method.

For efficiency reasons there is no need to solve the sub-systems to high accuracy. In doing so, we will get more than three eigenvalues for the complete system, leading to more *outer iterations*. This has to be balanced with the amount of inner iterations. Both the inner and outer iteration methods can be Krylov sub-space methods. However, a Krylov sub-space is generated by a constant matrix. This will not be the case if the inner iteration method is also a Krylov method. There exist, however, flexible variants of Krylov methods, FGMRES (Saad, 1993) and GMRESR (van der Vorst and Vuik, 1992), that can deal with a varying matrix, which will be employed in this case. An alternative would be to use a stationary method for the inner iteration with a fixed number of iterations and use a standard Krylov sub-space method in the outer loop.

Instead of taking just the block diagonal as a pre-conditioner, one could also combine it with one of the other two factors. For instance, if we take the factor on the left in (8.36), then we obtain the pre-conditioner

$$\begin{bmatrix} I & 0 \\ B^T A^{-1} & I \end{bmatrix} \begin{bmatrix} A & 0 \\ 0 & -B^T A^{-1} B \end{bmatrix} = \begin{bmatrix} A & 0 \\ B^T & -B^T A^{-1} B \end{bmatrix} \quad (8.40)$$

and we have to consider the generalized eigenvalue problem

$$
\begin{bmatrix}
(1 - \lambda)A & B \\
(1 - \lambda)B^T & \lambda B^T A^{-1} B
\end{bmatrix}
\mathbf{x} = 0. \tag{8.41}
$$

For eigenvalues other than one we can zero B^T and obtain, to our surprise, the Schur-complement $(1 - \lambda)B^T A^{-1} B$, which does not give us eigenvalues different from one. So, this is an example where the generalized eigenvalue problem has less eigenvalues than the order of the matrix. Another view on this is that, applying the pre-conditioner (8.40) to the left, that is, multiplying by the inverse of it, we observe that the pre-conditioned matrix is (8.36)

$$
U = \begin{bmatrix} I & A^{-1}B \\ 0 & I \end{bmatrix},
$$

and from this matrix we have to compute the eigenvalues. Now we find that the algebraic multiplicity of the eigenvalue 1 is of the order of the matrix, but the geometric multiplicity is only equal to the order of A. So, this matrix is equivalent to a Jordan matrix. In order to study the order of such blocks we have to find the smallest k such that $(U - I)^k = 0$, which is $k = 2$ in this case. This means that according to Property 3 in Section 8.1.1 only two iterations are needed for convergence, which makes it a good example of what can occur for a non-normal matrix. But of course, here also we will solve the systems only approximately, leading to inner iterations.

Farrell and colleagues (2019) show results of such a method for the 3D lid-driven cavity problem. The foregoing can be generalized to the case of multiple blocks, which occurs if tracer equations are added for temperature, salinity, or any species, as is worked out in the Teko package from Trilinos (Cyr *et al.*, 2012).

8.9.3 Structure-Preserving Multi-level Pre-conditioners

The indefiniteness of the saddle-point system (8.35) prohibits robustness of pre-conditioners. This can be alleviated if we build a pre-conditioner that iterates in the constraint space, that is, for every iterate $\mathbf{x}^{(k)}$ we have $[B^T 0]\mathbf{x}^{(k)} = 0$. One way to achieve this test space is by constructing a basis for the constraint space

$$
\begin{bmatrix} V & 0 \\ 0 & I \end{bmatrix}
$$

where $B^T V = 0$. Next, one makes a Galerkin projection of the linear problem on this space, where it appears that we get a linear problem with the matrix $V^T A V$. If V is a dense matrix, then also this system matrix will be dense. Fortunately, for the Navier–Stokes equation there exists a sparse basis for V. In fact, V will be a discretization of the rotation operator, because we know that for a vector field, having zero divergence, there exists a potential such that the vector field follows from the rotation of a scalar field in the 2D case. In the Navier–Stokes case, the potential is the stream function.

In this section, we follow another route achieving the same goal for the incompressible Navier–Stokes equations discretized by the finite-volume approach. We will not go into the details, but discuss the main ingredients for the discretization discussed in Section 4.2.3. The method can be seen as an incomplete variant of the nested dissection method treated in Section 6.4. In nested dissection the domain is split by a separator into two sub-domains. Next, each sub-domain is split up by a separator and this process is repeated until the sub-domains are very small. The reverse order in which we created the separators is used in the elimination process. As a direct method, it has optimal complexity; however, the computational work is still $O(N^{3/2})$ for a 2D problem and $O(N^2)$ for a 3D problem (see Table 6.2). Moreover, the work is dominated by the last step in the elimination process; for example, in the last step we have to solve a full system on the separator dividing the domain into two sub-domains. To counteract this increase of work we prune the separators in every step. For that we use a Householder transformation on each part of the separator. We will explain this in more detail since it is not commonly discussed in textbooks on iterative methods. The layout of the algorithm computing the incomplete factorization is given in Algorithm 11 and more details can be found in Wubs and Thies (2011) and Baars and colleagues (2021).

We restrict to a general positive definite matrix A (not necessarily symmetric) of order $m \times n$. Suppose we pre- and post-multiply this matrix by a block diagonal matrix $H = I_n \otimes \hat{H}$, where $\hat{H}$ is a Householder matrix (Quarteroni *et al.*, 2007, Section 5.6.1) of order m and $\otimes$ denotes the Kronecker product. Note that a Householder matrix is symmetric and hence H is symmetric. By construction HAH also will be positive definite. Recall that every principal sub-matrix of a positive definite matrix is also positive definite, hence every selection of a block diagonal part of HAH is positive definite. We will exploit this in the dropping procedure. An appropriate choice of $\hat{H}$ is one which maps a constant vector, say $\mathbf{e}$, to a vector in the direction of $\mathbf{e}_1$, that is, $\hat{H} = I - 2\mathbf{v}\mathbf{v}^T/||\mathbf{v}||^2$ with $\mathbf{v} = \mathbf{e} - ||\mathbf{e}||\mathbf{e}_1$. The matrix has the property that the first row and first column are constant.

Write $\hat{H} = [\hat{H}_1, \hat{H}_2]$ where $\hat{H}_1$ is the first column of $\hat{H}$. Next, use a permutation $P = [P_h, P_l]$ such that $HP = [H_h, H_l]$ where $H_h = [I \otimes \hat{H}_2]$ and $H_l = [I \otimes \hat{H}_1]$. The subscripts l and h stand for low and high frequencies, respectively, because H_l consists of a basis of smooth (here constant) vectors, whereas H_h contains vectors which are perpendicular to all of those. So, the problem $A\mathbf{x} = \mathbf{b}$ is transformed into

$$P^T HAHPP^T H\mathbf{x} = P^T H\mathbf{b},$$

Algorithm 11 Structure-preserving incomplete LU factorization

---

    function $[L, U, H, P]=$ factor(A)

        if (coarsest grid is not reached yet) **then**

            Determine separators

            Eliminate internal unknowns in all domains $\rightarrow$ L_i, U_i

            Create H and P

            $[L_h, U_h] = \mathrm{lu}(H_h A H_h)$

            $[L_l, U_l, H_l, P_l] =$ factor($H_l A H_l$)

            $L = [L_i, L_h, L_l]$

            $U = [U_i, U_h, U_l]$

            $H = [H_h, H_l]$

            $P = [P_h, P_l]$

        else

            H and P void

            $[L, U] = \mathrm{lu}(A)$

        end

        return $[L, U, H, P]$

    end

---

where

$$P^T HAHP = \begin{bmatrix} H_h^T A H_h & H_h^T A H_l \\ H_l^T A H_h & H_l^T A H_l \end{bmatrix} \tag{8.42}$$

and where the vector **x** turns into

$$P^T H\mathbf{x} = \begin{bmatrix} H_h^T \mathbf{x} \\ H_l^T \mathbf{x} \end{bmatrix}.$$

Now $H_l^T \mathbf{x}$ is particularly important because it can represent smooth parts in the solution, which are often difficult to deal with in iteration methods. In fact, one can see $H_l^T A H_l$ as a coarse-grid approximation of the original matrix A (Notay, 2012). Therefore, we approximate (8.42) by

$$\begin{bmatrix} H_h^T A H_h & O \\ O & H_l^T A H_l \end{bmatrix}, \tag{8.43}$$

which is positive definite as discussed earlier, and will use that as a pre-conditioner for (8.42). Note that if additionally A would be symmetric, then the pre-conditioned matrix will be similar to a symmetric positive definite matrix too, which makes this way of dropping robust for arbitrary symmetric positive definite matrices. Formally, the pre-conditioner for A is defined by

$$K = HP \begin{bmatrix} H_h^T A H_h & O \\ O & H_l^T A H_l \end{bmatrix} P^T H,$$

but we will never use it in this form. Of course, we can repeat the process on the bottom block $H_l^T A H_l$ of (8.43). We can also prune $H_h^T A H_h$ further because it is a positive definite matrix and we can again take block diagonal parts of it, for instance the diagonal. Note that, by our definition of the Householder matrix, $\mathbf{e} \in \mathrm{span}(H_l)$ and therefore

$$H_l^T A \mathbf{e} = H_l^T K \mathbf{e}, \tag{8.44}$$

and, if $\mathbf{e}$ is an eigenvector of A and K, then also

$$A \mathbf{e} = K \mathbf{e}. \tag{8.45}$$

The properties (8.44) and (8.45) assure that the pre-conditioner and the original matrix do about the same on smooth components in the solution; see also the discussion on probe/test vectors in Section 8.8.1. Modified incomplete factorizations bring the same property but might fail for some cases, even for M-matrices. The preceding works for a much wider class of definite matrices. Pre-conditioner (8.43) will be a good one for (8.42) if $H_h^T A H_l$ is small in the latter. This is often the case. For instance, suppose that A discretizes a Laplace operator; then $A H_l$ is the application of A to smooth low-frequency vectors in H_l and it is likely that A will map them almost completely into the space spanned by these vectors, say $A H_l = H_l S + R_h$, where R_h is small. Hence, $H_h^T A H_l = H_h R_h$ will be small. In Wubs and Thies (2011) it is shown for a similar case that the condition number of the pre-conditioned matrix is independent of the mesh size. Finally, we mention the property that if $H_h^T A H_h$ and/or $H_l^T A H_l$ is not positive definite, then A is not positive definite.

Now let us turn to our model problem (8.35) and suppose we have used the finite-volume method on a domain with closed walls. We just explain what the effect of elimination will be in the case of two sub-domains divided by one vertical separator. Then this separator consists of all the horizontal velocities and all the vertical velocities. If we eliminate the unknowns in each domain, then in each domain one pressure unknown remains. All the horizontal velocities on the separator are connected to both the pressures, but the vertical velocities are not. Moreover, the sub-matrix relating all the velocities can be shown to be positive definite. Now we can use a Householder transformation to decouple all but one horizontal velocity from pressure. This will also create a low-frequency part and a high-frequency part in the horizontal velocities. To the vertical velocities we then apply a Householder transformation as earlier. Finally, we drop connections between high frequencies and between high frequencies and low frequencies. This leads to a new system in which we have in fact only one horizontal and one vertical velocity and two pressure unknowns. For a domain with closed wall we can fix one of the pressures, which leads to a solvable problem. It can be shown that two discrete divergence relations which result after the elimination steps contain only the horizontal velocities on the separator, with exactly the same weights. Since we do not drop anything in the gradient and divergence part we are working exactly there.

8.9.4 Recent Developments and Software

For eigenvalue problems using Krylov sub-space methods the ArPack package (Lehoucq *et al.*, 1998) is often used. For example, it is used in MATLAB.

In the realm of iterative solvers for linear systems there is much more than the methods treated here. For instance, the Induced Dimension Reduction method, already introduced in 1980, is put into block form and appears to be an attractive alternative to GMRES and BiCGstab (Sonneveld and van Gijzen, 2008). Combinations of BiCGstab and IDR can be made that have a short recurrence (the number of vectors needed to compute an improved solution is low, like in the Lanczos and two-sided Lanczos method) while showing a convergence like GMRES, which needs to keep and store the whole Krylov space to proceed. For details and references, see van Gijzen and colleagues (2015). As mentioned already in the previous subsection, there exist also variants which allow a pre-conditioning that varies from step to step, for example FGMRES (Saad, 1993) and GMRESR (van der Vorst and Vuik, 1992).

Deflation is an important tool to speed up the solution process. If one has knowledge of those vectors retarding convergence, one can exploit that in the Krylov sub-space method (Gaul *et al.*, 2013; Ebadi *et al.*, 2016). Another development of interest, especially in the context of continuation, is recycling of the sub-space from the previous continuation step; see, for example, Daas *et al.* (2021).

There are many packages which have implemented pre-conditioners and Krylov sub-space methods. Important ones are Trilinos (Heroux *et al.*, 2005), PetsC (Balay *et al.*, 2021), and PHIST (Thies *et al.*, 2020) which allow one to work with a whole variety of pre-conditioners. These packages are developed for parallel computing and are continuously extended as new platforms arise.

8.10 Summary

- The most used generalization of stationary iterations is iteration in the Krylov sub-space. The essential difference with stationary iteration is that all iterates are stored. By this the Krylov sub-space is invariant under shifts of A, that is, one can replace A by $A - \alpha I$ with α arbitrary and get the same space. As with stationary methods, the convergence depends on the eigenvalues of A. However, due to the shift invariance property, one can for the convergence analysis shift those eigenvalues such that the best convergence results. Moreover, due to the shift invariance, the notion of smallest and largest eigenvalue is lost. For Krylov methods one uses the notion exterior eigenvalues for eigenvalues that can be made largest in modulus by an appropriate shift.

- The dimension of the Krylov space needed depends on the number of different eigenvalues of A, their geometric multiplicity, and on the number of eigenvectors represented in the initial vector $\mathbf{v}$. This determines also the number of iterations needed to solve $A\mathbf{x} = \mathbf{b}$ exactly using the Krylov sub-space if $\mathbf{v} = \mathbf{b}$ or to find the eigenvalues associated to the eigenvectors that constitute $\mathbf{v}$.

- During the generation of the Krylov sub-space one can immediately create an orthogonal basis for it. This leads to using the Arnoldi method and Lanczos method to compute eigenvalues for general and symmetric matrices, respectively. In contrast to the former method, the latter has the same complexity in each iteration. For general matrices one can also obtain this property by using two Krylov sub-spaces at the same time, which leads to the two-sided Lanczos method.

- Newton's method can be accelerated by storing all the Newton corrections and minimizing the function to be zeroed over the sub-space these build. Applying this approach to the eigenvalue problem leads to the Jacobi–Davidson method.

- Generalized eigenvalue problems can be transformed into a standard eigenvalue problem to which Krylov sub-space methods can be applied.

- The conjugate gradient process is the Galerkin projection of the problem $A\mathbf{x} = \mathbf{b}$, with A symmetric positive definite, on the Krylov sub-space $K^m(A, \mathbf{b})$. For general matrices one can make a least-squares approximation in the Krylov sub-space. This leads to the GMRES method.

- By pre-conditioning, the problem $A\mathbf{x} = \mathbf{b}$ is replaced by $K^{-1}A\mathbf{x} = K^{-1}\mathbf{b}$, leading to a new system matrix $K^{-1}A$. The goal of the pre-conditioner is to cluster the eigenvalues around a limited number of points in the complex plane and to decrease non-normality of the matrix. Standard approaches are: direct approximation of the inverse of A (sparse approximate inverse) and incomplete LU factorization of A. Moreover, this can all be combined and a pre-conditioner can also consist of a number of iteration schemes.

> • Using multi-grid ideas, multi-level pre-conditioners can be con-
> structed that lead to condition numbers of $K^{-1}A$ that are grid
> independent. Moreover, the complexity can be kept linear in the
> order of A.
> • For block systems arising from coupled PDEs like the Navier–
> Stokes equations one can reduce the problem to solving Poisson- and
> convection–diffusion-like problems for which off-the-shelf multi-
> grid-like methods exist. However, one can also develop special-
> purpose methods to solve the complete system.

8.11 Exercises

Exercise 8.1 *Consider the Krylov space.*

a. *Let $\mathbf{v}^{(1)} = \sum_{l=1}^{k} \alpha_l \mathbf{q}^{(l)}$ where $\mathbf{q}^{(l)}$ is an eigenvector and show that the maximum dimension of the space is k.*

b. *Suppose that for a defect eigenvalue μ of A we have a $\mathbf{q}_\mu$ such that $(A - \mu I)^3 \mathbf{q}_\mu = 0$ while $(A - \mu I)^2 \mathbf{q}_\mu \neq 0$. Let $\mathbf{v}^{(1)} = \mathbf{q}_\mu$; show that the maximum dimension is reached after two steps.*

Exercise 8.2 *Suppose that in Arnoldi's method the eigenvector $\mathbf{q}$ is part of V_m; show that $\mathbf{q}$ will be a Ritz vector.*

Exercise 8.3 *Let a positive definite matrix A have eigenvalues $\lambda^{(1)} = \epsilon$ and $\lambda^{(i)} = i^2$ for $i = 2, \ldots, N$, with eigenvectors $\mathbf{v}^{(i)}, i = 1, \ldots, N$ and $||\mathbf{v}^{(i)}|| = 1$. Let $\mathbf{b} = \epsilon \mathbf{v}^{(1)} + \sum_{i=2}^{N} \mathbf{v}^{(i)}$. Take $\epsilon = 1e-3$. Based on the Power method analogy discussed in Section 8.2, how many iterations will it take before the contribution of $\mathbf{v}^{(1)}$ is half that of its final contribution in the solution $\mathbf{x}$ of $A\mathbf{x} = \mathbf{b}$?*

Exercise 8.4 *A real symmetric matrix has real eigenvalues. Give a simple 2×2 example showing that for a symmetric generalized eigenvalue problem, such as in (8.9), this needs not be the case.*

Exercise 8.5

a. *Consider the generalized eigenvalue problem $A\mathbf{x} = B\mathbf{x}$ with B SPD. Let the Krylov subspace be $\mathcal{K}^m(B^{-1}A, \mathbf{v})$. What will the reduced eigenvalue problem look like if we create a B-orthogonal space?*

b. *For the incompressible Navier–Stokes equations we will encounter a generalized eigenvalue problem of the shape*

$$\begin{bmatrix} A & B \\ B^T & 0 \end{bmatrix} \mathbf{x} = \lambda \begin{bmatrix} M & 0 \\ 0 & 0 \end{bmatrix} \mathbf{x}.$$

Show that for $\lambda \neq 0$ one can replace this eigenvalue problem by

$$\begin{bmatrix} A & 0 \\ 0 & 0 \end{bmatrix} \mathbf{x} = \lambda \begin{bmatrix} M & B \\ B^T & 0 \end{bmatrix} \mathbf{x}.$$

c. *Next, we consider a Krylov space similar to the one in part a, and also similarly we orthogonalize this space with the right-hand-side matrix. What will the reduced eigenvalue problem look like now?*

Exercise 8.6 *If one considers the generalized eigenvalue problem associated to a flow in a lid-driven cavity, then one knows beforehand that due to diffusion there will be a lot of big negative eigenvalues close to the real axis. In this case we are interested in eigenvalues passing the imaginary axis. What is a minimal requirement on a real choice for σ and τ in order to avoid finding the big negative eigenvalues?*

Exercise 8.7

a. *Suppose we consider a real symmetric matrix A. Which eigenvectors, expressed in eigenvalues of A, will be filtered out by the transformation $q_{0,n}(x) = T_n(0)/T_n(x)$, with n large, when used in a Lanczos process? Here $T_n(x)$ is the Chebyshev polynomial of degree n.*
b. *Suppose we have a normal matrix. Which eigenvectors, expressed in eigenvalues of A, will be filtered out by the transformation $q_{0,n}(z) = 1/(1 - (z/r)^n) = \prod_{k=1}^{n} 1/(\exp(i2\pi k/n) - z/r)$, with n large and r a fixed positive number, when used in an Arnoldi process?*

Exercise 8.8

a. *Show that (8.23) is just a Galerkin projection of the problem find $\mathbf{y}^{\dagger} \in R^N$ such that $((\mathbf{x} - \mathbf{y}^{\dagger}), A\mathbf{y}) = 0$ for all $\mathbf{y}$ in R^N.*
b. *What is the relation between the functional defined in (3.18) and g($\mathbf{y}$) in Section 8.7.1?*

Exercise 8.9 *Show that the condition number κ in (8.26) is also equal to the ratio of the largest and smallest eigenvalues of A*

Exercise 8.10 *What is the Krylov space used in the GMRES algorithm and which if we solve the normal equations?*

Exercise 8.11 *Consider the convection-diffusion problem $u_t = -\bar{u}u_x + \mu u_{xx}$, $\mu, \bar{u} \geq 0$ on the interval $[0, 1]$, and $t > 0$ with $u(x, 0)$, $u(0, t)$ and $u(1, t)$ given.*

a. *Solve the eigenvalue problem associated to the continuous form and observe that for any $\mu/\bar{u} > 0$ the eigenvalues are real.*
b. *Assume now periodic boundary conditions at $x = 0$ and $x = 1$. Compute the eigenvalues for this case. Compare these to the ones found in the previous part. Draw a picture of these eigenvalues.*

Exercise 8.12

a. *In fluid dynamics problems, coefficients in the Jacobian may vary greatly over the domain. In order to get some insight into the influence of varying coefficients on eigenvalues and eigenfunctions, search for Bessel's differential equation on the Internet, define it on [0,1] and require $u(0) = u(1) = 0$, and compare the results to those of the constant coefficient case. As an alternative you could solve the eigenvalue problem with a computer algebra package.*

b. *Consider the operator $\mathcal{L}_{p,q}$ defined by $\mathcal{L}_{p,q}u \equiv -d/dx(p(x)du/dx) + q(x)u$ on [0,1] with $u(0) = u(1) = 1$ and $p(x) \geq 0$. Show that*

$$\lambda^{(1)}(\mathcal{L}_{p_-,q_-}) \leq \lambda^{(1)}(\mathcal{L}_{p,q}) \leq \lambda^{(1)}(\mathcal{L}_{p_+,q_+}),$$

where $p_- = \min_{x\in[0,1]} p(x)$ and $q_- = \min_{x\in[0,1]} q(x)$ and $p_+ = \max_{x\in[0,1]} p(x)$ and $q_+ = \max_{x\in[0,1]} q(x)$

c. *Search for Weyl's inequality on the Internet and observe that the previous is a special case of that. The conclusion of this is that eigenvalues may shift due to varying coefficients but all do so in more or less the same way.*

9

Matrix-Free Techniques

*There are clever alternatives
to reach one's goals.*

In the previous chapters, we have presented methods for numerical bifurcation analysis where it was assumed that the Jacobian matrix was available. In many applications, however, only a time-stepping code is available, where the state vector $\mathbf{x}_{n+1}$ at time t_{n+1} is determined from state vector $\mathbf{x}_n$ at time t_n, without giving information on the Jacobian matrix. In this chapter, we present alternative methods of the computation of fixed points and periodic orbits versus parameters. In these methods only the product of the Jacobian matrix times a vector needs to be evaluated, without requiring the Jacobian matrix itself.

9.1 Jacobian-Free Newton–Krylov Methods

After discretization of the model equations in space, any time-stepping application code can be cast into the following form:

$$\frac{d\mathbf{x}}{dt} = \Phi(\mathbf{x}, \lambda), \tag{9.1}$$

where $\mathbf{x} \in \mathbf{R}^d$ is the state vector containing all prognostic variables on all the grid points and $\Phi(\mathbf{x}, \lambda)$ is often called the 'right-hand side' or 'tendency term'. Note that the usual mass matrix M on the left-hand side of (9.1) is in most cases the identity matrix for time-stepping codes. Assume that such a model is in a regime where steady states exist. An equilibrium solution is reached then by integrating the application code forward in time until close enough to a steady state. This steady state is a solution to the set of non-linear equations

$$\Phi(\mathbf{x}, \lambda) = 0. \tag{9.2}$$

As we have seen in Chapter 6, an alternative method for time stepping is to use Newton's method for solving the system of equations (9.2). In this case, we start from an initial guess $\mathbf{x}_0$ and apply the iteration (with superscript k)

$$\mathbf{x}^{k+1} = \mathbf{x}^k + \Delta\mathbf{x}^{k+1} \tag{9.3}$$

with $\mathbf{x}^0 = \mathbf{x}_0$. In this case, $\Delta\mathbf{x}^{k+1}$ is satisfying

$$\Phi_{\mathbf{x}}(\mathbf{x}^k)\,\Delta\mathbf{x}^{k+1} = -\Phi(\mathbf{x}^k, \lambda), \tag{9.4}$$

where $\Phi_{\mathbf{x}}(\mathbf{x}^k)$ is the Jacobian matrix of Φ. If we want to apply this method to an existing time-stepping code, the right-hand side Φ is available, but the problem is that the Jacobian matrix $\Phi_{\mathbf{x}}(\mathbf{x}^k)$ is in general not easily extracted.

To deal with this problem, Jacobian-Free Newton–Krylov (JFNK) methods were introduced about two decades ago (Knoll and Keyes, 2004; Reisner *et al.*, 2000, 2003; Knoll *et al.*, 2005). In a JFNK method, the system (9.4) is solved using a Krylov method, for example, the GMRES (Saad, 2003) method (Chapter 8). This is an iterative method in which at the lth iteration, the approximation $\Delta\mathbf{x}^{k+1,l}$ of the solution of the system (9.4), more conveniently written as $A_k\,\Delta\mathbf{x}^{k+1} = b_k$, is such that $||A_k\Delta\mathbf{x}^{k+1,l} - b_k||_2$ is minimized over the Krylov sub-space given by

$$K_l = \langle \mathbf{r}_{k+1}, A_k\mathbf{r}_{k+1}, A_k^2\mathbf{r}_{k+1}, \ldots, \ldots, A_k^{l-1}\mathbf{r}_{k+1}\rangle \tag{9.5}$$

with $\mathbf{r}_{k+1} = A_k\Delta\mathbf{x}^{k+1,0} - b_k$.

Since the construction of these Krylov sub-spaces only requires matrix-vector products, the matrix A_k is not needed itself, but rather the matrix applied to a vector $A_k\mathbf{v}$. In (9.4), $A_k = \Phi_{\mathbf{x}}(\mathbf{x}^k)$ and hence the matrix vector product can be approximated using the following finite-difference approximation,

$$\Phi_{\mathbf{x}}(\mathbf{x}^k)\mathbf{v} \sim \frac{\Phi(\mathbf{x}^k + \epsilon\mathbf{v}) - \Phi(\mathbf{x}^k)}{\epsilon}, \tag{9.6}$$

with small ϵ and scaled depending on $||\mathbf{x}^k||_1 = \sum |x_i^k|$ and $||\mathbf{v}||$. For a fast convergence rate of the GMRES method, a pre-conditioner may be needed as explained in Chapter 8. The construction of a pre-conditioner is usually model dependent. One may also obtain such a pre-conditioner from the time-stepping code, as will be shown in the next section. *See example Ex. 9.1*

9.2 Time-Integration-Based Methods

Another set of methods that use a different approach to compute bifurcation diagrams from only time-stepping application codes was suggested in Tuckerman and Barkley (1988) and Mamun and Tuckerman (1995). Assume that such an application code can be modified easily to also integrate the linearized system around some chosen background state $(\bar{\mathbf{x}}, \bar{\lambda})$ in time. Hence, with $\mathbf{x} = \bar{\mathbf{x}} + \mathbf{y}$ and $\lambda = \bar{\lambda} + \mu$, this linearized system, the so-called variational equations, can be written as

$$\frac{d\mathbf{y}}{dt} = \Phi_{\mathbf{x}}(\bar{\mathbf{x}}, \bar{\lambda})\mathbf{y} + \Phi_{\lambda}(\bar{\mathbf{x}}, \bar{\lambda})\mu, \tag{9.7}$$

with initial conditions $\mathbf{y}(0) = \mathbf{y}_0$ and fixed μ.

The computation of fixed points using these time-stepping codes will be discussed in Section 9.2.1. In Section 9.2.2 the computation of periodic orbits is described. It is also possible to compute invariant tori (Sánchez $et\ al.$, 2010), but this will not be discussed here.

9.2.1 Continuation of Fixed Points

A time-discretization scheme can be seen as replacing (9.1) by a map (with iteration subscript i):

$$\mathbf{x}_{i+1} = G(\mathbf{x}_i, \lambda). \tag{9.8}$$

As an example, using an Euler forward method to (9.1) gives

$$\mathbf{x}_{i+1} = \mathbf{x}_i + \Delta t \Phi(\mathbf{x}_i, \lambda) \tag{9.9}$$

and hence

$$G(\mathbf{x}_i, \lambda) = \mathbf{x}_i + \Delta t \ \Phi(\mathbf{x}_i \lambda). \tag{9.10}$$

As another example, we write $\Phi(\mathbf{x}, \lambda) = L\mathbf{x} + N(\mathbf{x})$, with L a matrix and $N(\mathbf{x})$ representing the non-linear terms. Both L and N depend on λ, but this is omitted in the notation for clarity. A semi-implicit time-integration scheme (linear terms backward Euler, non-linear terms forward Euler, also called the BEFE scheme) can be written as

$$\mathbf{x}_{i+1} = \mathbf{x}_i + \Delta t (L\mathbf{x}_{i+1} + N(\mathbf{x}_i)) \tag{9.11}$$

and hence

$$G(\mathbf{x}_i, \lambda) = (I - \Delta t L)^{-1}(\mathbf{x}_i + \Delta t N(\mathbf{x}_i)). \tag{9.12}$$

When a continuation method is used to obtain fixed points of (9.1), systems of the form

$$\mathbf{x} - G(\mathbf{x}, \lambda) = 0 \tag{9.13a}$$

$$\Sigma(\mathbf{x}, \lambda) = 0. \tag{9.13b}$$

must be solved. In the case of pseudo-arclength continuation, with step size Δs, the normalization condition is

$$\Sigma(\mathbf{x}, \lambda) = \dot{\mathbf{x}}_0^T(\mathbf{x} - \mathbf{x}^0) + \dot{\lambda}_0(\lambda - \lambda^0), \tag{9.14}$$

where $(\mathbf{x}^0, \lambda^0)$, $(\mathbf{x}_0, \lambda_0)$ and $(\dot{\mathbf{x}}_0, \dot{\lambda}_0)$ are predictions of a new point on the curve of solutions, the known point, and the curve's tangent, respectively. Hence,

$$\mathbf{x}^0 = \mathbf{x}_0 + \Delta s\, \dot{\mathbf{x}}_0 \tag{9.15a}$$

$$\lambda^0 = \lambda_0 + \Delta s\, \dot{\lambda}_0. \tag{9.15b}$$

The system (9.13) is solved by Newton's method. At each step, the linear system (we use again a superscript k for the Newton iterates)

$$\begin{pmatrix} I - G_{\mathbf{x}}(\mathbf{x}^k, \lambda^k) & -G_{\lambda}(\mathbf{x}^k, \lambda^k) \\ \dot{\mathbf{x}}_0^T & \dot{\lambda}_0 \end{pmatrix} \begin{pmatrix} \Delta \mathbf{x}^{k+1} \\ \Delta \lambda^{k+1} \end{pmatrix} = \begin{pmatrix} -\mathbf{x}^k + G(\mathbf{x}^k, \lambda^k) \\ -\Sigma(\mathbf{x}^k, \lambda^k) \end{pmatrix} \tag{9.16}$$

is solved iteratively and the estimated values are updated via

$$(\mathbf{x}^{k+1}, \lambda^{k+1}) = (\mathbf{x}^k, \lambda^k) + (\Delta \mathbf{x}^{k+1}, \Delta \lambda^{k+1}). \tag{9.17}$$

Iterative methods such as GMRES or BiCGStab require only the computation of matrix-vector products, that is, the calculation of actions of the form

$$G_{\mathbf{x}}(\mathbf{x}^k, \lambda^k)\Delta \mathbf{x} + G_{\lambda}(\mathbf{x}^k, \lambda^k)\Delta \lambda \tag{9.18}$$

for given $\Delta \mathbf{x}$ and $\Delta \lambda$. The matrix products required to solve the linear systems are obtained by integrating (9.7) with initial condition $\mathbf{y}_0 = \Delta \mathbf{x}$, and $\mu = \Delta \lambda$. Each Newton iteration then requires the time integration of a system of $2d$ equations, i.e., (9.1) for $\mathbf{x}$ to calculate the right-hand side in (9.16) and (9.7) for $\mathbf{y}$ to calculate the product (9.18). *See example Ex. 9.2*

The solution of the linear systems is facilitated if the spectrum of $G_{\mathbf{x}}$ (evaluated at the fixed point) is clustered around the origin, more precisely if very few of its eigenvalues are located outside a disk centered at the origin with a radius of one. This is indeed the case for maps like (9.8) which involve the time integration of systems of elliptic-parabolic partial differential equations. In this case, the map G is a strong contraction to the fixed points, except along the unstable manifold if the fixed point is unstable. We assume that the dimension of this manifold is very small compared to the dimension d of the system. In this case, the system (9.16) needs no pre-conditioning and we can expect fast convergence (Sánchez et al., 2004).

To determine how to choose the time step Δt, consider again the semi-implicit BEFE scheme (9.12), where it is assumed that $I - \Delta t L$ can be inverted inexpensively. We seek solutions of

$$\begin{aligned} 0 &= G(\mathbf{x}, \lambda) - \mathbf{x} \\ &= (I - \Delta t L)^{-1} (\mathbf{x} + \Delta t N(\mathbf{x})) - \mathbf{x} \\ &= (I - \Delta t L)^{-1} (\mathbf{x} + \Delta t N(\mathbf{x}) - (I - \Delta t L)\mathbf{x}) \\ &= (I - \Delta t L)^{-1} \Delta t (N(\mathbf{x}) + L\mathbf{x}). \end{aligned} \tag{9.19}$$

This demonstrates that steady states of (9.1) are indeed fixed points of G for the time-stepping scheme (9.11), for any value of Δt.

The Newton scheme for determining fixed points leads in this case to linear systems of the form

$$((L + N_{\mathbf{x}}(\mathbf{x}^k))\Delta \mathbf{x} = -(L\mathbf{x}^k + N(\mathbf{x}^k)) \tag{9.20}$$

while the BEFE time-integration scheme of the variational equations (9.7), for $\Delta\lambda$ is

$$\mathbf{y}_{i+1} = (I - \Delta t L)^{-1}(I + \Delta t N_{\mathbf{x}})\mathbf{y}_i. \tag{9.21}$$

If we solve (9.20) with a pre-conditioner $\Delta t(I - \Delta t L)^{-1}$, then we obtain

$$((I - \Delta t L)^{-1}(I + \Delta t N_{\mathbf{x}}(\mathbf{x}^k)) - I)\Delta \mathbf{x} = ((I - \Delta t L)^{-1}(\mathbf{x}^k + N(\mathbf{x}^k)) - \mathbf{x}^k. \tag{9.22}$$

The right-hand side of (9.22) is the difference between two time steps $\mathbf{x}_{i+1} - \mathbf{x}_i$, while the left-hand side is the difference between the time steps $\mathbf{y}_{i+1} - \mathbf{y}_i$ of the variational equation, and hence any Newton correction $\Delta\mathbf{x}$ can be determined by performing time integrations. In this case, the time-integration scheme effectively provides the pre-conditioner.

A few numerical experiments are usually required to determine the optimal value of the integration interval T. Increasing T concentrates the spectrum of $G_{\mathbf{x}}$ at the origin, reducing the number of matrix evaluations required to solve (9.16). However, increasing T (for fixed time step Δt) also increases the computation time to carry out each matrix evaluation. Alternatively, a large single time step T can be used and so no trade-off is necessary to select the value of T. In this case, G no longer represents accurate time stepping of the original equations, but its fixed points can nonetheless be steady states. In this case, it can be shown that L^{-1} can act as an adequate pre-conditioner of $N_{\mathbf{x}} + L$ (Carey *et al.*, 1989; Brown and Saad, 1990; Tuckerman and Barkley, 2000). This method has been used to calculate bifurcation diagrams in many physical systems, such as spherical Couette flow (Mamun and Tuckerman, 1995), cylindrical Rayleigh–Bénard convection (Tuckerman and Barkley, 1988; Boronska and Tuckerman, 2010), Bose–Einstein condensation (Huepe *et al.*, 2003), and binary fluid convection (Batiste *et al.*, 2006; Assemat *et al.*, 2008). *See example Ex. 9.3*

Far away from the fixed point $\bar{\mathbf{x}}$, the linear solver may fail to converge, and other pre-conditioners of $N_{\mathbf{x}} + L$ may be sought. If finite-difference, finite-volume, or finite-element methods are employed, incomplete LU decomposition or other techniques can be used as pre-conditioners to accelerate the convergence (Molemaker and Dijkstra, 2000; Sánchez *et al.*, 2002). For spectral methods, it is not easy to find good pre-conditioners. Finite-difference or finite-element versions of the problem have been successfully used as pre-conditioners (Canuto *et al.*, 2006, 2007), but the coding becomes much more complicated.

To compute the stability of a fixed point, $\bar{\mathbf{x}}$, any variant of the power method (sub-space iteration, Arnoldi methods, etc.) can be used to compute the leading eigenvalues (largest in modulus) of $G_{\mathbf{x}}(\bar{\mathbf{x}})$ (Barkley and Henderson, 1996; Net et al., 2008). If the fixed points are stable, the method can be seen as an acceleration of the time evolution towards the steady state. If they are unstable, the method is similar to a stabilization method like the Recursive Projection Method of Shroff and Keller (1993), which is discussed in Section 9.3. This method was used in Net and colleagues (2003) to compute solution branches of flows in a two-dimensional annular region subject to gravity and differential heating.

9.2.2 Continuation of Periodic Orbits

For computing periodic orbits, two possibilities are available for the map G in (9.8). The first consists of a Poincaré map on a certain manifold. A hyperplane Σ^+ is again defined by

$$g(\mathbf{x}) = \mathbf{n}^T(\mathbf{x} - \mathbf{x}_{\Sigma^+}) = 0. \tag{9.23}$$

In this case, G in (9.8) is the Poincaré map $\mathcal{P}(\mathbf{x}, \lambda) = \phi(t(\mathbf{x}), \mathbf{x}, \lambda)$ as defined in Section 5.4.2. Hence, $g(\phi(p, \mathbf{x}, \lambda)) = 0$, is the first intersection close to $\mathbf{x}$ of the trajectory starting at $\mathbf{x} \in \Sigma^+$ with a certain arrival time. When $\mathbf{x} = \phi(p, \mathbf{x}, \lambda)$ there is a periodic orbit of period p. The details on how to parameterize Σ^+, which has dimension $d - 1$, are given in Sánchez and colleagues (2004). The action of the Jacobian can now be computed by the formula (Simó, 1990):

$$G_{\mathbf{x}}(\mathbf{x}, \lambda)\Delta\mathbf{x} + G_{\lambda}(\mathbf{x}, \lambda)\Delta\lambda = \mathbf{y} - \frac{\mathbf{v}^T\mathbf{y}}{\mathbf{v}^T\mathbf{z}}\,\mathbf{z}, \tag{9.24}$$

where $\mathbf{v}^T\Delta\mathbf{x} = 0$, $\mathbf{z} = \Phi(G(\mathbf{x}, \lambda), \lambda)$, and $\mathbf{y}$ is the solution of the first variational equation (9.7) with initial condition $\mathbf{y}_0 = \Delta\mathbf{x}$, and $\mu = \Delta\lambda$. Again, each matrix product requires the time integration of a system of $2d$ equations. *See example Ex. 9.4*

This formulation, together with Newton–Krylov methods, was used in thermal convection problems (Sánchez et al., 2004; Puigjaner et al., 2011), and in Sánchez and colleagues (2006) to obtain a normal form about a periodic orbit with symmetries. The latter study also required the eigenfunctions of an adjoint operator at the critical periodic orbit. The computations of segments of two-dimensional invariant manifolds in large-scale systems have been described in van Veen and colleagues (2011), generalizing the ideas of Krauskopf and Osinga (2007) and of Krauskopf and colleagues (2005).

An approach without Poincaré maps is used, for instance, in AUTO (Doedel, 1986), CONTENT (Kuznetsov and Levitin, 1996), or MatCont (Dhooge et al., 2003) and consists of solving the bordered system

$$\mathbf{x} - \phi(p, \mathbf{x}, \lambda) = 0$$
$$\Pi(\mathbf{x}) = 0 \qquad (9.25)$$
$$\Sigma(\mathbf{x}, \lambda) = 0$$

for $(\mathbf{x}, \lambda, p)$ by Newton–Krylov methods. In this case the map depends explicitly on T, that is, $G(\mathbf{x}, p, \lambda) = \phi(T, \mathbf{x}, \lambda)$. In (9.25), $\Pi(\mathbf{x}) = 0$ is a phase condition which selects a single point on the periodic orbit, and $\Sigma(\mathbf{x}, \lambda) = 0$ is again the pseudo-arclength condition (9.13). The phase condition can be the equation of a hyperplane (9.23) as before, or an integral constraint, as is done in AUTO. It is possible, but not strictly necessary, to scale the time as $t = p\tau$, with p the unknown period of the periodic orbit (so that now the period is $\tau = 1$), and to rewrite the original system (9.1) as

$$\frac{d\mathbf{x}}{d\tau} = p\Phi(\mathbf{x}, \lambda). \qquad (9.26)$$

This is done when the periodic orbits are computed by collocation in time, in order to fix the endpoints of the time interval.

The action of the Jacobian can be computed through (9.24). The only difference is that the derivative with respect to p must be included. These algorithms have been used to find periodic orbits for plane channel and pipe flows (Duguet *et al.*, 2008; Gibson *et al.*, 2009). In some cases the authors did not have a good initial condition to start the continuation and therefore used a globalized version of Newton's method to increase the size of the basin of attraction (Pawlowski *et al.*, 2006).

Additional Material

- In Gelfgat (2019a), there are four contributions (Tuckerman *et al.*, 2019; Loiseau *et al.*, 2019; Sánchez Umbría and Net, 2019; Gelfgat, 2019b) which provide an excellent overview of the state-of-the-art of time-stepping Krylov methods for numerical bifurcation analysis.

9.3 Hybrid Methods

In this section the so-called Newton–Picard method will be described. The underlying idea of the method can already be found in Jarausch and Mackens (1987), where it is applied to fixed-point computations. Shroff and Keller (1993) called their method the Recursive Projection Method (RPM). This method is at the basis of the PDECONT software package; for an extensive treatment, see Lust and Roose (2000).

If a time-integration method is used to compute a fixed point, then eventually the error will be dominated by a set of slowly decaying (periodic) components. It is possible to make a basis for this set of slowly decaying components which form an invariant sub-space of limited dimension. The crux is to compute the update in the invariant sub-space and its orthogonal complement separately. In the former sub-space, Newton's method is used, and in the latter the time integration is still carried out; hence, the method is of hybrid type. To explain the Newton–Picard method, first a description of the RPM will be given.

9.3.1 Recursive Projection Method

Consider again the fixed-point problem

$$\mathbf{x}_{i+1} = G(\mathbf{x}_i, \lambda).$$

The fixed-point iteration may converge slowly or even diverge when the eigenvalues of the Jacobian of G are in magnitude close to 1 or larger than 1.

Note that there is a relation between the stability of the time-stepping scheme and the stability of the fixed points, determined by the eigenvalues of the Jacobian matrix of the dynamical system. Consider, for example, the Euler forward case (9.9) for which $G(\mathbf{x}_i, \lambda) = \mathbf{x}_i + \Delta t \Phi(\mathbf{x}_i, \lambda)$; then the relation between the eigenvalues α of the Jacobian $\Phi_{\mathbf{x}}$ and the ones determining the stability of the Euler forward scheme, and so the eigenvalues μ of G_x, is

$$\mu = 1 + \alpha \Delta t. \tag{9.27}$$

Hence, when $\alpha = \rho + i\tau$, then $|\mu| < 1$ if and only if $\rho < -\Delta t |\alpha|^2/2$. So, the iteration (9.27) converges when, for all eigenvalues, $\rho < 0$ and the time step is restricted by $\Delta t = \theta \min_\rho 2|\rho|/|\alpha|^2$, where $0 < \theta < 1$.

To stabilize this iteration procedure, the space is split into two orthogonal sub-spaces H and $H^\perp$ with projectors P and $Q = I - P$, respectively (Fig. 9.1).

Then the fixed-point iteration is re-written as

$$\mathbf{p}_{i+1} = \mathbf{p}_i + C_1(PG(\mathbf{x}_i, \lambda) - \mathbf{p}_i),$$
$$\mathbf{q}_{i+1} = \mathbf{q}_i + C_2(QG(\mathbf{x}_i, \lambda) - \mathbf{q}_i),$$
$$\mathbf{x}_{i+1} = \mathbf{p}_{i+1} + \mathbf{q}_{i+1},$$

where $\mathbf{p}_0 = P\mathbf{x}_0$ and $\mathbf{q}_0 = Q\mathbf{x}_0$. Here C_1 and C_2 should be chosen such that good convergence is obtained in both sub-spaces. In Shroff and Keller (1993), C_2 is just the identity matrix, yielding a Picard iteration in $H^\perp$, and C_1 is minus the inverse of the Jacobian of $PG(\mathbf{p}_i, \lambda) - \mathbf{p}_i$ yielding a Newton step in H.

More generally, let the columns of V be the orthogonal vectors spanning H; then every $\mathbf{p} \in H$ can be written as $\mathbf{p} = V\hat{\mathbf{p}}$. Since operators restricted to H are easily

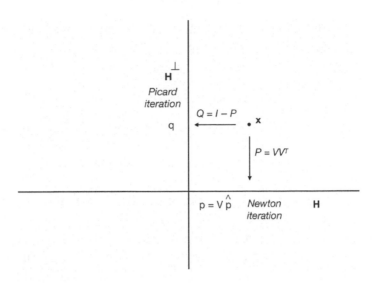

Figure 9.1 Sketch of the Recursive Projection (and Newton–Picard) approach.

formed, contrary to those restricted to $H^\perp$, one prefers to work with $\hat{\mathbf{p}}$ instead of $\mathbf{p}$. Using $P = VV^T$ we can re-write the update equation for $\hat{\mathbf{p}}$ as

$$V\hat{\mathbf{p}}_{i+1} = V\hat{\mathbf{p}}_i + C_1(VV^T G(V\hat{\mathbf{p}}_i + \mathbf{q}_i, \lambda) - V\hat{\mathbf{p}}_i).$$

Pre-multiplying this equation by V^T yields (using $V^T V = I$)

$$\hat{\mathbf{p}}_{i+1} = \hat{\mathbf{p}}_i + V^T C_1 V(V^T G(V\hat{\mathbf{p}}_i + \mathbf{q}_i, \lambda) - \hat{\mathbf{p}}_i). \qquad (9.28a)$$

To solve the equation

$$\mathbf{p} - PG(\mathbf{p} + \mathbf{q}, \lambda) = 0 \quad \rightarrow \quad \hat{\mathbf{p}} - V^T G(V\hat{\mathbf{p}} + \mathbf{q}, \lambda) = 0$$

with Newton's method, the Jacobian is given by $I - V^T G_{\mathbf{x}} V$. Hence, (9.28a) provides such a Newton step in H if we choose

$$V^T C_1 V = (I - V^T G_{\mathbf{x}} V)^{-1}.$$

Here, the directional derivatives $G_{\mathbf{x}} \mathbf{v}_j$, where $\mathbf{v}_j$ is the jth column of V, can be found by numerical differentiation:

$$G_{\mathbf{x}}(\mathbf{x}_i)\mathbf{v}_j \approx \frac{G(\mathbf{x}_i + \varepsilon \mathbf{v}_j, \lambda) - G(\mathbf{x}_i, \lambda)}{\varepsilon}. \qquad (9.28b)$$

Note that this iteration is in general more contractive than the original fixed-point problem. Even if the fixed-point problem is mildly unstable, with only a few eigenvalues of $G_{\mathbf{x}}$ in magnitude larger than one, then the RPM is converging if the eigenvectors corresponding to those eigenvalues are in H. The remaining problem is to find V. Its columns build the invariant sub-space H which usually is the space corresponding to the dominant eigenvalues of G_x. V can be found in a variety of ways which we discuss in the next section.

9.3.2 Newton–Picard Method

Lust and Roose (2000) re-considered and generalized the RPM for the computation of periodic orbits starting from a linear algebra point of view and called their method the Newton–Picard method. Note that in the periodic orbit case, G_x in the previous subsection is the monodromy matrix and the matrix $V^T G_x V$ is the monodromy matrix restricted to H; in this subsection we will therefore indicate G_x by M.

In the Newton–Pickard method, the invariant sub-space V is found by a variant of orthogonal iteration with Ritz acceleration (Golub and van Loan, 1996, page 423) or sub-space iteration with projection, as it is called in Saad (2011). The iteration process is given by Algorithm 12. If steps 2 to 4 are omitted and for $V = W$

Algorithm 12 Construction of invariant sub-space

Give some V_0 with $V_0^T V_0 = I$ and $M = G_x$.

Set $k = 1$ and perform the following steps:

1. Apply M to V_{k-1}: $W_k = M V_{k-1}$.
2. Compute the restriction of M to the space spanned by V_{k-1}: $M_k = V_{k-1}^T W_k (\equiv V_{k-1}^T M V_{k-1})$.
3. Convert M_k to its real Schur normal form where the eigenvalues on the diagonal of R are in decreasing order of magnitude: $U^T M_k U = R$.
4. Combine columns in W_k such that they correspond to the order of the eigenvalues: $V_k = W_k U$.
5. Orthogonalize V_k.
6. Increment k and go to step 1, or stop if converged.

---

the algorithm just boils down to the standard orthogonal iteration process (a generalization of the power method). The influence of steps 2 to 4 is subtle, as these acceleration steps do not improve the convergence of the sub-space as a whole and therefore they also do not improve the convergence of the eigenvalues. In these steps, a basis is constructed for the sub-space such that the eigenvector (eigensubspace) corresponding to the largest eigenvalue (eigenvalue pair) in magnitude is on the first position, repeating itself for the remaining eigenvalues. This basis will eventually also appear if standard orthogonal iteration is used.

To show the favourable influence of this action, consider, for example, the extreme case in which the construction of the invariant sub-space is started with the eigenvectors corresponding to the two largest eigenvalues in magnitude but in reverse order (the vector with the smaller eigenvalue is at the first position). Then orthogonal iteration builds the eigenvector with the largest eigenvalue from rounding error in the first position, which may take many iteration steps, while the acceleration will reverse the order in the first step and exit the routine.

In the standard acceleration (Golub and van Loan, 1996, page 423) step 2 is replaced by first orthogonalizing W, which allows one to omit step 5, and then computing $M_k = \tilde{W}_k^T M \tilde{W}_k$. This requires a new application of M, which will slow down the construction of V by almost a factor of two. The difference is that in the algorithm used, the re-ordering is based on the old V, which has a limited effect only (see the paragraph on the convergence behaviour in what follows). Eventually, when the respective sub-spaces are converging, U tends to the identity matrix, and the eigenvalues on the diagonal of R, that is, the eigenvalues of the matrix M restricted to the invariant sub-space, are the amplification factors mentioned in Section 3.2.

In exact arithmetic the construction can be simplified to the iteration $V_k = MV_{k-1}$, requiring only that the distance (Golub and van Loan, 1996, page 76) between V_0 and the dominant eigensubspace is less than one. If h is the dimension of the space spanned by V_k then the ith Ritz value σ_i converges as $(\sigma_{h+1}/\sigma_i)^k$. Here the Ritz value can be obtained by orthogonalizing V_k, restricting M to that space as in step 2, and computing the eigenvalues of that restriction in decreasing order (the ith eigenvalue of this matrix is the ith Ritz value).

In order to get the same result in the presence of rounding errors, V_k has to be orthogonalized now and then (usually at each step), resulting in orthogonal iteration. In this iteration the eigenvectors will converge in order of decreasing magnitude of the eigenvalues. To accelerate this process one recombines the columns in V_k when the Ritz values are computed (here at every step), yielding the same convergence in the eigenvectors as in the Ritz values. For more considerations we refer to chapter 5 of Saad (2011).

The computation of MV is well parallelizable. Since this is the most time-consuming operation (over 90 per cent) of the code, this is very attractive. For the stopping criterion in Algorithm 12, the matrix

$$(I - VV^T)MV$$

is considered. Here $(I - VV^T)$ is the projector on the orthogonal complement of V. Hence, if V is an invariant sub-space then this expression is zero. It is not necessary that all columns of V have converged. The user specifies a lower bound for the magnitude of the eigenvalues on the diagonal of R that must converge and a tolerance which is used as follows. If the eigenvalues corresponding to the space spanned by the first k columns of V are larger than the lower bound and if the norm of the first k columns of $(I - VV^T)MV$ is less than the tolerance, then these k columns are assumed to have converged. Of course, the space V must be taken large enough such that it contains all desired eigenvectors. A larger space than strictly necessary is even beneficial, as we have seen in the previous paragraph on convergence behaviour. As mentioned before, the eigenvectors converge in order. This can be exploited by freezing (or locking) a converged vector. In this case step 1 of Algorithm 12 is only performed for the non-converged part. The other steps are applied to all vectors in V.

Additional Material ∎

- Many details on the Newton–Picard method and the applications to a diverse set of problems can be found in Lust (1997).

9.3.3 Continuation of Periodic Orbits Using Newton–Picard

Again, the solution of the dynamical system (9.1) can be described in terms of the flow ϕ, that is,

$$\mathbf{x}(t) = \phi(t, \mathbf{x}(0), \lambda).$$

So, the solution depends on the initial solution $\mathbf{x}(0)$ and the parameter λ. If the system has a periodic solution with period p, then the flow must have a fixed point, that is,

$$\mathbf{x} = \phi(p, \mathbf{x}, \lambda). \tag{9.29}$$

As explained in Chapter 5, the fixed point is not unique and can be made unique by imposing a phase condition, which formally can be written as

$$\Pi(\mathbf{x}, p, \lambda) = 0. \tag{9.30}$$

Of course, one would like to solve the preceding two coupled equations (9.29) and (9.30) with the Newton method. Unfortunately, the Jacobian matrix is full and hence, for large systems, this road has a dead end.

With the RPM ideas, a rapidly linearly converging method is obtained. Writing down the Newton-correction equation as

$$\begin{bmatrix} \phi_{\mathbf{x}} - I & \phi_p \\ \Pi_{\mathbf{x}} & \Pi_p \end{bmatrix} \begin{bmatrix} \Delta\mathbf{x} \\ \Delta p \end{bmatrix} = - \begin{bmatrix} \mathbf{r} \\ \Pi \end{bmatrix},$$

where the matrix depends on the best solution so far, and $\mathbf{r}$ is the residual of the fixed-point equation. The matrix ϕ_x is full, and since it is an approximation to the monodromy matrix, it is denoted again by M. Furthermore, ϕ_p is the derivative of ϕ with respect to t at the time p. Since ϕ is nearly periodic, ϕ_p is approximated by the derivative of ϕ with respect to t at $t = 0$.

In the Newton–Picard approach, $\Delta\mathbf{x}$ is written as $\Delta\mathbf{x} = V\Delta\hat{\mathbf{p}} + \Delta\mathbf{q}$ and the correction equation can be rewritten as

$$
\begin{bmatrix}
Q(M-I)Q & Q(M-I)V & Q\phi_p \\
V^T(M-I)Q & V^T(M-I)V & V^T\phi_p \\
\Pi_x Q & \Pi_x V & \Pi_p
\end{bmatrix}
\begin{bmatrix}
\Delta\mathbf{q} \\
\Delta\hat{\mathbf{p}} \\
\Delta p
\end{bmatrix}
= -
\begin{bmatrix}
Q\mathbf{r} \\
V^T\mathbf{r} \\
\Pi
\end{bmatrix}, \quad (9.31)
$$

where Q is the orthogonal projector on the complement of V, $Q = I - VV^T$. If V is an invariant sub-space of M, then $Q(M - I)V = 0$; we will assume that this is the case. Moreover, since for a periodic solution ϕ_p is the eigenvector of M with eigenvalue 1 (this can be deduced from (9.29)) and hence is in the invariant subspace V, $Q\phi_p$ will be negligible during the Newton process (9.31). This results in a Gauss–Seidel-like iteration; first, $\Delta\mathbf{q}$ has to be computed and then the other unknowns follow from the reduced system (note that $Q\Delta\mathbf{q} = \Delta\mathbf{q}$):

$$
\begin{bmatrix}
V^T(M-I)V & V^T\phi_p \\
\Pi_x V & \Pi_p
\end{bmatrix}
\begin{bmatrix}
\Delta\hat{\mathbf{p}} \\
\Delta p
\end{bmatrix}
= -
\begin{bmatrix}
V^T\mathbf{r} + V^T M\Delta\mathbf{q} \\
\Pi + \Pi_x\Delta\mathbf{q}
\end{bmatrix}. \quad (9.32)
$$

In order to compute $\Delta\mathbf{q}$, a system with matrix $Q(M - I)Q$ has to be solved. This forms the bottleneck in the computations since we do not want to compute this full matrix explicitly. However, a multiplication of M with a vector can be computed by using numerical differentiation as in (9.28b). Therefore, the inverse of $Q(M - I)Q$ will be approximated by a polynomial. In PDECONT, a truncated Neumann series is used,

$$
[Q(M-I)Q]^+ \approx Q + QMQ + (QMQ)^2 + \cdots + (QMQ)^k, \quad (9.33)
$$

where '+' denotes the pseudo-inverse; this yields exactly the inverse of $M - I$ restricted to the orthogonal complement of V; the value of k is specified by the user. Once the reduced system has been computed, its solution can be found by a direct method (as in Section 6.1) since the system has a small dimension.

In order to avoid giving a phase condition a variant based on least-squares is available. Here, the reduced system to be solved is

$$
\begin{bmatrix} V^T(M-I)V & V^T\phi_p \end{bmatrix}
\begin{bmatrix}
\Delta\hat{\mathbf{p}} \\
\Delta p
\end{bmatrix}
= -\begin{bmatrix} V^T\mathbf{r} + V^T M\Delta\mathbf{q} \end{bmatrix}. \quad (9.34)
$$

As the number of unknowns exceeds the number of equations by one and hence the solution is not unique. A way to make it unique is to find the solution with minimal

Euclidean norm. This solution is found by employing the pseudo-inverse, which implicitly defines a phase condition.

In case of a periodic solution, ϕ_p is the eigenvector of M with eigenvalue 1. In general, this is a simple eigenvalue and due to the construction in the Newton–Picard process it is in the invariant sub-space V. Hence, if $[\widehat{\Delta\hat{\mathbf{p}}}, \widehat{\Delta p}]^T$ is a solution of the under-determined system (9.34), then $[\widehat{\Delta\hat{\mathbf{p}}} + \mu V^T \phi_p, \widehat{\Delta p}]^T$ is also a solution. Minimization of the length of this vector leads to the condition that the inner product $(V^T \phi_p, \widehat{\Delta\hat{\mathbf{p}}} + \mu V^T \phi_p) = 0$. In other words, the unique solution of the system has to be perpendicular to $[V^T \phi_p, 0]^T$. This is the implicit phase condition when the system is solved using the pseudo-inverse (Lust and Roose, 2000). As usual in the application of the pseudo-inverse, the user has to specify a drop tolerance for the singular values. Here, a singular value is dropped if it is less than the user-specified drop tolerance times the maximum singular value.

The stopping criterion of the Newton–Picard process is based on the residual and the corrections. More precisely, the root mean square of each of the residuals, $\Delta\mathbf{p}$, $\Delta\mathbf{q}$, and the magnitude of Δp should be less than a user-specified tolerance. After a periodic solution has been computed accurately, the Floquet multipliers must be computed in order to study the stability of the solution. Therefore, the current dominant sub-space V is updated according to Algorithm 12 and the Floquet multipliers can be read from the diagonal of the matrix R.

Finally, the Newton–Picard method can also be used to compute a fixed-point solution as a degenerate case of a periodic solution. *See example Ex. 9.5*

9.4 Summary

- Several numerical methods exist to determine the fixed points and periodic orbits of autonomous dynamical systems versus parameters using only time-stepping codes, hence without the availability of the Jacobian matrix.
- In the Jacobian-free Newton–Krylov methods, only products of the Jacobian and a vector are used and the Jacobian is approximated using finite differences.
- In the time-stepping techniques, the product of the Jacobian and a vector is computed by taking differences of state vectors at different time steps.
- With both Jacobian-free Newton–Krylov methods and time-stepping methods, periodic orbits also can be computed and followed in one of the parameters.

> • The Newton–Picard iteration is a hybrid method. The key aspect
> of the method is to compute the update in the invariant sub-space
> and its orthogonal complement separately. In the former sub-space,
> Newton's method is used and in the latter the time integration is
> carried out.

9.5 Exercises

Exercise 9.1 *For* $\mathbf{x} \in \mathbb{R}^d$, *derive that the error in the left-hand side of (9.6) due to the following choice of* ϵ *in (9.6):*

$$\epsilon = \epsilon_0 \frac{\left(1 + \frac{\|\mathbf{x}^k\|_1}{d}\right)}{\|\mathbf{v}\|}$$

is $\mathcal{O}(\epsilon_0)$.

Exercise 9.2 *Write a skeleton of a Python program that performs pseudo-arclength continuation using only time-stepping codes (9.1) and (9.7), and the actions (9.18).*

Exercise 9.3 *a. Show that for large time step T, (9.19) can be approximated as*

$$G(\mathbf{x}) - \mathbf{x} \approx L^{-1}(N + L)\mathbf{x}.$$

b. Show that L^{-1} *plays the role of a pre-conditioner for* $N_{\mathbf{x}} + L$, *and that* $G_{\mathbf{x}} - I$ *is essentially a pre-conditioned version of* $N_{\mathbf{x}} + L$.

Exercise 9.4 *a. Derive the equation (9.24).*

b. Estimate the computational costs to obtain one matrix-vector multiplication using (9.24).

Exercise 9.5 *The Newton–Picard method can be used to compute a fixed-point solution as a degenerate case of a periodic solution.*

a. Show that for a fixed-point $V^T(M - I)V$ *is in general non-singular.*

b. Show that $\phi_p = 0$ *and hence* $V^T \phi_p = 0$.

c. Show that the solution with minimal Euclidian norm is the one with $\Delta p = 0$.

10

Benchmark Results for Canonical Problems

Only through practise
can one become a master.

In this last chapter, we provide explicit results for the bifurcation analysis of flows introduced in Chapter 1 using the concepts of Chapter 2 and the numerical techniques as discussed in Chapters 3 through 9. The aim of this chapter is not to give an extensive overview of results in the literature but to provide cases for which computations can be relatively easily done. Since the results described in this chapter are all two-dimensional, they can be reproduced on a local computer without the need for parallel implementations. The computations have been performed using Python implementations of the finite volume method (FVM)[1] and the Jacobi–Davidson method (JaDaPy).[2] We also provide optional interfaces to parallel solvers, which allow for the computation of results for three-dimensional problems.

Additional Material

- Methods of numerical bifurcation theory have been applied to many more flow problems than the ones discussed in this chapter. Much studied systems are the spherical Couette flow (Mamun and Tuckerman, 1995), cylindrical Rayleigh–Bénard convection (Tuckerman and Barkley, 1988; Boronska and Tuckerman, 2010), coating flows (Ly *et al.*, 2020), and binary fluid convection (Batiste *et al.*, 2006; Assemat *et al.*, 2008). An overview of the different flows studied can be obtained by going through Doedel and Tuckermann (2000), Henry and Bergeon (2000), Gelfgat (2019a), and Dijkstra *et al.* (2014).

> • Other publicly available codes (than the FVM code mentioned earlier) to perform numerical bifurcation analyses on fluid-flow problems are LOCA (Salinger *et al.*, 2002); Oomph-lib (Heil and Hazel, 2006), see https://oomph-lib.github.io/oomph-lib/doc/html/; PDECONT Lust and Roose (2000); and pde2path (Uecker, 2021), see www.staff.uni-oldenburg.de/hannes.uecker/pde2path/.

10.1 Two-Dimensional Lid-Driven Cavity Flow

The governing equations (1.4) in Cartesian coordinates can be put into dimensionless form by using scales U for velocity, L for length, and L/U for time. Moreover, a dimensionless pressure p is introduced by $p_* = -\rho g z_* + \rho U^2 p$, where the $*$ subscript indicates the dimensional variables. In this way, the governing equations for the dimensionless velocity vector $\mathbf{u} = (u, w)$ and pressure p become

$$\frac{\partial u}{\partial t} = -\frac{\partial (uu)}{\partial x} - \frac{\partial (wu)}{\partial z} - \frac{\partial p}{\partial x} + Re^{-1} \nabla^2 u, \tag{10.1a}$$

$$\frac{\partial w}{\partial t} = -\frac{\partial (uw)}{\partial x} - \frac{\partial (ww)}{\partial z} - \frac{\partial p}{\partial z} + Re^{-1} \nabla^2 w, \tag{10.1b}$$

$$0 = \frac{\partial u}{\partial x} + \frac{\partial w}{\partial z}, \tag{10.1c}$$

where $Re = UL/\nu$ is the Reynolds number which indicates the ratio between inertial and viscous forces. The dimensionless boundary conditions become

$$x = 0, 1: \quad u = w = 0 \tag{10.2a}$$

$$z = 0: \quad u = w = 0 \tag{10.2b}$$

$$z = D/L = 1/A_x: \quad u = 1, w = 0 \tag{10.2c}$$

In this way, the dimensionless equations contain two parameters: the aspect ratio A_x and the Reynolds number Re.

The system (10.1) is first discretized in space, where we here use central differences on a space-staggered non-uniform grid. Various choices can be made for a discretization on such a grid, but the proper treatment of the convective terms is crucial (Verstappen and Veldman, 1998). The discretization is as discussed in Section 4.2.3 where also explicit formulas are given, and it can still have a decoupling of odd and even grid points at high values of Re. This can be taken care of by using grid refinements towards the walls in order to capture the boundary layers. The refinement used here is determined by the mapping

$$x = \frac{1 + \tanh(2\eta_s(\xi - \frac{1}{2}))}{2 \tanh \eta_s}. \tag{10.3}$$

This expression maps grid points from a uniform grid in terms of ξ to those of a non-equidistant stretched grid in terms of x (see Section 4.1.1). In the results shown in this section, a stretching factor of $\eta_s = 1.5$ is used in both horizontal and vertical directions. The number of grid points in the x- and z-directions is indicated by n_x and n_z, respectively.

The spatial discretization results in a large system of ordinary differential equations of the form

$$\frac{d\mathbf{x}}{dt} = \Phi(\mathbf{x}, \lambda), \quad \mathbf{x}(0) = \mathbf{x}_0 \tag{10.4}$$

where the state vector $\mathbf{x}$ consists of all unknowns (velocity components, pressure) at the grid points and $\lambda = Re$. For the time discretization of equation (10.4) we will use the θ-method with time step Δt, which is given by

$$\mathbf{x}_{n+1} = \mathbf{x}_n + \Delta t[(1 - \theta)\, \Phi(\mathbf{x}_n, \lambda) + \theta\, \Phi(\mathbf{x}_{n+1}, \lambda)]. \tag{10.5}$$

For $\theta = 1/2$ and $\theta = 1$ this is the trapezoidal method (Crank–Nicolson) and backward Euler method, respectively.

As a standard parameter value we take $A_x = 1$. The volume-averaged kinetic energy of the flow, that is,

$$E = \frac{1}{2} \int_0^1 \int_0^{1/A_x} (u^2 + w^2)\, dx dz \tag{10.6}$$

for steady state flows versus Re is plotted in Fig. 10.1 for different values of the number of grid points $n_x = n_z$. It can be seen that a resolution of $n_x = n_z = 256$ is needed to obtain a reasonably converged (partial) bifurcation diagram.

The linear stability of each steady state was determined by solving the associated generalized eigenvalue problem. The path of the eigenvalues $\sigma = \sigma_r \pm i\sigma_i$ with the largest real parts is plotted versus Re (for $n_x = n_z = 256$) in Fig. 10.2. The eigenvalues for $Re = 7,000$ are the left end points of the curves. Note that the eigenvalue with $\sigma_i \sim 2.8$ is the first to cross the imaginary axis and a Hopf bifurcation occurs at $Re_c = 8,089$, with a dimensionless period of oscillation equal to $p = 2\pi/\sigma_i = 2.24$.

Patterns of the flow at $Re = Re_c = 8,089$ can be seen in Fig. 10.3 where the stream function and speed $(u^2 + w^2)^{1/2}$ are plotted. For this value of Re no substantial corner cells can be seen yet. Using the transient integration method (10.5), we next determined a periodic solution for $Re = 8,200$ for $n_x = n_z = 256$. The time series of the volume-averaged kinetic energy is plotted in Fig. 10.4a. Patterns of the stream function of the perturbed flow (actual flow minus the steady state at $Re = 8,200$) are plotted for three phases along the periodic orbit in Figs. 10.4b through 10.4d. The period $p \sim 2.23$ is close to that determined at criticality ($p = 2\pi/\sigma_i$). For each grid size, the location of the first Hopf bifurcation

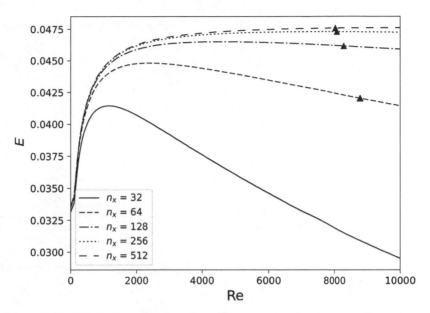

Figure 10.1 Bifurcation diagram for the case $A_x = 1$ showing the volume-averaged kinetic energy E as in (10.6) for various values of $n_x = n_z$. The position of the first Hopf bifurcation is denoted by a triangle.

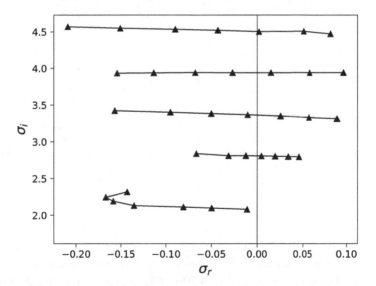

Figure 10.2 Eigenvalues in the complex plane for Reynolds numbers 7,000, 7,500, 8,000, 8,500, 9,000, 9,500, and 10,000 for $n_x = n_z = 256$ and $A_x = 1$.

is indicated by a triangle in Fig. 10.1. Due to the second-order convergence of the location of the bifurcation, we can extrapolate the results and estimate that the first Hopf bifurcation for $A_x = 1$ is located at approximately $Re_c = 8,023$.

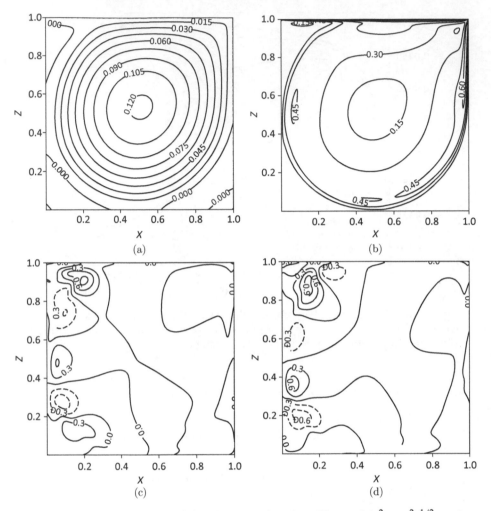

Figure 10.3 Contour plots of the (a) stream function, (b) speed $(u^2 + w^2)^{1/2}$, and (c–d) stream function of the real and imaginary parts of the least stable normal mode of the lid-driven cavity flow for $n_x = n_z = 256$ at $Re = Re_c = 8,089$.

To compare with results in Tiesinga and colleagues (2002), we provide in Table 10.1 the values of the two eigenvalues with the largest real part versus Re for $n_x = n_z = 128$. The first Hopf bifurcation at this resolution occurs at $Re_c = 8,375$ and the second one at about $Re = 8,700$. These results are in agreement with those in Tiesinga and colleagues (2002), who used the PDECONT software (see Section 9.3) with $n_x = n_z = 128$.

Tiesinga and colleagues (2002) also computed the periodic orbit arising at $Re_c = 8,375$ for $n_x = n_z = 128$ by continuation and determined its linear stability. The path of the Floquet multipliers versus Re along this branch of periodic

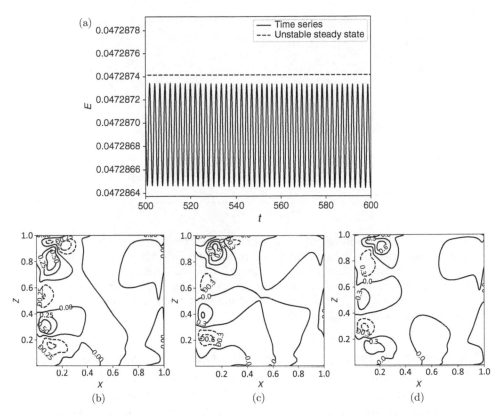

Figure 10.4 (a) Transient of the perturbed solution at $Re = 8,200$ and patterns of the stream function of the perturbed flow at (b) $t = 599$, (c) $t = 599.5$, and (d) $t = 600$ normalized with the maximum value of the perturbation of $6.73 \cdot 10^{-5}$.

Table 10.1 *Eigenvalues and corresponding periods and frequencies for $A_x = 1$ and $n_x = n_z = 128$.*

Re	σ_1		p	$1/p$	σ_2		p	$1/p$
7,750	$-0.0163\pm$	2.7813i	2.2590	0.4427				
8,000	$-0.0068\pm$	2.7762i	2.2633	0.4418				
8,375	$0.0009\pm$	2.7640i	2.2732	0.4399	$-0.0312\pm$	3.7956i	1.6554	0.6041
8,875	$0.0189\pm$	2.7518i	2.2833	0.4380	$0.0275\pm$	3.8058i	1.6509	0.6057
9,375	$0.0313\pm$	2.7354i	2.2970	0.4353	$0.0583\pm$	3.7903i	1.6577	0.6032
9,875	$0.0473\pm$	2.7247i	2.3060	0.4337	$0.0871\pm$	3.7818i	1.6614	0.6019
10,375	$0.0608\pm$	2.7148i	2.3144	0.4321	$0.1136\pm$	3.7755i	1.6642	0.6009

orbits is shown in Fig. 10.5a, where the markers follow the sequence $\circ, \times, +, *, \blacksquare, \diamond$ (cyclically) for increasing values of Re. The periodic orbit is stable for $Re = 8,375$ up to about $Re = 9,150$, where a pair of complex conjugate Floquet multipliers leave the unit disc and hence the periodic solution becomes unstable. The periodic

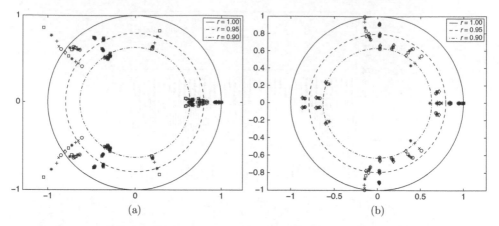

Figure 10.5 (a) The dominant Floquet multipliers of the periodic solution emerging at the first Hopf bifurcation for Re in the interval 8,550–10,000. Note that the fastest moving multipliers move outward. (b) The dominant Floquet multipliers of the periodic solution emerging at the second Hopf bifurcation for Re in the range 8,700–10,000. The fastest moving multipliers move inward. Results from Tiesinga and colleagues (2002).

solution remains unstable up to $Re = 10,000$. At the second Hopf bifurcation point (at $Re = 8,700$), the steady state as well as the emerging periodic solution are unstable. The path of the Floquet multipliers versus Re along this branch of periodic orbits is shown in Fig. 10.5b. Already close to $Re = 8,700$, the unstable periodic orbit stabilizes as the associated Floquet multiplier moves into the unit circle. The periodic solution remains stable up to $Re = 10,000$ (Fig. 10.5b).

Other studies have used (only) transient integration to determine the first Hopf bifurcations and periodic orbits. In Cazemier and colleagues (1998), a smaller value of $Re_c = 7,972$ is found due to the higher resolution and higher-order discretization used. They found 0.60 as the frequency ($1/p$) of the periodic orbit at the first bifurcation point, whereas we find a frequency of 0.44. However, the frequency of the second mode at criticality (causing the second Hopf bifurcation) is 0.6, and hence it appears that in Cazemier and colleagues (1998), they found the stable periodic orbit arising from the second Hopf bifurcation.

Additional Material

- More details on the bifurcation diagrams of lid-driven cavity flows (with the relevant literature) can be found in the relatively recent papers of Kuhlmann and Romanò (2019) and Gelfgat (2019c).

10.2 Axisymmetric Taylor–Couette Flow

The two-dimensional version of the Taylor–Couette flow is that of axisymmetric flow where the velocities (u, v, w) do not depend on the azimuthal coordinate θ.

To obtain non-dimensional equations for the case where $\omega_o = 0$, length is scaled with r_i, velocity with $v_i = \omega_i r_i$, time with r_i/v_i, and pressure (after subtracting the hydrostatic component) with ρv_i^2. This leads to the following equations:

$$\frac{\partial u}{\partial t} + u\frac{\partial u}{\partial r} + w\frac{\partial u}{\partial z} - \frac{v^2}{r} = \tag{10.7a}$$
$$\frac{1}{\sqrt{Ta}}\left(\frac{1}{r}\frac{\partial}{\partial r}\left(r\frac{\partial u}{\partial r}\right) - \frac{u}{r^2} + \frac{\partial^2 u}{\partial z^2}\right) - \frac{\partial p}{\partial r}$$

$$\frac{\partial v}{\partial t} + u\frac{\partial v}{\partial r} + w\frac{\partial v}{\partial z} + \frac{uv}{r} = \tag{10.7b}$$
$$\frac{1}{\sqrt{Ta}}\left(\frac{1}{r}\frac{\partial}{\partial r}\left(r\frac{\partial v}{\partial r}\right) - \frac{v}{r^2} + \frac{\partial^2 v}{\partial z^2}\right) - \frac{1}{r}\frac{\partial p}{\partial \theta}$$

$$\frac{\partial w}{\partial t} + u\frac{\partial w}{\partial r} + w\frac{\partial w}{\partial z} = \tag{10.7c}$$
$$\frac{1}{\sqrt{Ta}}\left(\frac{1}{r}\frac{\partial}{\partial r}\left(r\frac{\partial w}{\partial r}\right) + \frac{\partial^2 w}{\partial z^2}\right) - \frac{\partial p}{\partial z}$$

$$\frac{1}{r}\frac{\partial(ru)}{\partial r} + \frac{\partial w}{\partial z} = 0. \tag{10.7d}$$

Here the Taylor number Ta is related to Re_i in (1.60) by

$$\sqrt{Ta} = \frac{r_i v_i}{\nu} = Re_i\frac{r_i}{d} = Re_i\frac{r_i}{r_o - r_i} = Re_i(\eta^{-1} - 1)^{-1}$$

where $d = r_o - r_i$ is the gap width and $\eta = r_i/r_o$ is the radius ratio. The equations are discretized using central differences on the domain $[1, \eta^{-1}] \times [0, L/r_i]$, and as standard values of the parameters, we choose $\Gamma = 30$ and $\eta = 0.833$; periodic boundary conditions are applied in the z-direction, no-slip in the r-direction, with $\omega_i = 1$ and $\omega_o = 0$.

The parallel Couette flow solution (see Section 1.2) is a solution for all values of the parameters. The linear stability of the Couette flow, while increasing Re_i is determined by the eigenvalues shown for $n_r = n_z = 256$ in Table 10.2. This indicates that there is a sign change in the real part of the first eigenvalue between $Re_i = 98$ and $Re_i = 99$. The convergence of this value of $Re_{i,c}$ at the first bifurcation point with grid size is shown in Table 10.3.

In Fig. 10.6a we show the partial bifurcation diagram over the interval $Re_i \in [96, 100]$ for $n_r = n_z = 256$. At the y-axis, now the value of the vertical velocity w at the location $(1.1, 1.5)$ is plotted, as it is better to distinguish the different solutions. At $Re_i^c = 98.09$, the Couette flow becomes unstable at a pitchfork bifurcation

Table 10.2 *The first five eigenvalues along the Couette flow solution branch versus Re_i between 96 and 100 for $n_r = n_z = 256$; note that $\Gamma = 30$ and $\eta = 0.833$.*

Re_i	σ_1	σ_2	σ_3	σ_4	σ_5
96	$-1.46 \cdot 10^{-1}$	$-1.46 \cdot 10^{-1}$	$-1.62 \cdot 10^{-1}$	$-1.62 \cdot 10^{-1}$	$-1.71 \cdot 10^{-1}$
97	$-7.52 \cdot 10^{-2}$	$-7.52 \cdot 10^{-2}$	$-9.49 \cdot 10^{-2}$	$-9.49 \cdot 10^{-2}$	$-9.69 \cdot 10^{-2}$
98	$-6.15 \cdot 10^{-3}$	$-6.15 \cdot 10^{-3}$	$-2.46 \cdot 10^{-2}$	$-2.46 \cdot 10^{-2}$	$-2.90 \cdot 10^{-2}$
99	$6.16 \cdot 10^{-2}$	$6.16 \cdot 10^{-2}$	$4.65 \cdot 10^{-2}$	$4.65 \cdot 10^{-2}$	$3.56 \cdot 10^{-2}$
100	$1.28 \cdot 10^{-1}$	$1.28 \cdot 10^{-1}$	$1.16 \cdot 10^{-1}$	$1.16 \cdot 10^{-1}$	$9.91 \cdot 10^{-2}$

Table 10.3 *Convergence of the first bifurcation point of the axisymmetric Taylor–Couette flow with grid size for the standard values $\Gamma = 30$ and $\eta = 0.833$ and periodic boundary conditions in z.*

n_x	n_z	$Re_{i,c}$
32	32	107.3454
64	64	98.0332
128	128	98.1072
256	256	98.0911
512	512	98.0909

and then two symmetry-related Taylor vortex patterns become stable solutions; one of these (on the lower branch) is plotted for $Re_i = 100$ in Figs. 10.6b through 10.6c. The value of $Re_{i,c}$ is close to that determined from experimental results as shown in Fig. 1.3, where a value $Re_{i,c} = 94.7$ is given (Recktenwald *et al.*, 1993). The difference is due to the finite length of container and the slightly different value of η.

Additional Material

- First applications of numerical bifurcation methods on the Taylor–Couette flow were by Andrew Cliffe (Cliffe, 1983, 1988) using the finite-element code ENTWIFE. Other results can be found in Pfister et al. (1988) and Mullin et al. (2017).

10.3 Two-Dimensional Rayleigh–Bénard Flow

The general non-dimensional two-dimensional equations (using the same scaling as in Section 2.1.2), with $Ma = 0$, are given by

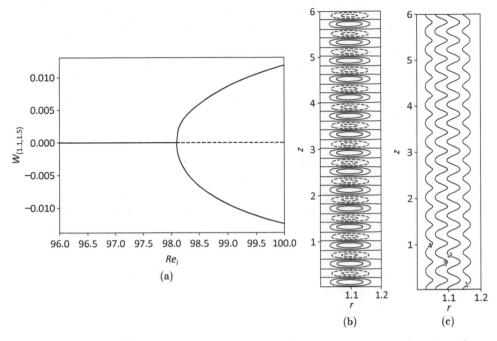

Figure 10.6 (a) Bifurcation diagram axisymmetric case; $w_{(1.1,1.5)}$ as a function of Re_i for $n_r = n_z = 256$. (b–c) Stream function and speed $(u^2 + w^2)^{1/2}$ at $Re_i = 100$ at the lower branch.

$$Pr^{-1} \left(\frac{\partial u}{\partial t} + u\frac{\partial u}{\partial x} + w\frac{\partial u}{\partial z} \right) = -\frac{\partial p}{\partial x} + \nabla^2 u \qquad (10.8a)$$

$$Pr^{-1} \left(\frac{\partial w}{\partial t} + u\frac{\partial w}{\partial x} + w\frac{\partial w}{\partial z} \right) = -\frac{\partial p}{\partial z} + \nabla^2 w + Ra\, T \qquad (10.8b)$$

$$\frac{\partial u}{\partial x} + \frac{\partial w}{\partial z} = 0 \qquad (10.8c)$$

$$\frac{\partial T}{\partial t} + u\frac{\partial T}{\partial x} + w\frac{\partial T}{\partial z} = \nabla^2 T \qquad (10.8d)$$

with boundary conditions

$$z = 0: \quad T = 1\,; \quad \frac{\partial u}{\partial z} = w = 0 \qquad (10.9a)$$

$$z = 1: \quad \frac{\partial u}{\partial z} = 0\,; w = 0;\ \frac{\partial T}{\partial z} = Bi\, T \qquad (10.9b)$$

$$x = 0, A_x: \quad u = w = \frac{\partial T}{\partial x} = 0. \qquad (10.9c)$$

Standard values of the parameters are chosen as $Pr = 10$, $A_x = 10$, and $Bi = 1$. The motionless solution $\bar{u} = \bar{w} = 0, \bar{T}(z) = 1 - Bi\, z/(Bi + 1)$ is a steady solution for all values of Ra (see Section 1.3).

Table 10.4 *The first five eigenvalues for various values of Ra along the motionless solution branch for $n_x = n_z = 256$ and standard parameter values $Bi = 1$, $A_x = 10$, and $Pr = 10$.*

Ra	σ_1	σ_2	σ_3	σ_4	σ_5
1500	$-6.75 \cdot 10^{-3}$	$-8.53 \cdot 10^{-3}$	$-1.45 \cdot 10^{-2}$	$-1.81 \cdot 10^{-2}$	$-2.85 \cdot 10^{-2}$
1520	$-4.77 \cdot 10^{-3}$	$-6.22 \cdot 10^{-3}$	$-1.23 \cdot 10^{-2}$	$-1.60 \cdot 10^{-2}$	$-2.58 \cdot 10^{-2}$
1540	$-2.92 \cdot 10^{-3}$	$-4.33 \cdot 10^{-3}$	$-1.04 \cdot 10^{-2}$	$-1.40 \cdot 10^{-2}$	$-2.40 \cdot 10^{-2}$
1560	$-1.13 \cdot 10^{-3}$	$-2.49 \cdot 10^{-3}$	$-8.65 \cdot 10^{-3}$	$-1.20 \cdot 10^{-2}$	$-2.22 \cdot 10^{-2}$
1580	$6.28 \cdot 10^{-4}$	$-6.94 \cdot 10^{-4}$	$-6.91 \cdot 10^{-3}$	$-1.02 \cdot 10^{-2}$	$-2.05 \cdot 10^{-2}$
1600	$2.28 \cdot 10^{-3}$	$2.07 \cdot 10^{-3}$	$-4.29 \cdot 10^{-3}$	$-7.77 \cdot 10^{-3}$	$-1.88 \cdot 10^{-2}$

Table 10.5 *Convergence of the value of Ra for the first bifurcation point for $n_x = n_z = 256$.*

n_x	n_z	Ra	Factor
16	16	1540.377	
32	32	1560.737	
64	64	1564.305	5.706
128	128	1565.516	2.946
256	256	1565.851	3.615
512	512	1565.940	3.764

The eigenvalues along the motionless flow solution for $n_x = n_z = 256$ are given in Table 10.4. The first eigenvalue crosses the imaginary axis between $Ra = 1,560$ and $Ra = 1,580$. The convergence of this first bifurcation point with grid resolution is shown in Table 10.5. This value of Ra_c is very close to that computed with a very different technique and listed in section 4.8 of Dijkstra (2005), where for a 512×16 grid, a value of $Ra_c = 1563.78$ was found.

Because of the reflection symmetry through the mid-container axis ($x = A_x/2$), a pitchfork bifurcation is expected to occur. The bifurcation structure for the standard parameter values is plotted in the weakly non-linear regime in Fig. 10.7a for $n_x = n_z = 256$. On the vertical axis, the vertical velocity at the grid point $(3, 12)$ (near the upper left corner) is plotted. At the first primary bifurcation point ($Ra = 1,566$), the motionless solution becomes unstable to the roll-cell pattern (Fig. 10.7b) which stabilizes in a super-critical pitchfork bifurcation. Also its symmetry-related pattern stabilizes (Fig. 10.7c), and both patterns are stable up to the end of the computational domain.

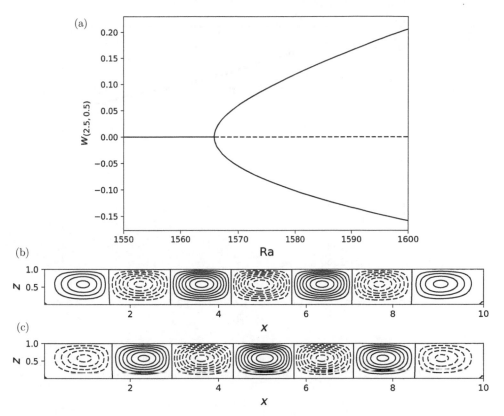

Figure 10.7 (a) Bifurcation diagram and (b–c) symmetry-related cellular solutions for the stream function arising at the first pitchfork bifurcation for $n_x = n_z = 256$, on the lower and upper branches, respectively.

Additional Material

- There is a rich literature on the bifurcation analysis of Rayleigh–Bénard–Marangoni flows, with nice examples in Gelfgat (1999), Puigjaner and colleagues (2011), and Dauby and Lebon (1996).

10.4 Two-Dimensional Differentially Heated Cavity Flow

Using the same scaling as in the Rayleigh–Bénard problem, the dimensionless equations for the Differentially Heated Cavity flow are given by

$$Pr^{-1}\left(\frac{\partial u}{\partial t} + u\frac{\partial u}{\partial x} + w\frac{\partial u}{\partial z}\right) = -\frac{\partial p}{\partial x} + \nabla^2 u \qquad (10.10a)$$

$$Pr^{-1}\left(\frac{\partial w}{\partial t} + u\frac{\partial w}{\partial x} + w\frac{\partial w}{\partial z}\right) = -\frac{\partial p}{\partial z} + \nabla^2 w + Ra\, T \qquad (10.10b)$$

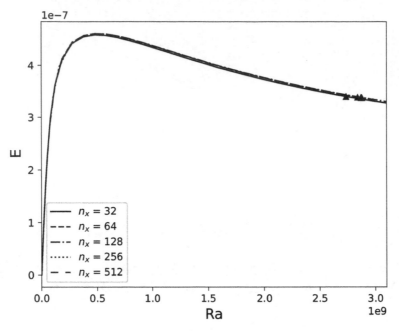

Figure 10.8 Bifurcation diagram showing the volume-averaged kinetic energy E as in (10.6) for various values of $n_x = n_z$. The first Hopf bifurcation is denoted by a triangle; standard parameter values are $Pr = 1,000$ and $A_x = 0.051$.

$$\frac{\partial u}{\partial x} + \frac{\partial w}{\partial z} = 0 \qquad (10.10c)$$

$$\frac{\partial T}{\partial t} + u\frac{\partial T}{\partial x} + w\frac{\partial T}{\partial z} = \nabla^2 T \qquad (10.10d)$$

with boundary conditions

$$z = 0, 1: \frac{\partial T}{\partial z} = u = w = 0 \qquad (10.11a)$$

$$x = 0: \quad u = w = T - 1 = 0 \qquad (10.11b)$$

$$x = A_x: \quad u = w = T = 0 \qquad (10.11c)$$

where the Rayleigh number is now defined as

$$Ra = \frac{\alpha_T g(T_H - T_L)D^3}{\nu\kappa}$$

where D is the layer depth. With the aspect ratio $A_x = L/D$ and the Prandtl number Pr, a total of three parameters results. Motivated by the laboratory experiments in Elder (1965), we choose $Pr = 1,000$ and $A_x = 0.051$ as standard parameter values. The bifurcation diagram, showing the volume averaged kinetic energy versus Ra, is shown in Fig. 10.8 for different mesh sizes. There is already a quite good convergence of this diagram for the coarser grid resolutions.

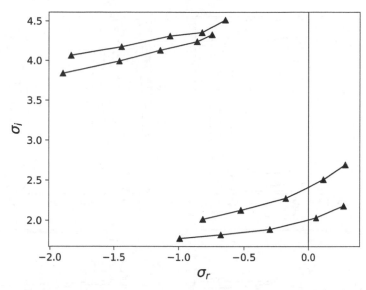

Figure 10.9 Eigenvalues in the complex plane for Rayleigh numbers $2.816 \cdot 10^9$ (left points), $2.842 \cdot 10^9$, $2.867 \cdot 10^9$, $2.893 \cdot 10^9$, and $2.918 \cdot 10^9$ (right points) for $n_x = n_z = 256$.

The linear stability of each steady state was determined by solving the associated generalized eigenvalue problem. The path of the eigenvalues with the largest real parts is plotted versus Ra (for $n_x = n_z = 256$) in Fig. 10.9. The eigenvalues for $Ra = 2.816 \cdot 10^9$ are the left end points of the curves. Note that the eigenvalue with $\sigma_i \sim 2.4$ is the first to cross the imaginary axis and hence a Hopf bifurcation occurs.

Patterns of the steady flow at $Ra = Ra_c = 2.88 \cdot 10^9$ can be seen in Figs. 10.10a through 10.10b, where the Stream function and speed are plotted. The stream function patterns of the real and imaginary parts of the eigenvector at criticality indicate fine-scale instabilities with largest amplitudes in the center of the cavity (Figs. 10.10c–10.10d). Using the transient solution method (10.5), we next determined a periodic solution for $Ra = 2.89 \cdot 10^9$, and the time series of the volume-averaged kinetic energy is plotted in Fig. 10.11a. Patterns of the stream function of the perturbed flow (actual flow minus the unstable steady state at $Ra = 2.88 \cdot 10^9$) are plotted for three phases along the periodic orbit in Figs. 10.11b through 10.11d. We find again that the period p is related to the eigenvalue corresponding to that at the bifurcation by $p = 2\pi/\sigma_i \sim 2.58$.

The location of the first Hopf bifurcation is indicated by a triangle in Fig. 10.8. Due to the second-order convergence of the location of the bifurcation, we can extrapolate the results and estimate that the first Hopf bifurcation is located at approximately $Ra_c = 2.88 \cdot 10^9$. As there have been few studies on the bifurcation behaviour of this flow, this result is difficult to compare with other results and even with the very qualitative experimental results (Elder, 1965).

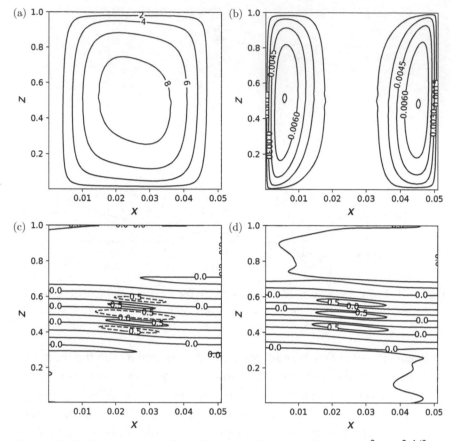

Figure 10.10 Contour plots of the (a) stream function, (b) speed $(u^2 + w^2)^{1/2}$ of the steady-state flow for $n_x = n_z = 256$ at $Ra = Ra_c = 2.88 \cdot 10^9$. (c–d) Stream function of the real and imaginary parts of the least stable eigenmode of the flow for $n_x = n_z = 256$ at $Ra = Ra_c = 2.88 \cdot 10^9$.

Additional Material

- Much work has been done on the bifurcation behaviour of natural convection flows in rectangular cavities with (partially or fully) heated sidewalls (Erenburg *et al.*, 2003; Gelfgat *et al.*, 1999a,b; Gelfgat, 2007).

10.5 Summary

- For the two-dimensional Lid-Driven Cavity flow, the first bifurcation is a Hopf bifurcation and it occurs at $Re_c = 8{,}023$ for $A_x = 1$.
- For the axisymmetric Taylor–Couette flow, the first bifurcation is a pitchfork bifurcation and it occurs, for $\Gamma = 30$ and $\eta^{-1} = 0.833$, at $Re_{i,c} = 98.09$.

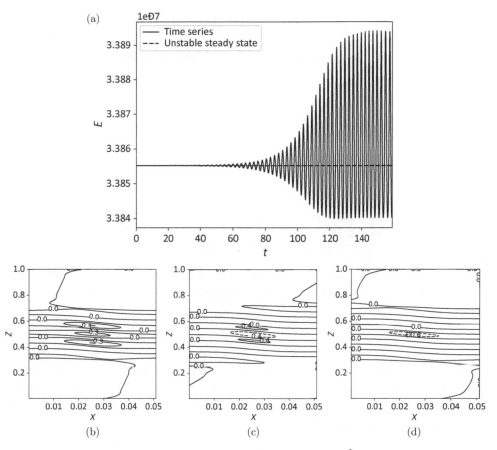

Figure 10.11 (a) Time series generated for $Ra = 2.89 \cdot 10^9$ and patterns of the stream function of the equilibrium perturbed flow at (b) $t = 158.5$, (c) $t = 159$, and (d) $t = 159.5$ normalized with the maximum value of the perturbation of 0.21; again $n_x = n_z = 256$.

- For the two-dimensional Rayleigh–Bénard flow, the first bifurcation is a pitchfork bifurcation and it occurs at $Ra_c = 1{,}566$ for $A_x = 10$, $Pr = 10$.
- For the two-dimensional Differentially Heated Cavity flow, the first bifurcation is a Hopf bifurcation and it occurs at $Ra_c = 2.88 \cdot 10^9$ for $A_x = 0.051$, $Pr = 1{,}000$.
- The Python code to compute all results in this chapter is available on https://github.com/BIMAU/fvm/.

Appendix A

Proofs Related to Chapter 3

In this appendix we just list the proofs related to theorems in Chapter 3.

Proof Lemma 3.1 We first consider the case where there is no viscous term and neither a source or sink. In that case, the contribution of $(\psi, -\mathcal{C}\phi)_\Omega$ to (3.54) is given by

$$- (\phi, \mathcal{C}\phi)_\Omega = -\frac{1}{2}(\phi, (\mathcal{C} + \mathcal{C}^*)\phi)_\Omega. \tag{A.1}$$

So, we have to show that the self-adjoint part of $\mathcal{C}$ with respect to the used inner product is zero inside the domain and that only boundary terms remain. First, we will determine the adjoint of $\mathcal{C}$ defined by $\mathcal{C}\phi = \nabla \cdot (\phi\mathbf{u})$. By replacing ω and $\mathbf{v}$ in (3.36) by ψ and $\phi\mathbf{u}$, respectively, we get

$$\int_\Gamma \phi\psi\mathbf{u} \cdot \mathbf{n}d\Gamma = \int_\Omega \psi\nabla \cdot (\phi\mathbf{u})d\Omega + \int_\Omega \phi\mathbf{u} \cdot \nabla\psi d\Omega = (\psi, \mathcal{C}\phi)_\Omega + \int_\Omega \phi\mathbf{u} \cdot \nabla\psi d\Omega.$$

So, the adjoint $\mathcal{C}^*$ has a domain and a boundary part and is given implicitly by

$$(\mathcal{C}^*\psi, \phi)_\Omega = -\int_\Omega \phi\mathbf{u} \cdot \nabla\psi d\Omega + \int_\Gamma \phi\psi\mathbf{u} \cdot \mathbf{n}d\Gamma. \tag{A.2}$$

Now observe, using (3.35), that $\nabla \cdot (\phi\mathbf{u}) - \mathbf{u} \cdot \nabla\phi = \phi\nabla \cdot \mathbf{u}$ which is zero if $\nabla \cdot \mathbf{u} = 0$. Hence, (A.1) turns into

$$-\frac{1}{2}\int_\Gamma \phi^2\mathbf{u} \cdot \mathbf{n}d\Gamma, \tag{A.3}$$

which, using that $\phi = 0$ on Γ/Γ_R, gives us the associated contribution to the right-hand side of (3.55).

Next, we consider the contribution of diffusion to (3.54) which will add a term $2\mu(\phi, \Delta\phi)$ to (A.1). By replacing ω and $\mathbf{v}$ in (3.36) by ψ and $\nabla\phi$, respectively, we get

$$\int_\Gamma \psi(\nabla\phi \cdot \mathbf{n})d\Gamma = \int_\Omega \psi\Delta\phi d\Omega + \int_\Omega \nabla\psi\nabla\phi d\Omega = (\psi, \Delta\phi)_\Omega + (\nabla\psi, \nabla\phi)_\Omega.$$

So, using again that $\phi = 0$ on Γ / Γ_R, diffusion will add the term

$$\mu\Big(\int_{\Gamma_R} \phi(\nabla\phi \cdot \mathbf{n})d\Gamma - (\nabla\phi, \nabla\phi)_\Omega\Big)$$

to the right-hand side of (3.55).

The remaining part is just the Robin boundary condition, namely $(\phi, \alpha\phi - \beta\frac{\partial\phi}{\partial n})_{\Gamma_R}$. ∎

Proof Lemma 3.2 We have $\frac{d}{dt}(\phi, \phi)_\Omega = 2(\phi, \frac{\partial\phi}{\partial t})_\Omega$. Substitution of the right-hand side of (3.52) and adding the term with the Robin condition, namely $(\phi, \alpha\phi - \beta\frac{\partial\phi}{\partial n} - b)_{\Gamma_R}$, leads to the desired result. ∎

Proof Theorem 3.4 From (3.55) we have

$$\frac{d}{dt}(\phi, \phi)_\Omega \leq -2c_P\mu(\phi, \phi)_\Omega + 2(f, \phi)_\Omega. \tag{A.4}$$

Note that

$$\frac{d}{dt}(\phi, \phi)_\Omega = \frac{d}{dt}||\phi||_\Omega^2 = 2||\phi||_\Omega \frac{d}{dt}||\phi||_\Omega.$$

and by the Cauchy–Schwarz inequality $(f, \phi)_\Omega \leq ||f||_\Omega ||\phi||_\Omega$. By substituting these expressions into (A.4) and dividing by $2||\phi||_\Omega$, we obtain the result. ∎

Proof Theorem 3.5 We can bound (3.55) from Theorem 3.2 by

$$\frac{d}{dt}||\phi||_\Omega^2 \leq -2a(\phi, \phi) + 2c_{b,0}||\phi||_\Omega + 2c_{b,1}||\nabla\phi||_\Omega + 2||f||_\Omega||\phi||_\Omega.$$

The boundary conditions given by Corollary 3.2 are such that $||\nabla\phi||_\Omega^2 \leq a(\phi, \phi)/\mu$. Hence, we can bound $||\nabla\phi||_\Omega$ by the square root of $a(\phi, \phi)$ giving

$$\frac{d}{dt}||\phi||_\Omega^2 \leq -2a(\phi, \phi) + 2c_{b,1}\sqrt{a(\phi, \phi)/\mu} + 2(c_{b,0} + ||f||_\Omega)||\phi||_\Omega.$$

Again, performing the differentiation as in the proof of Theorem 3.5 earlier in the left-hand side and dividing the inequality by $2||\phi||_\Omega$ yields

$$\frac{d}{dt}||\phi||_\Omega \leq -\frac{a(\phi, \phi)}{||\phi||_\Omega} + \frac{c_{b,1}}{\sqrt{\mu}}\sqrt{\frac{a(\phi, \phi)}{||\phi||_\Omega^2}} + ||f||_\Omega + c_{b,0}$$

$$= \sqrt{\frac{a(\phi, \phi)}{||\phi||_\Omega^2}}(-\sqrt{a(\phi, \phi)} + \frac{c_{b,1}}{\sqrt{\mu}}) + ||f||_\Omega + c_{b,0}$$

$$\leq \sqrt{\frac{a(\phi, \phi)}{||\phi||_\Omega^2}}(-\sqrt{c_c}||\phi||_\Omega + \frac{c_{b,1}}{\sqrt{\mu}}) + ||f||_\Omega + c_{b,0}.$$

If in this expression $||\phi||_\Omega \geq \frac{c_{b,1}}{\sqrt{\mu c_c}}$, then the right-hand side is less than

$$\sqrt{c_c}(-\sqrt{c_c}||\phi||_\Omega + \frac{c_{b,1}}{\sqrt{\mu}}) + ||f||_\Omega + c_{b,0}.$$

This is negative if $||\phi||_\Omega > (c_{b,1}\sqrt{\frac{c_c}{\mu}} + ||f||_\Omega + c_{b,0})/c_c$. The maximum of the two criteria on ϕ, which make the right-hand side negative, gives us the sphere in which the solution will dwell for t to infinity. ∎

Proof Lemma 3.3 Using (3.36) as in convection diffusion in incompressible flows, described in Section 3.6.4, one can rewrite $a(\mathbf{u}, \mathbf{u})$ from (3.58a) to

$$a(\mathbf{u}, \mathbf{u}) = (|\mathbf{u}|^2, \mathbf{u} \cdot \mathbf{n})_\Gamma/2 + \mu||(I \otimes \nabla)\mathbf{u}||_\Omega^2 - \mu((\frac{\partial}{\partial n}\mathbf{u}, \mathbf{u})_\Gamma + (\nabla p, \mathbf{u})_\Omega$$
$$- (B(\mathbf{u}, p), \mathbf{u})_\Gamma.$$

Now we employ (3.36) again to find that $(\nabla p, \mathbf{u})_\Omega = (1, \nabla \cdot (p\mathbf{u}))_\Omega = (p, \mathbf{u} \cdot \mathbf{n})_\Gamma$, where we have used that $\nabla \cdot \mathbf{u} = \mathcal{C}1 = 0$. So, $a(\mathbf{u}, \mathbf{u})$ transforms into the result (3.59). ∎

Proof Theorem 3.6 The Dirichlet boundary condition $\mathbf{u} = 0$ makes all boundary contributions zero. Hence, this is a valid boundary condition.

Next, we only set the normal velocity to the wall to zero, that is, $\mathbf{u} \cdot \mathbf{n} = 0$. Then the part $-\mu(\frac{\partial}{\partial n}\mathbf{u}, \mathbf{u})_\Gamma - (B(\mathbf{u}, p), \mathbf{u})_\Gamma$ remains. As before, we can let these terms cancel each other. Hence, $B(\mathbf{u}, p) = -\mu\frac{\partial}{\partial n}\mathbf{u} = -\mu\frac{\partial}{\partial n}(I - \mathbf{nn}^T)\mathbf{u}$, where the last step holds because the normal velocity is zero. This leads to the condition (3.60b).

Finally, suppose the normal velocity is not zero. For that, we combine the terms on the boundary

$$(|\mathbf{u}|^2/2 + p, \mathbf{u} \cdot \mathbf{n})_\Gamma - \mu(\frac{\partial}{\partial n}\mathbf{u}, \mathbf{u})_\Gamma - (B(\mathbf{u}, p), \mathbf{u})_\Gamma =$$
$$((|\mathbf{u}|^2/2 + p - \mu\nabla\mathbf{u}) \cdot \mathbf{n} - B(\mathbf{u}, p), \mathbf{u})_\Gamma. \tag{A.5}$$

So, choosing $B(\mathbf{u}, p) = (p - \mu\nabla\mathbf{u}) \cdot \mathbf{n}$ can be done on an outflow boundary, since then the dropped square has a positive contribution. The expression suggests we should not do that at the inflow boundary. However, this appears to work also. This is due to the fact that we demand monotonic decrease of the energy. If one allows some temporary increase, one can get the same condition to hold at inflow. We will not go into this detail; see Nordström (2022) for a more general exposition. ∎

In order to see what prescribing $(p - \mu\nabla\mathbf{u}) \cdot \mathbf{n}$ means, we use the splitting $\mathbf{u} = (\mathbf{u} \cdot \mathbf{n})\mathbf{n} + (I - \mathbf{nn}^T)\mathbf{u}$ to obtain

$$\nabla\mathbf{u}^T = \mathbf{n}^T\nabla(\mathbf{u} \cdot \mathbf{n}) + \nabla((I - \mathbf{nn}^T)\mathbf{u})^T. \tag{A.6}$$

Consequently, we have

$$\nabla \mathbf{u}^T \cdot \mathbf{n} = \frac{\partial}{\partial n}(\mathbf{u} \cdot \mathbf{n})\mathbf{n} + \frac{\partial}{\partial n}(I - \mathbf{n}\mathbf{n}^T)\mathbf{u}. \tag{A.7}$$

So, we prescribe $(p - \mu\frac{\partial}{\partial n}(\mathbf{u} \cdot \mathbf{n}))\mathbf{n}$ and $\frac{\partial}{\partial n}(I - \mathbf{n}\mathbf{n}^T)\mathbf{u}$, where the latter is a slip condition if it is set to zero.

Proof Theorem 3.7 The energy equation related to (3.62) becomes

$$\frac{d}{dt}||\tilde{\mathbf{u}}||_\Omega^2 = -2\mu||(I\otimes\nabla)\tilde{\mathbf{u}}||_\Omega^2 - 2(\tilde{\mathbf{u}}, (I\otimes\mathcal{C}(\bar{\mathbf{u}}+\tilde{\mathbf{u}}))\tilde{\mathbf{u}})_\Omega - 2(\tilde{\mathbf{u}}, (I\otimes\mathcal{C}(\tilde{\mathbf{u}}))\bar{\mathbf{u}})_\Omega - 2(\tilde{\mathbf{u}}, \mathbf{g})_\Omega,$$

where the whole term with p cancels out due to the fact that normal velocities are zero at the walls; see the contribution it gives in the energy expression (3.59) in Lemma 3.3. The second term is just (A.1) for each component of $\tilde{\mathbf{u}}$ and leads to (A.3) for each component and hence, since the normal velocity is zero, it is zero. The third term is not zero, and we need to bound it. Using the adjoint of $\mathcal{C}$ (see (A.2)), we can write it as

$$((I \otimes \mathcal{C}(\tilde{\mathbf{u}}))\bar{\mathbf{u}}, \tilde{\mathbf{u}})_\Omega = -(I \otimes \tilde{\mathbf{u}} \cdot \nabla)\bar{\mathbf{u}}, \tilde{\mathbf{u}})_\Omega.$$

Now, by Cauchy Schwarz on the vector, where the length of a vector is indicated by $|\cdot|$, this will be less than

$$(|(I \otimes \tilde{\mathbf{u}} \cdot \nabla)\bar{\mathbf{u}}|, |\bar{\mathbf{u}}|)_\Omega \leq \max_\Omega |\bar{\mathbf{u}}|(|(I \otimes \tilde{\mathbf{u}} \cdot \nabla)\bar{\mathbf{u}}|, 1)_\Omega.$$

Furthermore, $|(I \otimes \tilde{\mathbf{u}} \cdot \nabla)\bar{\mathbf{u}}|^2 = (\tilde{\mathbf{u}}, \nabla u)^2 + (\tilde{\mathbf{u}}, \nabla v)^2 + (\tilde{\mathbf{u}}, \nabla w)^2$. This is less than $|\tilde{\mathbf{u}}|^2(|\nabla u|^2 + |\nabla v|^2 + |\nabla w|^2) = |\tilde{\mathbf{u}}|^2|(I \otimes \nabla)\bar{\mathbf{u}}|^2$. So the energy equation can be bounded by

$$\frac{d}{dt}||\tilde{\mathbf{u}}||_\Omega^2 < -2\mu||(I \otimes \nabla)\tilde{\mathbf{u}}||_\Omega^2 + 2U||\tilde{\mathbf{u}}||_\Omega||(I \otimes \nabla)\tilde{\mathbf{u}}||_\Omega + 2||\tilde{\mathbf{u}}||_\Omega||\mathbf{g}||_\Omega. \tag{A.8}$$

As in the proof of Theorem 3.5 earlier, performing the differentiation in the left-hand side and dividing by $2||\tilde{\mathbf{u}}||_\Omega$ yields the result (3.63). ∎

Proof Theorem 3.8 Similar to (3.62), we get now

$$\frac{\partial}{\partial t}\tilde{\mathbf{u}} = -(I \otimes \mathcal{C}(\tilde{\mathbf{u}}))\tilde{\mathbf{u}} - (I \otimes \mathcal{C}(\tilde{\mathbf{u}}))\bar{\mathbf{u}} - (I \otimes \mathcal{C}(\bar{\mathbf{u}}))\tilde{\mathbf{u}} + \mu(I \otimes \Delta)\tilde{\mathbf{u}}$$
$$+ \nabla\tilde{p} - \alpha g\tilde{T}\mathbf{e}_z + \mathbf{g}, \tag{A.9a}$$
$$0 = \mathcal{C}(\tilde{\mathbf{u}})1, \tag{A.9b}$$

where

$$\mathbf{g} = [I \otimes (-\mathcal{C}(\bar{\mathbf{u}}) + \mu\Delta)]\bar{\mathbf{u}} + \nabla\bar{p} + \alpha g\bar{T}\mathbf{e}_z,$$

and

$$\frac{\partial \tilde{T}}{\partial t} = -\mathcal{C}(\tilde{\mathbf{u}})\bar{T} - \mathcal{C}(\bar{\mathbf{u}})\tilde{T} - \mathcal{C}(\tilde{\mathbf{u}})\tilde{T} + \kappa \Delta \tilde{T} + g_T,$$

where $g_T = -\mathcal{C}(\bar{\mathbf{u}})\bar{T} + \kappa \Delta \bar{T}$. When we now consider the energy $||\tilde{\mathbf{u}}||_\Omega^2 + \gamma^2 ||\tilde{T}||_\Omega^2$, then for the $\tilde{T}$ part we have, since all normal velocities are zero and $\tilde{T} = 0$ at the lateral walls and at the other walls we have insulation,

$$\frac{1}{2}\frac{d}{dt}||\tilde{T}||_\Omega^2 = -\kappa ||\nabla \tilde{T}||_\Omega^2 + (\tilde{T}, g_T)_\Omega \leq -\kappa ||\nabla \tilde{T}||_\Omega^2 + ||\tilde{T}||_\Omega ||g_T||_\Omega.$$

If we combine this with (A.8), then we have

$$\begin{aligned}
\frac{1}{2}\frac{d}{dt} \, ||[\tilde{\mathbf{u}}; \gamma \tilde{T}]||_\Omega^2 &= \frac{1}{2}\frac{d}{dt}||\tilde{\mathbf{u}}||_\Omega^2 + \gamma^2 ||\tilde{T}||_\Omega^2 \\
&\leq -\mu ||(I \otimes \nabla)\tilde{\mathbf{u}}||_\Omega^2 + U||\tilde{\mathbf{u}}||_\Omega ||(I \otimes \nabla)\tilde{\mathbf{u}}||_\Omega + \alpha g ||\tilde{\mathbf{u}}||_\Omega ||\tilde{T}||_\Omega \\
&\quad + ||\tilde{\mathbf{u}}||_\Omega ||\mathbf{g}||_\Omega - \gamma^2 \kappa ||\nabla \tilde{T}||_\Omega^2 + \gamma^2 ||\tilde{T}||_\Omega ||g_T||_\Omega \qquad\qquad \text{(A.10)} \\
&\leq -\min(\mu, \kappa)||(I \otimes \nabla)[\tilde{\mathbf{u}}; \gamma \tilde{T}]||_\Omega^2 + U||[\tilde{\mathbf{u}}; \gamma \tilde{T}]||_\Omega ||(I \otimes \nabla)[\tilde{\mathbf{u}}; \gamma \tilde{T}]||_\Omega \\
&\quad + \frac{\alpha g}{|\gamma|}||[\tilde{\mathbf{u}}; \gamma \tilde{T}]||_\Omega^2 + ||[\tilde{\mathbf{u}}; \gamma \tilde{T}]||_\Omega \max(||g_T||_\Omega, ||\mathbf{g}||_\Omega).
\end{aligned}$$

This leads to

$$\begin{aligned}
\frac{d}{dt}||[\tilde{\mathbf{u}}; \gamma \tilde{T}]||_\Omega &\leq -\nu ||(I \otimes \nabla)[\tilde{\mathbf{u}}; \gamma \tilde{T}]||_\Omega \left(\frac{||(I \otimes \nabla)[\tilde{\mathbf{u}}; \gamma \tilde{T}]||_\Omega}{||[\tilde{\mathbf{u}}, \gamma \tilde{T}||_\Omega} - \frac{U}{\nu} \right) \\
&\quad + \frac{\alpha g}{|\gamma|}||[\tilde{\mathbf{u}}; \gamma \tilde{T}]||_\Omega + \max(||g_T||_\Omega, ||\mathbf{g}||_\Omega) \\
&\leq -\nu ||(I \otimes \nabla)[\tilde{\mathbf{u}}; \gamma \tilde{T}]||_\Omega \left(||(\sqrt{c_P} - \frac{U}{\nu}) \right) \\
&\quad + \frac{\alpha g}{|\gamma|}||[\tilde{\mathbf{u}}; \gamma \tilde{T}]||_\Omega + \max(||g_T||_\Omega, ||\mathbf{g}||_\Omega),
\end{aligned}$$

where $\nu = \min(\mu, \kappa)$. If $Re_{c_P} \equiv U/(\nu \sqrt{c_P}) < 1$, we can bound this by

$$\begin{aligned}
\frac{d}{dt}||[\tilde{\mathbf{u}}; \gamma \tilde{T}]||_\Omega &\leq -\nu c_P ||[\tilde{\mathbf{u}}; \gamma \tilde{T}]||_\Omega (1 - Re_{c_P}) \\
&\quad + \frac{\alpha g}{|\gamma|}||[\tilde{\mathbf{u}}; \gamma \tilde{T}]||_\Omega + \max(||g_T||_\Omega, ||\mathbf{g}||_\Omega) \\
&\leq \left(-\nu c_P (1 - Re_{c_P}) + \frac{\alpha g}{|\gamma|} \right) ||[\tilde{\mathbf{u}}; \gamma \tilde{T}]||_\Omega + \max(||g_T||_\Omega, ||\mathbf{g}||_\Omega).
\end{aligned}$$

■

Appendix B
Relevant Linear Algebra

To know whether a matrix is non-singular is important when it comes to solving a linear system with that matrix. Knowing in which area of the domain eigenvalues of a Jacobian matrix are located makes it possible to show the stability of an ODE. Theorems on these subjects will be treated in the first section of this appendix. In the second, we will discuss bounds on solutions in iterative processes and solutions of ODEs; it sheds light on the question of when we may expect monotonic decrease of energy. Finally, in the last section we will consider conditions under which solutions of iterations and ODEs remain positive, which is relevant when computing positive physical quantities like concentrations and masses. We give most of the theorems without proof; for proofs we refer to the references in Hogben (2006).

B.1 Localization of Eigenvalues

There exist a few useful tools to determine whether a matrix is non-singular or to localize eigenvalues of a matrix. We start off with a few definitions (see also Burden and Faires, 2001, Sections 7.1–2, 9.1).

Definition B.1 (Spectrum) *The spectrum is the set of all eigenvalues of a matrix A and denoted by $\sigma(A)$.*

It is obvious that all eigenvalues are contained in a disc in the complex plane with the origin as center and a radius as big as the eigenvalue biggest in absolute value. This leads to the following definition.

Definition B.2 (Spectral radius) *The spectral radius $\rho(A)$ is $\max_{\lambda \in \sigma(A)} |\lambda|$.*

Definition B.3 (Vector-induced matrix norm) *Any vector p-norm defined by $||\mathbf{x}||_p \equiv (\sum_{i=1}^{n} |x_i|^p)^{1/p}$ induces a matrix norm defined by $||A||_p = \max_{\mathbf{x} \in C^n} ||A\mathbf{x}||_p / ||\mathbf{x}||_p$.*

Theorem B.1 *It holds that $\rho(A) \leq ||A||_p$ for any p.*

This theorem is particularly useful for $p = 1$ and $p = \infty$, because then the matrix norms are easy to compute.

Theorem B.2 *It holds that $||A||_1 = \max_{1 \leq j \leq N} \sum_{i=1}^{N} |a_{ij}|$ and $||A||_\infty = ||A^T||_1$.*

Definition B.4 (Similarity) *Two matrices A and B are called similar if there exists a non-singular matrix Q such that $B = Q^{-1}AQ$.*

The operation with Q on A is called a *similarity transformation*.

Theorem B.3 (Similar matrices) *Two similar matrices have the same spectrum and spectral properties, that is, the same algebraic and geometric multiplicities.*

Definition B.5 *A complex matrix A is a normal matrix if $AA^* = A^*A$.*

Theorem B.4 *Normal matrices are similar to a diagonal matrix through a unitary transformation.*

Definition B.6 (Permutation matrix) *A matrix P is an identity matrix in which the rows are permuted.*

An example of a permutation matrix is

$$P = \begin{bmatrix} 0 & 0 & 1 \\ 1 & 0 & 0 \\ 0 & 1 & 0 \end{bmatrix}.$$

Pre-multiplication of A with P enforces a permutation of the rows of A. In the case of the example, all rows shift down one and the last row becomes the first. Similarly, post-multiplication enforces a permutation of the columns.

The transformation PAP^T results in a mutual permutation where the rows and columns are permuted in the same way. This ensures that a diagonal of the original is also a diagonal element of the permuted matrix. It also holds that $P^{-1} = P^T$, hence the special transformation is a similarity transformation.

Definition B.7 (Irreducibility) *A matrix A is called irreducible if there is no permutation matrix P such that*

$$PAP^T = \begin{bmatrix} B_{11} & B_{12} \\ O & B_{22} \end{bmatrix}, \tag{B.1}$$

where B_{11} and B_{22} are square matrices and O a zero matrix.

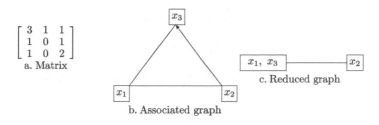

Figure B.1 Matrix with associated graph and reduced graph.

A is *reducible* if it is not irreducible. In fact, for a matrix to be irreducible, for every unknown there should be a connection to every other unknown. This connection may run via other unknowns. So, a reducible matrix exists if some unknown, and the set of all unknowns it is connected to, is not connected to another unknown outside this set.

This can also be interpreted by using graphs. The nodes (or vertices) of the graph are the unknowns, and there is an arrow (or edge) from x_i to x_j if $A_{i,j} \neq 0$. If also $A_{j,i} \neq 0$, which is always the case for a symmetric matrix, there is also an arrow from x_j to x_i. The resulting two arrows between x_i and x_j are replaced by one arc indicating that the connection is two-sided. In Fig. B.1a, one finds a matrix and in part b its associated graph. In order to find whether a matrix is reducible one repeatedly groups nodes that are connected by an arc. In Fig. B.1b, one can group x_1 and x_3 in this way and find a new graph as in Fig. B.1c. This graph shows again an arc. Hence, one can also add x_2 to the group, which means that the matrix is irreducible. In general, in the end this will lead to a graph with arrows only, or all nodes are in one set. By a proper ordering of the nodes, for example by putting the nodes with least incoming arrows last in the vector, one then can find the block form of the matrix.

An important property of reducible matrices is the following.

Theorem B.5 *It holds that the set of eigenvalues of a reducible matrix A are the union of the sets of eigenvalues of the diagonal blocks.*

For the matrix (B.1) we have $\sigma(A) = \sigma(B_{11}) \cup \sigma(B_{22})$. One should be aware that there can be many blocks, for example a triangular matrix has N blocks of size 1×1.

Theorem B.6 (Gershgorin) *Let A be an arbitrary (complex) matrix of order N and let every row define a radius by*

$$r_i = \sum_{\substack{j=1 \\ j \neq i}}^{N} |a_{ij}|, \qquad i = 1, 2, \dots, N.$$

Then the eigenvalues λ of A are in the union of the Gershgorin discs

$$|z - a_{ii}| \leq r_i, \qquad i = 1, 2, \ldots, N.$$

Moreover, if m discs are disjunct from the remaining $N - m$ circles, then these m discs contain m eigenvalues.

Proof We only prove the first part here. Let λ be an eigenvalue of A with corresponding eigenvector x. We normalize x such that for the biggest element it holds that $|x_k| = 1$. From the row corresponding to this element in $Ax = \lambda x$ it follows that

$$a_{kk}x_k + \sum_{\substack{j=1 \\ j \neq k}}^{N} a_{kj}x_j = \lambda x_k.$$

After rearranging and taking absolutes, we have

$$|\lambda - a_{kk}| = |\sum_{\substack{j=1 \\ j \neq k}}^{N} a_{kj}x_j| \leq \sum_{\substack{j=1 \\ j \neq k}}^{N} |a_{kj}||x_j| \leq \sum_{\substack{j=1 \\ j \neq k}}^{N} |a_{kj}| = r_k. \qquad (B.2)$$

Hence, λ is in the disc corresponding to the row where the eigenvector is biggest. From this it follows that any λ is inside a Gershgorin disc. ∎

Since the theorem also holds for A^T one can also base the radii of the Gershgorin circles on the columns. Moreover, this gives anoher set containing the eigenvalues and we can take the intersection of the two which must contain all the eigenvalues.

Theorem B.7 (Taussky) *For an irreducible matrix A of order N it holds that a point λ on the boundary of the union of all Gershgorin discs can only be an eigenvalue of A if it is on the boundary of each disc (and the corresponding eigenvector has components of equal magnitude).*

Proof Suppose that the eigenvalue λ of A is on the boundary of the union of Gershgorin discs. This means that for each disc, λ is on its boundary or outside of it. Hence, using the notation of the previous theorem,

$$|\lambda - a_{ii}| \geq r_i, \; i = 1, 2, \ldots, N.$$

Combining with (B.2), we see that this is only possible if we have equality, hence

$$\sum_{\substack{j=1 \\ j \neq k}}^{N} |a_{kj}||x_j| = \sum_{\substack{j=1 \\ j \neq k}}^{N} |a_{kj}|.$$

This in itself is only possible if $|x_l| = 1$ for all l for which $a_{kl} \neq 0$. So, the eigenvalue must be on the boundary of the disc where the eigenvector is biggest and moreover the eigenvector must have values of equal magnitude for all its components that are used on this row. The last means that we could also have taken the row l instead of the row k in (B.2), and that also here we will find equality as above and that the elements of the eigenvector used in this row should be of equal magnitude 1. Due to the fact that the matrix is irreducible, we can reach every row in the matrix, herewith proving that the eigenvalue should be on the boundary of all the discs and that the magnitude of all the components of the corresponding eigenvector should be equal. ∎

Usually we apply this theorem to show that a matrix is non-singular. However, if we have to deal with a symmetric real matrix, then we know that all eigenvalues and eigenvectors are real. So, if the circles go through one point on the boundary of the union, then this point is an eigenvalue if the components of the eigenvector can be chosen plus or minus one. If not one vector of such a type is an eigenvector, the point is still not an eigenvalue.

We can use this to show that the class of weakly diagonally dominant matrices are non-singular. This is defined as follows

Definition B.8 (Weak diagonal dominance) *A matrix A of order N is called weakly diagonally dominant when*

$$|a_{ii}| \geq \sum_{\substack{j=1 \\ j \neq i}}^{N} |a_{ij}| \ \textit{for} \ i = 1, 2, \ldots, N$$

and the $>$ sign should hold for at least one of the equations.

Theorem B.8 *If A is an irreducible, weakly diagonally dominant matrix of order N, then A is non-singular.*

Proof There is at least one disc which doesn't contain 0. Hence, according to Taussky's theorem, 0 cannot be an eigenvalue. Hence, the matrix is non-singular. ∎

We remark that the Gershgorin theorem only indicates that a matrix is non-singular. In the case where the matrix originates from a discretization of an elliptic equation also the discrete Poincaré inequality (cf. Theorem 4.4) shows that the matrix is non-singular. In fact, it gives a lower bound for the smallest eigenvalue of the symmetric part of the matrix.

Before, we mentioned that the coercivity condition (3.24) is a kind of positive definiteness condition, and in Example 4.9b we already referred to the following definition.

Definition B.9 (Positive definiteness) *A matrix is A is called positive definite if* $(x, Ax) > 0$
for all nonzero complex x.

Theorem B.9 *If A is positive definite, then it has positive eigenvalues. The reverse holds if
A is real symmetric or Hermitian.*

In numerical analysis it is quite important to know that a matrix is positive definite. For instance, if we also add time integration, then A may be the Jacobian
of the problem. With positive eigenvalues we know that perturbations will grow
exponentially, while with negative eigenvalues perturbations damp out. For some
special cases we can read positive definiteness immediately from the matrix. The
following theorem gives an example.

Theorem B.10 *If A has positive diagonal elements and moreover is symmetric, irreducible,
and weakly diagonally dominant, then it is positive definite.*

Proof The eigenvalues of a symmetric matrix are real, and from the location of the
Gershgorin discs we know that all eigenvalues are on the positive real axis.   ∎

Also the *Field of Values* of a matrix can help to locate the eigenvalues of a matrix
in the complex plane.

Definition B.10 (Field of Values) *The field of values of a matrix A is the region in the
complex plane given by*

$$F(A) = \{\frac{(x, Ax)}{(x, x)} | x \in \mathbb{C}^n, x \neq 0\}.$$

Definition B.11 (Generalized Field of Values) *The generalized field of values of a matrix
A is the region in the complex plane given by*

$$F_H(A) = \{\frac{(x, HAx)}{(x, Hx)} | x \in \mathbb{C}^n, x \neq 0\}$$

where H is positive definite, but not necessarily Hermitian.

Theorem B.11 $\sigma(A) \subset F(A)$ *and* $\sigma(A) \subset F_H(A)$.

Theorem B.12 *Let* $A = \alpha I + \beta B$; *then* $F(A) = \alpha + \beta F(B)$.

Theorem B.13 *If A is a principle submatrix of B, then* $F(A) \subseteq F(B)$.

Theorem B.14 $F(A)$ *is contained in the disc around zero with radius the largest singular
value of A.*

Proof From the Cauchy–Schwarz inequality, we have $|(x, Ax)| \leq ||x||_2||Ax||_2 \leq ||x||^2||A||_2 = ||x||_2^2\sigma_{max}$. ∎

Theorem B.15 *If A or $P^{-1}AP$ is normal, then the convex hull of $\sigma(A)$ is $F(A)$ or convex hull of $\sigma(A)$ is $F_{(PP^T)^{-1}}(A)$, respectively.*

For Toeplitz matrices one can find a relation between its field of values and that obtained from a Fourier analysis on the associated difference stencil, that is, the regular part of the matrix.

Definition B.12 (Toeplitz matrix) *A Toeplitz matrix A of order n is determined by a sequence of coefficients a_i, $i = 0, 1, \ldots, 2n - 1$ through $A_{ij} = a_{\{j-i \bmod 2n-1\}}$.*

Theorem B.16 *If, for the sequence of coefficients a_i, $i = 0, 1, \ldots, 2n - 1$, $a_i = 0$ for $i = k + 1, \ldots, 2n - l - 1$, $-1 \leq k \leq n - 1$, and $1 \leq l \leq n - 1$. For the Toeplitz matrix A it determines by Definition B.12, it holds that $F(A)$ is contained in the convex hull of the Fourier eigenvalues of the associated difference operator defined by $\sum_{i=-l}^{k} a_{\{i \bmod 2n-1\}} u_i$.*

Proof A Toeplitz matrix A of order n, can be extended to a circulant matrix B of order $2n - 1$, with coefficients $B_{ij} = a_{\{j-i \bmod 2n-1\}}$, and we can apply Th. B.13. The circulant matrix is normal and hence we can apply Theorem B.15, where the eigenvalues are given by $q(\exp(im2\pi/(2n - 1)))$, $m = 1, \ldots, 2n - 1$ where $q(x) = \sum_{i=-l}^{k} a_{\{i \bmod 2n-1\}} x^i$. The convex hull of these eigenvalues will be contained in that of $q(\exp(i\phi))$ for $\phi \in [0, 2\pi]$, which will result from applying the Fourier analysis to the given difference scheme. ∎

Theorem B.17 *Any matrix can be split into a Hermitian and a skew Hermitian part: $A = (A + A^*)/2 + (A - A^*)/2$, where the first term is Hermitian and the second skew Hermitian.*

Theorem B.18 *$F_{(A+A^*)/2}$ is real and $\min(\sigma((A + A^*)/2)) \leq F_{(A+A^*)/2} \leq \max(\sigma((A + A^*)/2))$. Likewise, the Field of Values of the skew symmetric part is purely imaginary and it holds that $\min(imag(\sigma((A - A^*)/2))) \leq imag(F_{(A-A^*)/2}) \leq \max(imag(\sigma((A - A^*)/2)))$.*

Theorem B.19 (Bendixon) *$F_A \subseteq (F_{(A+A^*)/2} + F_{(A-A^*)/2})$, hence $\sigma(A)$ in the rectangle is defined by the two corners $(\min(\sigma((A + A^*)/2)), i\min(imag(\sigma((A - A^*)/2)))$ and $(\max(\sigma((A + A^*)/2)), i\max(imag(\sigma((A - A^*)/2)))$*

Theorem B.20 *If A is real, then if (λ, x) is an eigenpair, then $(\bar{\lambda}, \bar{x})$ also is an eigenpair. Hence, $\min(imag(\sigma((A - A^*)/2))) = -\max(imag(\sigma((A - A^*)/2)))$.*

Theorem B.21 *If A is positive definite, then also $S^T A S$ is positive definite for any non-singular. S, and moreover any matrix that is similar to $S^T A S$ has its spectrum in the positive half plane.*

Corollary In particular $\sigma(SS^T A) > 0$, hence $\sigma(HA) > 0$ for any SPD matrix H.

Loisel and Maxwell (2018) presented a method to compute the boundary of the field of values of a matrix.

Theorem B.22 *Any real irreducible tridiagonal matrix A is similar to a tridiagonal matrix B with $a_{i,i-1}a_{i-1,i} = b_{i,i-1}b_{i-1,i}$ for $i = 2, \ldots, n$ such that $|b_{i,i-1}| = |b_{i-1,i}|$. The transformation matrix is real and diagonal.*

Corollary B.1

i. *If additionally the matrix A has positive off-diagonal elements, then it is similar to a symmetric matrix B.*
ii. *If the matrix A has a zero diagonal part and $sign(a_{i,i-1}) = -sign(a_{i-1,i})$ for $i = 1, \ldots, n$, then it is similar to a skew symmetric matrix.*

Theorem B.23 *If λ is an eigenvalue of the generalized eigenvalue problem*

$$
\begin{pmatrix}
\lambda B_1 & & & & A_m \\
A_1 & \lambda B_2 & & & \\
& A_2 & \ddots & & \\
& & & A_{m-1} & \lambda B_m
\end{pmatrix}
\begin{pmatrix}
x_1 \\ x_2 \\ \vdots \\ x_m
\end{pmatrix} = 0, \tag{B.3}
$$

then if the B_i $i = 1, \ldots, m$ are non-singular, then λ^m is an eigenvalue of $A_1 B_1^{-1} A_2 B_2^{-1} \ldots A_m B_m^{-1}$. If A_i $i = 1, \ldots, m$ are non-singular, then λ^{-m} is an eigenvalue of $B_1 A_1^{-1} B_2 A_2^{-1} \ldots B_m A_m^{-1}$.

Theorem B.24 (Jordan normal form) *Any square matrix A is similar to its Jordan normal form J given by*

$$
J = \begin{bmatrix}
J_1 & & & 0 \\
& J_2 & & \\
& & \ddots & \\
0 & & & J_m
\end{bmatrix}
$$

in which every Jordan block J_i of order p_i, with $\sum_{i=1}^{m} p_i = N$, is of the form

$$J_i = \begin{bmatrix} \lambda_i & 1 & 0 & \cdots & 0 \\ 0 & \lambda_i & 1 & & \\ \vdots & & \ddots & & \\ 0 & & & \lambda_i & 1 \\ 0 & . & . & . & \lambda_i \end{bmatrix} \tag{B.4}$$

where λ_i is an eigenvalue of A. See example Ex. 7.2

Theorem B.25 (Schur normal form) *Any square matrix A is similar to an upper triangular matrix through a unitary transformation matrix. If A is real, it can be transformed to a real Schur form by orthogonal matrices. The diagonal of the latter Schur form may contain 2×2 blocks related to complex eigenvalues.*

B.2 Growth Bounds

In stability analysis it is important to be able to bound the growth of a perturbation. We give a few theorems on these.

Theorem B.26 *If J is a Jordan block of order p, then for $\lambda \neq 0$ it holds for $n \to \infty$*

$$\|J^n\| \approx cn^{p-1}|\lambda|^n,$$

where c is a constant that depends on the choice of the norm.

Proof Write $J = \lambda I + E$ where E is zero except for the first upper diagonal which has ones. With this,

$$J^n = (\lambda I + E)^n = \sum_{k=0}^{n} \binom{n}{k} \lambda^{n-k} E^k = \sum_{k=0}^{p-1} \binom{n}{k} \lambda^{n-k} E^k$$

where in the last step we used the fact that E is an idempotent matrix, that is, for any power bigger than $p - 1$ it becomes zero. Consequently,

$$\|J^n\| \leq \sum_{k=0}^{p-1} \binom{n}{k} |\lambda|^{n-k} \|E^k\| \leq cn^{p-1}|\lambda|^n$$

where $c = \max_{k \in \{0,1,\ldots,p-1\}} |\lambda|^{-k} \|E^k\|$. ∎

Theorem B.27 *Consider the square matrix $A = \lambda I + \alpha E$ where E is a matrix with ones on the first superdiagonal and zeros elsewhere. Then*

i. $A^n \to 0$ for $n \to \infty$ if $|\lambda| < 1$.
ii. $\|A^n\|_\infty \leq (|\lambda| + |\alpha|)^n$ for any n.

Theorem B.28 *If $u_n = A u_{n-1}$ and $P^{-1} A P$ is normal, then $||u_n||_2 \leq \kappa_2(P) \rho^n(A) ||u_0||_2$, where $\kappa_2(P) \equiv ||P||_2 ||P^{-1}||_2$ is the condition number of P.*

Theorem B.29

i. *For general real A, if $u_n - u_{n-1} = A(u_n + \alpha u_{n-1})$ with u_n real, and $\sigma((P^{-1}AP)^T + (P^{-1}AP)) \leq 2\mu$ with $\mu < \frac{1}{2}$, then for all $\alpha \in [0,1]$ $||u_n||_2 \leq \beta^n \kappa_2(P) ||u_0||_2$, with $\beta = \alpha$ if $\alpha \geq 1/(1-2\mu)$ and $\beta = (1 + \mu\alpha)/(1-\mu)$ if $\alpha \leq 1/(1-2\mu)$.*

ii. *For general real A, if $M(u_n - u_{n-1}) = A(u_n + \alpha u_{n-1})$ with u_n real, and $(u, Q^T A P u) \leq \mu(u, Q^T M P u)$ with $Q^T M P$ SPD and with $\mu < \frac{1}{2}$, then for all $\alpha \in [0,1]$ $||P^{-1} u_n||_{Q^T M P} \leq \beta^n |||P^{-1} u_0||_{Q^T M P}$, with $\beta = \alpha$ if $\alpha \geq 1/(1-2\mu)$ and $\beta = (1 + \mu\alpha)/(1-\mu)$ if $\alpha \leq 1/(1-2\mu)$.*

Proof First, the proof of part i. Suppose first that $\sigma(A + A^T) \leq 2\mu$. Take the inner product of the equation with $u_n + \alpha u_{n-1}$ and we find that

$$(u_n + \alpha u_{n-1}, u_n - u_{n-1}) \leq \mu(u_n + \alpha u_{n-1}, u_n + \alpha u_{n-1}). \tag{B.5}$$

From this it follows that $(1 - \mu)||u_n||^2 - (1 + \mu\alpha)\alpha||u_{n-1}||^2 \leq (2\mu\alpha + 1 - \alpha)(u_n, u_{n-1}) \leq |2\mu\alpha + 1 - \alpha| ||u_n|| ||u_{n-1}||$. (In this proof all norms are two norms.) Division by $||u_{n-1}||^2$ and defining a new positive unknown $x = ||u_n||/||u_{n-1}||$ yields the inequality $(1 - \mu)x^2 - |2\mu\alpha + 1 - \alpha|x - (1 + \mu\alpha)\alpha \leq 0$. Since $1 - \mu > \frac{1}{2}$, this holds between the two roots of the quadratic form. Now the discriminant is composed of the square of $2\mu\alpha + 1 - \alpha$ and the product $(1-\mu)(1+\mu\alpha)\alpha$. Note that the former is $2(1+\mu\alpha)-(1+\alpha)$ and in the latter $(1-\mu)\alpha = (1+\alpha)-(1+\mu\alpha)$. Hence, the discriminant is $(2(1+\mu\alpha)-(1+\alpha))^2 + 4((1+\alpha)-(1+\mu\alpha))(1+\mu\alpha) = (1+\alpha)^2$. So we find the zeros

$$x_\pm = \frac{|2(1 + \mu\alpha) - (1 + \alpha)| \pm (1 + \alpha)}{2(1 - \mu)}$$

For $\alpha \leq 1/(1 - 2\mu)$ we have the roots $-\alpha$ and $(1 + \mu\alpha)/(1 - \mu)$. Now β is the biggest positive value that the ratio x can take, which is the latter zero. For $\alpha \geq 1/(1-2\mu)$ we have the roots $-(1 + \mu\alpha)/(1 - \mu)$ and α where the latter is the biggest one. The expression with P follows if we pre-multiply the difference equation by P^{-1} and replace the unknown by a transformed one $P^{-1}u$. Now we get the matrix $P^{-1}AP$ playing the role of A. One finds in the end $||P^{-1}u_n|| \leq \beta ||P^{-1}u_{n-1}||$. Now $||u_n|| = ||PP^{-1}u_n|| \leq ||P|| ||P^{-1}u_n|| \leq ||P|| \beta^n ||P^{-1}u_0|| \leq \kappa_2(P)\beta^n||u_0||$.

For part ii, we take the inner product of the equation with $QP^{-1}(u_n + \alpha u_{n-1})$ which results in the same expression as in B.5 but now in the $Q^T M P$ inner product and u_n replaced by $P^{-1}u_n$ and similar for u_{n-1}. This leads straighforwardly to the statement. ∎

Theorem B.30 *If $\frac{d}{dt}u = Au$, $\sigma(A) \leq \mu$, and $P^{-1}AP$ is normal, then it holds that $\|u(t)\|_2 \leq \kappa_2(P)\exp(\mu t)\|u(0)\|_2$, where $\kappa_2(P) \equiv \|P\|_2\|P^{-1}\|_2$ is the condition number of P.*

Theorem B.31 *For general A, if $\frac{d}{dt}u = Au$, $\sigma((P^{-1}AP)^T + (P^{-1}AP)) \leq 2\mu$, then $\|u(t)\|_2 \leq \kappa_2(P)\exp(\mu t)\|u_0\|_2$.*

Crouzeix (2016) presented a general theorem bounding the norm of a general analytic function of a matrix by the maximum of that function on the field of values of the matrix.

Theorem B.32 (Crouzeix) *There exists a universal constant $C \in [2, 11.5]$ such that for any matrix A and for any function f analytic in the field of values $F(A)$ it holds that $\|f(A)\|_2 \leq C \sup_{z \in F(A)} |f(z)|$.*

The exponential function is an example of an analytic function. Moreover, rational functions in general are analytic on a certain domain if they do not have poles in that domain, that is, points where the function tends to infinity. The function $1/(1-z)$ has a singularity at $z = 1$, but nevertheless it is analytic on every domain not containing this point. Note that for normal matrices the inequality holds with $C = 1$.

B.3 Positivity of Solutions

In simulations, nonnegativeness of solutions also is an issue. For instance, we would like to keep a concentration positive during the computations. It is clear that, when we multiply a vector with positive elements by a matrix with positive elements, then the outcome is also positive. But what if we want to solve a system where the right-hand side is positive? For which matrices is the solution positive? Let us first define the class of nonnegative matrices.

Definition B.13 (Non-negativeness) *A matrix A is nonnegative ($A \geq 0$) if each element is non negative, and A is positive ($A > 0$) if each element is positive.*

Definition B.14 (Monotony) *A matrix A is monotone if its inverse exists and is nonnegative.*

Theorem B.33 *A matrix A is monotone if and only if from $Ax \geq 0$ it follows that $x \geq 0$.*

Proof Let A be monotone and $y = Ax \geq 0$. Then it follows straightforwardly that $x = A^{-1}y \geq 0$. The other direction is less trivial. First, we show non-singularity. Let $Ax \geq 0 \Rightarrow x \geq 0$ and suppose z is a non-zero singular vector so $Az = 0$. Then

from the outset, z must have nonnegative elements. However, $-z$ is also a singular
vector and hence also the elements of this vector should be nonnegative. From this
it follows that $z = 0$, so the matrix is non-singular. Furthermore the inverse of A
is the solution of the system $AX = I$, which just means that the columns of the
inverse follow from the solution of $Ax = e_i$ for $i = 1, \ldots, N$ where e_i is a unit
vector. Since the unit vector is nonnegative, the columns of the inverse of A are
nonnegative and hence $A^{-1} \geq 0$.   ∎

So if we want to solve a system $Ax = b$ where $b \geq 0$ and A is monotone, then
$x \geq 0$.

Definition B.15 (M-matrix) *A matrix A is an M-matrix if A is monotone and $a_{ij} \leq 0$*
($\forall i \neq j$).

Theorem B.34 *An M-matrix has positive diagonal elements.*

Proof Let $A^{-1} = (b_{ij})$, then all $b_{ij} \geq 0$. From $AA^{-1} = I$ it follows that

$$a_{ii}b_{ii} + \sum_{\substack{j=1 \\ j \neq i}}^{N} a_{ij}b_{ji} = 1, \qquad i = 1, 2, \ldots, N.$$

In the sum all $a_{ij} \leq 0$ and all $b_{ji} \geq 0$, hence the sum is nonpositive and therefore

$$a_{ii}b_{ii} = 1 - \sum_{\substack{j=1 \\ j \neq i}}^{N} a_{ij}b_{ji} \geq 1, \qquad i = 1, 2, \ldots, N.$$

Since $b_{ii} \geq 0$, it follows from this that $a_{ii} > 0$, $i = 1, 2, \ldots, N$.   ∎

Theorem B.35 *An irreducible, weakly diagonally dominant matrix A with positive*
diagonal elements and $a_{ij} \leq 0$ ($\forall i \neq j$) is an M-matrix.

Proof From Theorem B.8 we know that A is non-singular, hence it only remains to
show that $A^{-1} \geq 0$. We write $A = D - B$ with D a diagonal matrix and the diagonal
of B equal to zero. From application of Taussky's theorem to $D^{-1}B$ it follows that
$\rho(D^{-1}B) < 1$. Hence, the sum

$$S = \sum_{k=0}^{\infty} (D^{-1}B)^k$$

converges and

$$S = (I - D^{-1}B)^{-1}.$$

From $D^{-1} \geq 0$ and $B \geq 0$ it follows that $D^{-1}B \geq 0$, hence, also $S \geq 0$. From

$$A^{-1} = (I - D^{-1}B)^{-1}D^{-1} = SD^{-1}$$

it finally follows that also $A^{-1} \geq 0$. ∎

Theorem B.36 *If $-A$ is an M-matrix, then* $\exp(At) \geq 0$.

Proof Let $\mu = \max_i |a_{ii}|$; then $A + \mu I \geq 0$ and we can write $\exp(At) = \exp(-\mu t)\exp((A+\mu I)t)$. The first factor is obviously positive for any μ. Expanding the $\exp((A+\mu I)t)$ in a Taylor series will show a sum of nonnegative matrices, hence that one will be nonnegative too, which proves that $\exp(At) \geq 0$ for $t \geq 0$. ∎

Appendix C

Proof of inf-sup Condition for Stokes Flow

In Section 4.4.2 it is shown that for the Stokes equations the occurring bi-linear form is not able to show that the problem is well-posed for the pressure part. In this appendix we show that the bi-linear form satisfies an inf-sup condition. For a slightly different approach see Auricchio *et al.* (2004).

To do so, we write the system in terms of operators (no curly letters are used here):

$$\begin{bmatrix} A & B \\ B^T & 0 \end{bmatrix} \begin{bmatrix} u \\ p \end{bmatrix} = \begin{bmatrix} \mathrm{f} \\ \mathrm{g} \end{bmatrix}. \tag{C.1}$$

Here A represents the negative of the Laplace operator including the natural boundary conditions. The Dirichlet conditions are all homogeneous. Similarly, B signifies the gradient operator. We will assume that

$$c_A||u||_1^2 \le (u, Au) \le M_A||u||_1^2, \tag{C.2a}$$

$$\beta \le \inf_{p \in H_0} \sup_{u \in H_1} (Bp, u)/||p||_0||u||_1, \tag{C.2b}$$

$$|(Bp, u)| \le M_B||p||_0||u||_1. \tag{C.2c}$$

Now we have to show that

$$\inf_{u \in H_1, p \in H_0} \sup_{v \in H_1, q \in H_0} \frac{\left(\begin{bmatrix} v \\ q \end{bmatrix}, \begin{bmatrix} A & B \\ B^T & 0 \end{bmatrix} \begin{bmatrix} u \\ p \end{bmatrix} \right)}{\sqrt{||v||_1^2 + ||q||_0^2} \sqrt{||u||_1^2 + ||p||_0^2}} \ge \gamma > 0. \tag{C.3}$$

First, we consider the problem of finding a lower bound for the supremum given u and p. The aim is to turn the system matrix into a block diagonal matrix by a suitable transformation, which makes it easier to make suitable choices for the variables v and q. We can achieve that by an LDL^T transformation of the system matrix (as in (8.36)). So, for given u, p we want to bound the supremum over v, q of the following expression from below:

$$\frac{\left(\begin{bmatrix} v \\ q \end{bmatrix}, \begin{bmatrix} A & B \\ B^T & 0 \end{bmatrix} \begin{bmatrix} u \\ p \end{bmatrix}\right)}{\sqrt{||v||_1^2 + ||q||_0^2}\sqrt{||u||_1^2 + ||p||_0^2}} = \frac{\left(L^T \begin{bmatrix} v \\ q \end{bmatrix}, \begin{bmatrix} A & 0 \\ 0 & -B^T A^{-1} B \end{bmatrix} L^T \begin{bmatrix} u \\ p \end{bmatrix}\right)}{\sqrt{||v||_1^2 + ||q||_0^2}\sqrt{||u||_1^2 + ||p||_0^2}}$$

(C.4)

where

$$L^T = \begin{bmatrix} I & A^{-1}B \\ 0 & I \end{bmatrix}.$$

(C.5)

Now we set

$$\begin{bmatrix} \tilde{u} \\ \tilde{p} \end{bmatrix} = L^T \begin{bmatrix} u \\ p \end{bmatrix}.$$

(C.6)

This introduces a new velocity variable $\tilde{u} = u + A^{-1}Bp$, while p remains the same. Similar for $\tilde{v}$. Next, we choose

$$\tilde{v} = \tilde{u}, \quad q = -p.$$

(C.7)

With this choice the enumerator of the right-hand side of (C.4) becomes

$$(\tilde{u}, A\tilde{u}) + (p, B^T A^{-1} B p).$$

(C.8)

This expression also reveals why we like to strive to a block diagonal form; we can choose $\tilde{v}$ and q independently from each other and such that we get positive definiteness in the enumerator. We want to bound (C.8) from below by $c_c \left(||\tilde{u}||_1^2 + ||p||_0^2\right)$. Moreover, we want to bound the denominator in (C.4) from above by $c_d \left(||\tilde{u}||_1^2 + ||p||_0^2\right)$, which will give in the end that the supremum of (C.4) will be larger than c_c/c_d for any $\tilde{u}$ and p. For the first term in (C.8) we have the standard coercivity, so the difficulty is in the second term. We write this term as $(p, B^T \tilde{w})$ where $\tilde{w}$ is the solution of $A\tilde{w} = Bp$. Then we have from (C.2b) that there is a $\hat{w}$ such that

$$\beta ||\hat{w}||_1 ||p||_0 \le (Bp, \hat{w}).$$

(C.9)

Moreover, using (C.2a) we have that

$$(Bp, \hat{w}) = (A\tilde{w}, \hat{w}) = (\tilde{w}, A\hat{w}) \le \sqrt{(\tilde{w}, A\tilde{w})}\sqrt{(\hat{w}, A\hat{w})} \le \sqrt{(\tilde{w}, Bp)}\sqrt{M_A}||\hat{w}||_1.$$

Combining this result with (C.9) and squaring gives finally

$$(p, B^T A^{-1} B p) = (Bp, \tilde{w}) \ge \frac{\beta^2}{M_A}||p||_0^2.$$

(C.10)

Hence, we find (C.8) is bounded from below by

$$c_A||\tilde{u}||_1^2 + \frac{\beta^2}{M_A}||p||_0^2 \geq \min\left(c_A, \frac{\beta^2}{M_A}\right)(||\tilde{u}||_1^2 + ||p||_0^2) \qquad (C.11)$$

and consequently $c_c = \min\left(c_A, \frac{\beta^2}{M_A}\right)$.

Next, we turn to the denominator (C.4). The argument of the second square root is

$$||u||_1^2 + ||p||_0^2 = ||\tilde{u} - A^{-1}Bp||_1^2 + ||p||_0^2. \qquad (C.12)$$

Now we write, by using the triangle inequality,

$$||\tilde{u} - A^{-1}Bp||_1^2 \leq (||\tilde{u}||_1 + ||A^{-1}Bp||_1)^2, \qquad (C.13)$$

which shows that we have to find a bound for $||A^{-1}Bp||_1$. We replace this by $||\tilde{w}||_1$ where $\tilde{w}$ is the solution of $A\tilde{w} = Bp$. Using (C.2a, c) we have

$$c_A||\tilde{w}||_1^2 \leq (\tilde{w}, A\tilde{w}) = (\tilde{w}, Bp) \leq M_B||\tilde{w}||_1||p||_0$$

and therefore

$$||A^{-1}Bp||_1 = ||\tilde{w}||_1 \leq \frac{M_B}{c_A}||p||_0.$$

So

$$||u||_1^2 + ||p||_0^2 \leq ||\tilde{u}||_1^2 + \frac{M_B}{c_A}||\tilde{u}||_1||p||_0 + \left(1 + \left(\frac{M_B}{c_A}\right)^2\right)||p||_0^2. \qquad (C.14)$$

Using the relation $2ab \leq a^2 + b^2$, one can find a constant c such that the last expression is less than $c(||\tilde{u}||_1^2 + ||p||_0^2)$. The argument of the other square root is

$$||v||_1^2 + ||q||_0^2 = ||\tilde{v} - A^{-1}Bq||_1^2 + ||q||_0^2$$

which by our choice (C.7) becomes

$$||\tilde{u} + A^{-1}Bp||_1^2 + ||p||_0^2.$$

Again using the triangle equality on the first norm, we observe that we get the same result as in the right-hand side of (C.13) and hence this argument of the square root also is less than $c\left(||\tilde{u}||_1^2 + ||p||_0^2\right)$. Combining the two shows that the denominator of (C.4) is less than $c\left(||\tilde{u}||_1^2 + ||p||_0^2\right)$ and hence $c_d = c$. As a consequence we can take $\gamma = c_c/c_d$.

Notes

3 Well-Posed Problems

[1] It should be emphasized that the precise meaning of the word stability is context dependent. In the previous chapter, it was connected to the growth of initial perturbations from, for example, a steady state, but here we will use it in the sense of Hademard defined in Definition 3.1.

[2] An exception here is the case with only one space variable, as previously, where we can switch the roles of x and t.

6 Matrix-Based Techniques

[1] https://github.com/BIMAU/BifAnFF/

7 Stationary Iterative Methods

[1] https://github.com/BIMAU/BifAnFF/

8 Non-stationary Iterative Methods

[1] https://github.com/BIMAU/BifAnFF/

10 Benchmark Results for Canonical Problems

[1] https://github.com/BIMAU/fvm/
[2] https://github.com/BIMAU/jadapy/

References

Ahmed, N., Bartsch, C., John, V., and Wilbrandt, U. (2018). An assessment of some solvers for saddle point problems emerging from the incompressible Navier–Stokes equations. *Computer Methods in Applied Mechanics and Engineering*, **331**, 492–513.

Aidun, C. K., Triantafillopoulos, N. G., and Benson, J. D. (1991). Global stability of a lid-driven cavity with throughflow: Flow visualization studies. *Physics of Fluids A: Fluid Dynamics*, **3**, 2081–91.

Allgower, E. L., and Georg, K. (2003). *Introduction to Numerical Continuation Methods*. SIAM.

Amestoy, P., Duff, I., and L'Excellent, J.-Y. (2000). Multifrontal parallel distributed symmetric and unsymmetric solvers. *Computer Methods in Applied Mechanics and Engineering*, **184**, 501–20.

Andereck, C. D., Liu, S. S., and Swinney, H. L. (1986). Flow regimes in a circular Couette system with independently rotating cylinders. *Journal of Fluid Mechanics*, **164**, 155–83.

Argyros, I. K. (2008). *Convergence and Applications of Newton-Type Iterations*. Springer.

Arnoldi, W. E. (1951). The principle of minimized iteration in the solution of the matrix eigenproblem. *Quarterly of Applied Mathematics*, **9**, 17–29.

Assemat, P., Bergeon, A., and Knobloch, E. (2008). Spatially localized states in Marangoni convection in binary mixtures. *Fluid Dynamics Research*, **40**, 852–76.

Auricchio, F., Brezzi, F., and Lovadina, C. (2004). Mixed finite element methods. In E. Stein, R. de Borst, and T. R. J. Hughes, editors, *Encyclopedia of Computational Mechanics*, chapter 9. Wiley.

Baars, S., van der Klok, M., Thies, J., and Wubs, F. (2021). A staggered-grid multilevel incomplete LU for steady incompressible flows. *International Journal for Numerical Methods in Fluids*, **93**, 909–26.

Balay, S., Abhyankar, S., Adams, M. F., et al. (2021). PETSc Web page. `www.mcs.anl.gov/petsc`.

Barkley, D., and Henderson, R. D. (1996). Floquet stability analysis of the periodic wake of a circular cylinder. *Journal of Fluid Mechanics*, **322**, 215–41.

Batchelor, G. K. (1956). On steady laminar flow with closed streamlines at large Reynolds number. *Journal of Fluid Mechanics*, **1**, 177–90.

Batiste, O., Knobloch, E., Alonso, A., and Mercader, I. (2006). Spatially localized binary fluid convection. *Journal of Fluid Mechanics*, **560**, 149–58.

Bénard, H. (1901). Les tourbillons cellulaires dans une nappe liquide. Méthodes optiques d'observation et d'enregistrement. *Journal of Physics Theories and Applications*, **10**, 254–66.

Benzi, M., Golub, G. H., and Liesen, J. (2005). Numerical solution of saddle point problems. *Acta Numerica*, **14**, 1–137.

Bollhöfer, M. (2001). A robust ILU with pivoting based on monitoring the growth of the inverse factors. *Linear Algebra and Its Applications*, **338**, 515–26.

Bollhöfer, M., Eftekhari, A., Scheidegger, S., and Schenk, O. (2019). Large-scale sparse inverse covariance matrix estimation. *SIAM Journal on Scientific Computing*, **41**, A380–A401.

Boronska, K., and Tuckerman, L. (2010). Extreme multiplicity in cylindrical Rayleigh-Bénard convection. II. Bifurcation diagram and symmetry classification. *Physical Review*, **E81**, 036321.

Botta, E., and Wubs, F. (1999). MRILU: An effective algebraic multi-level ilu-preconditioner for sparse matrices. *SIAM Journal on Matrix Analysis and Applications*, pages 1007–26.

Boullé, N., Dallas, V., and Farrell, P. E. (2022). Bifurcation analysis of two-dimensional Rayleigh–Bénard convection using deflation. *Physical Review E*, **105**, 055106.

Braess, D. (2007). *Finite Elements – Theory, Fast Solvers, and Applications in Solid Mechanics*. Cambridge University Press.

Brand, C. (1992). An incomplete-factorization preconditioning using red-black ordering. *Numerische Mathematik*, **61**, 433–54.

Brenner, S. C., and Scott, L. R. (2008). *The Mathematical Theory of Finite Element Methods*, volume 15 of Texts in Applied Mathematics. Springer.

Brooks, A. N., and Hughes, T. J. (1982). Streamline upwind/Petrov–Galerkin formulations for convection dominated flows with particular emphasis on the incompressible Navier–Stokes equations. *Computer Methods in Applied Mechanics and Engineering*, **32**, 199–259.

Brown, P. N., and Saad, Y. (1990). Hybrid Krylov methods for nonlinear systems of equations. *SIAM Journal on Scientific Computing*, **11**, 450–81.

Burden, R., and Faires, J. (2001). *Numerical Analysis*. Brooks/Cole.

Canuto, C., Hussaini, M. Y., Quarteroni, A., and Zang, T. A. (2006). *Spectral Methods. Fundamentals in Single Domains*. Springer.

Canuto, C., Hussaini, M. Y., Quarteroni, A., and Zang, T. A. (2007). *Spectral Methods. Evolution to Complex Geometries and Applications to Fluid Dynamics*. Springer.

Carey, G. F., Wang, K., and Joubert, W. (1989). Performance of iterative methods for Newtonian and generalized Newtonian flows. *International Journal for Numerical Methods in Fluids*, **9**, 127–50.

Cazemier, W., Verstappen, R., and Veldman, A. (1998). POD and low-dimensional models for driven cavity flows. *Physics of Fluids*, **10**, 1685–99.

Chandrasekhar, S. (1961). *Hydrodynamic and Hydromagnetic Stability*. Dover Books on Physics Series. Dover.

Chen, K. (2005). *Matrix Preconditioning Techniques and Applications*. Cambridge Monographs on Applied and Computational Mathematics. Cambridge University Press.

Chossat, P., and Iooss, G. (2012). *The Couette–Taylor Problem*, volume 102 of Applied Mathematical Sciences. Springer Science & Business Media.

Chun, J. (2011). *Computational Fluid Dynamics*. Cambridge University Press. Second edition.

Cliffe, K. A. (1983). Numerical calculations of two-cell and single-cell Taylor flows. *Journal of Fluid Mechanics*, **135**, 219–33.

Cliffe, K. A. (1988). Numerical calculations of the primary-flow exchange process in the Taylor problem. *Journal of Fluid Mechanics*, **197**, 57–79.

Coleman, T. F., Garbow, B. S., and Moré, J. J. (1984). Software for estimating sparse Jacobian matrices. *ACM Transactions on Mathematical Software*, **10**, 329–45.

Couette, M. M. (1890). Études sur le frottement des liquides. *Annales de chimie et de physique*, **6**, 433–510.

Crawford, J. D., and Knobloch, E. (1991). Symmetry and symmetry-breaking bifurcations in fluid dynamics. *Annual Review of Fluid Mechanics*, **23**, 341–87.

Crouzeix, M. (2016). Some constants related to numerical ranges. *SIAM Journal on Matrix Analysis and Applications*, **37**, 420–42.

Curry, J. H., Herring, J. R., Loncaric, J., and Orszag, S. A. (1984). Order and disorder in two- and three-dimensional Bénard convection. *Journal of Fluid Mechanics*, **147**, 1–38.

Cyr, E. C., Shadid, J. N., and Tuminaro, R. S. (2012). Stabilization and scalable block preconditioning for the Navier–Stokes equations. *Journal of Computational Physics*, **231**, 345–63.

Daas, H. A., Grigori, L., Hénon, P., and Ricoux, P. (2021). Recycling Krylov subspaces and truncating deflation subspaces for solving sequence of linear systems. *ACM Transactions on Mathematical Software*, **47**, 1–30.

Dauby, P. C., and Lebon, G. (1996). Bénard–Marangoni instability in rigid rectangular containers. *Journal of Fluid Mechanics*, **329**, 25–64.

Davis, T. A. (2004). Algorithm 832: Umfpack v4.3 – an unsymmetric-pattern multifrontal method. *ACM Transactions on Mathematical Software*, **30**, 196–9.

Davis, T. A. (2006). *Direct Methods for Sparse Linear Systems (Fundamentals of Algorithms 2)*. Society for Industrial and Applied Mathematics.

Davis, T. A., Rajamanickam, S., and Sid-Lakhdar, W. M. (2016). A survey of direct methods for sparse linear systems. *Acta Numerica*, **25**, 383–566.

Dhooge, A., Govaerts, W. J. F., and Kuznetsov, Y. A. (2003). MatCont: A MATLAB package for numerical bifurcation analysis of ODEs. *ACM Transactions on Mathematical Software*, **29**, 141–64.

Dijkstra, H., Wubs, F., Cliffe, A., et al. (2014). Numerical bifurcation methods and their application to fluid dynamics: Analysis beyond simulation. *Communications in Computational Physics (CiCP)*, **15**, 1–45.

Dijkstra, H. A. (2005). *Nonlinear Physical Oceanography: A Dynamical Systems Approach to the Large Scale Ocean Circulation and El Niño, 2nd Revised and Enlarged Edition*. Springer.

Doedel, E. (1986). *AUTO: Software for Continuation and Bifurcation Problems in Ordinary Differential Equations*. Report Applied Mathematics, California Institute of Technology, Pasadena, USA.

Doedel, E. J. (1980). AUTO: A program for the automatic bifurcation analysis of autonomous systems. In *Proceedings of the Tenth Manitoba Conference on Numerical Mathematics and Computing*, volume 30, pages 265–74.

Doedel, E., Keller, H. B., and Kernevez, J. P. (1991). Numerical analysis and control of bifurcation problems. (i) Bifurcation in finite dimensions. *International Journal of Bifurcation and Chaos*, **1**, 493–520.

Doedel, E. J., and Tuckermann, L. S. (2000). *Numerical Methods for Bifurcation Problems and Large-Scale Dynamical Systems*. Springer.

Doedel, E. J., Paffenroth, R. C., Champneys, A. C., et al. (2007). Lecture Notes on Numerical Analysis of Nonlinear Equations. In B. Krauskopf, H. M. Osinga, and J. Galán-Vioque, editors. *Numerical Continuation Methods for Dynamical Systems. Understanding Complex Systems*. Springer. DOI: https://doi.org/10.1007/978-1-4020-6356-5_1.

Drazin, P., and Reid, W. (2004). *Hydrodynamic Stability*. Cambridge Mathematical Library. Cambridge University Press.

Drzisga, D., John, L., Rüde, U., Wohlmuth, B., and Zulehner, W. (2018). On the analysis of block smoothers for saddle point problems. *SIAM Journal on Matrix Analysis and Applications*, **39**, 932–60.

Duff, I., Erisman, A., and Reid, J. (1986). *Direct Methods for Sparse Matrices*. Monographs on Numerical Analysis. Oxford Science Publications.

Duff, I. S., Erisman, A. M., and Reid, J. K. (2017). *Direct Methods for Sparse Matrices*. Numerical Mathematics and Scientific Computation. Oxford University Press. Second edition.

Duguet, Y., Pringle, C. C. T., and Kerswell, R. R. (2008). Relative periodic orbits in transitional pipe flow. *Physics of Fluids*, **20**, 114102.

Ebadi, G., Alipour, N., and Vuik, C. (2016). Deflated and augmented global Krylov subspace methods for the matrix equations. *Applied Numerical Mathematics*, **99**, 137–50.

Eckert, E., and Carlson, W. O. (1961). Natural convection in an air layer enclosed between two vertical plates with different temperature. *International Journal of Heat and Mass Transfer*, **2**, 106–20.

Eckert, M. (2019). *The Turbulence Problem*. Springer Briefs in the History of Science and Technology. Springer. DOI: https://doi.org/10.1007/978-3-030-31863-5.

Elder, J. W. (1965). Laminar free convection in a vertical slot. *Journal of Fluid Mechanics*, pages 77–98.

Erenburg, V., Gelfgat, A., Kit, E., Bar-Yoseph, P., and Solan, A. (2003). Multiple states, stability and bifurcations of natural convection in rectangular cavity with partially heated vertical walls. *Journal of Fluid Mechanics*, **492**, 63–89.

Ern, A., and Guermond, J.-L. (2002). *Éléments finis: théorie, applications, mise en oeuvre*. Volume 36 of Mathématiques & Applications. Springer.

Farrell, P. E., Birkisson, A., and Funke, S. W. (2015). Deflation techniques for finding distinct solutions of nonlinear partial differential equations. *SIAM Journal on Scientific Computing*, **37**, A2026–A2045.

Farrell, P. E., Mitchell, L., and Wechsung, F. (2019). An augmented Lagrangian preconditioner for the 3d stationary incompressible Navier–Stokes equations at high Reynolds number. *SIAM Journal on Scientific Computing*, **41**, A3073–A3096.

Fokkema, D. R., Sleijpen, G. L. G., and van der Vorst, H. A. (1998). Jacobi–Davidson style QR and QZ algorithms for the reduction of matrix pencils. *SIAM Journal on Scientific Computing*, **20**, 94–125.

Freitas, C. J., Street, R. L., Findikakis, A. N., and Koseff, J. R. (1985). Numerical simulation of three-dimensional flow in a cavity. *International Journal for Numerical Methods in Fluids*, **5**, 561–75.

Freund, R. W., and Nachtigal, N. M. (1990). An implementation of the look-ahead Lanczos algorithm for non-Hermitian matrices, part 2. Technical Report 90.46, RIACS, NASA Ames Research Center.

Gaul, A., Gutknecht, M. H., Liesen, J., and Nabben, R. (2013). A framework for deflated and augmented Krylov subspace methods. *SIAM Journal on Matrix Analysis and Applications*, **34**, 495–518.

Gee, M., Siefert, C., Hu, J., Tuminaro, R., and Sala, M. (2006). Ml 5.0 smoothed aggregation user's guide. Technical Report SAND2006-2649, Sandia National Laboratories.

Gelfand, I., and Fomin, S. (2000). *Calculus of Variations*. Dover.

Gelfgat, A. Y. (1999). Different modes of Rayleigh–Bénard instability in two- and three dimensional rectangular enclosures. *Journal of Computational Physics*, **156**, 300–24.

Gelfgat, A. (2007). Stability of convective flows in cavities: Solution of benchmark problems by a low-order finite volume method. *International Journal for Numerical Methods in Fluids*, **53**, 485–506.

Gelfgat, A. (2019a). *Computational Modelling of Bifurcations and Instabilities in Fluid Dynamics*. Springer.

Gelfgat, A. (2019b). On acceleration of Krylov-subspace-based Newton and Arnoldi iterations for incompressible cfd: Replacing time steppers and generation of initial guess. In A. Gelfgat, editor, *Computational Modelling of Bifurcations and Instabilities in Fluid Dynamics*, pages 147–67. Springer International Publishing.

Gelfgat, A. Y. (2019c). Linear instability of the lid-driven flow in a cubic cavity. *Theoretical and Computational Fluid Dynamics*, **33**, 59–82.

Gelfgat, A., Bar-Yoseph, P., Solan, A., and Kowalewski, T. (1999a). An axisymmetry-breaking instability in axially symmetric natural convection. *International Journal of Transport Phenomena*, **1**, 173–90.

Gelfgat, A., Bar-Yoseph, P., and Yarin, A. (1999b). Stability of multiple steady states of convection in laterally heated cavities. *Journal of Fluid Mechanics*, **388**, 315–34.

George, A. (1973). Nested dissection of a regular finite-element mesh. *SIAM Journal on Numerical Analysis*, **10**, 345–63.

Getling, A. V. (1998). *Bénard–Rayleigh Convection: Structures and Dynamics*. World Scientific.

Gibson, J. F., Halcrow, J., and Cvitanovic, P. (2009). Equilibrium and traveling-wave solutions of plane Couette flow. *Journal of Fluid Mechanics*, **638**, 243–66.

Gladwell, M. (2000). *The Tipping Point*. Little Brown.

Gockenbach, M. (2002). *Partial Differential Equations, Analytical and Numerical Methods*. SIAM.

Golub, G., and van Loan, C. (1996). *Matrix Computations*. Johns Hopkins University Press. Third edition.

Govaerts, W. (1991). Stable solvers and block elimination for bordered systems. *SIAM Journal on Matrix Analysis and Applications*, **12**, 469–83.

Govaerts, W. J. F. (2000). *Numerical Methods for Bifurcations of Dynamical Equilibria*. Society for Industrial and Applied Mathematics.

Gresho, P. M., and Lee, R. L. (1981). Don't suppress the wiggles: They're telling you something! *Computers & Fluids*, **9**, 223–53.

Grossmann, C., Roos, H.-G., and Stynes, M. (2007). *Numerical Treatment of Partial Differential Equations*. Springer.

Grossmann, S., Lohse, D., and Sun, C. (2016). High–Reynolds number Taylor–Couette turbulence. *Annual Review of Fluid Mechanics*, **48**, 53–80.

Grote, M. J., and Huckle, T. (1997). Parallel preconditionings with sparse approximate inverses. *SIAM Journal on Scientific Computing*, **18**, 838–53.

Guckenheimer, J., and Holmes, P. (1990). *Nonlinear Oscillations, Dynamical Systems and Bifurcations of Vector Fields*. Springer. Second edition.

Hager, G., and Wellein, G. (2011). *Introduction to High Performance Computing for Scientists and Engineers*. Chapman and Hall/CRC Computational Science Series. CRC Press.

Hart, J. E. (1971). Stability of the flow in a differentially heated inclined box. *Journal of Fluid Mechanics*, **47**, 547–76.

Heil, M., and Hazel, A. L. (2006). oomph-lib – an object-oriented multi-physics finite-element library. In H.-J. Bungartz and M. Schäfer, editors, *Fluid-Structure Interaction*, pages 19–49. Springer.

Heinlein, A., Hochmuth, C., and Klawonn, A. (2020). Reduced dimension GDSW coarse spaces for monolithic Schwarz domain decomposition methods for incompressible fluid flow problems. *International Journal for Numerical Methods in Engineering*, **121**, 1101–19.

Henry, D., and Bergeon, A. (2000). *Continuation Methods in Fluid Dynamics. Contributions to the ERCOFTAC/EUROMECH Colloquium 383, Aussois, France, 6-9 September 1998*. Volume 74 of Notes on Numerical Fluid Mechanics. Vieweg+Teubner Verlag Wiesbaden.

Heroux, M. A., Bartlett, R. A., Howle, V. E., et al. (2005). An overview of the trilinos project. *ACM Transactions on Mathematical Software*, **31**, 397–423.

Hestenes, M., and Stiefel, E. (1954). Methods of conjugate gradients for solving linear systems. *Journal of Research of the National Bureau of Standards*, **49**, 409–36.

Hirsch, C. (1994). *Numerical Computation of Internal and External Flows*. Volume 1. Wiley.

Hogben, L., editor (2006). *Handbook of Linear Algebra*. CRC Press.

Horn, R., and Johnson, C. (1985). *Matrix Analysis*. Cambridge Universiy Press.

Hoyle, R. (2006). *Pattern Formation: An Introduction to Methods*. Cambridge University Press.

Huepe, C., Tuckerman, L. S., Métens, S., and Brachet, M. E. (2003). Stability and decay rates of non-isotropic attractive Bose–Einstein condensates. *Physical Review A*, **68**, 023609.

Hughes, T. (2000). *The Finite Element Method: Linear Static and Dynamic Finite Element Analysis*. Dover Civil and Mechanical Engineering. Dover.

Imberger, J. (1974). Natural convection in a shallow cavity with differentially heated end walls. Part 3. Experimental results. *Journal of Fluid Mechanics*, 247–60.

Jannot, M., and Mazeas, C. (1973). Étude expérimentale de la convection naturelle dans des cellules rectangulaires verticales. *International Journal of Heat and Mass Transfer*, **16**, 81–100.

Jarausch, H., and Mackens, W. (1987). Solving large nonlinear systems of equations by an adaptive condensation process. *Numerische Mathematik*, **50**, 633–53.

John, V., and Tobiska, L. (2000). Numerical performance of smoothers in coupled multigrid methods for the parallel solution of the incompressible Navier–Stokes equations. *International Journal for Numerical Methods in Fluids*, **33**, 453–73.

Joseph, D. D. (1976). *Stability of Fluid Motions: Vol. I. and Vol. II*. Springer.

Keller, H. B. (1977). Numerical solution of bifurcation and nonlinear eigenvalue problems. In P. H. Rabinowitz, editor, *Applications of Bifurcation Theory*, pages 359–384. Academic Press.

Kelley, C. T. (2007). *Solving Nonlinear Equations with Newton's Method*. SIAM.

Knoll, D., and Keyes, D. (2004). Jacobian-free Newton–Krylov methods: A survey of approaches and applications. *Journal of Computational Physics*, **193**, 357–97.

Knoll, D., Mousseau, V., Chacón, L., and Reisner, J. (2005). Jacobian-free Newton-Krylov methods for accurate time integration of stiff wave systems. *Journal of Scientific Computing*, **25**, 213–29.

Koschmieder, E. L. (1993). *Bénard Cells and Taylor Vortices*. Cambridge University Press.

Koseff, J. R., and Street, R. L. (1984). The lid-driven cavity flow: A synthesis of qualitative and quantitative observations. *Journal of Fluids Engineering*, **106**, 390–8.

Krantz, S., and Parks, H. R. (2002). *The Implicit Function Theorem*. Springer.

Krauskopf, B., and Osinga, H. (2007). Computing invariant manifolds via the continuation of orbit segments. In B. Krauskopf, H. Osinga, and J. Galán-Vioque, editors, *Numerical Continuation Methods for Dynamical Systems: Path Following and Boundary Value Problems*, Understanding Complex Systems, pages 117–54. Springer.

Krauskopf, B., Osinga, H. M., Doedel, E. J., et al. (2005). A survey of methods for computing (un)stable manifolds of vector fields. *International Journal of Bifurcation and Chaos in Applied Sciences and Engineering*, **15**, 763–91.

Krauskopf, B., Osinga, H. M., and Galán-Vioque, J. (2007). *Numerical Continuation Methods for Dynamical Systems: Path Following and Boundary Value Problems*. Springer.

Krishnamurti, R. (1973). Some further studies on the transition to turbulent convection. *Journal of Fluid Mechanics*, **60**, 285–303.

Kuhlmann, H. C., and Albensoeder, S. (2014). Stability of the steady three-dimensional lid-driven flow in a cube and the supercritical flow dynamics. *Physics of Fluids*, **26**.

Kuhlmann, H. C., and Romanò, F. (2019). *The Lid-Driven Cavity*, pages 233–309. Springer International.

Kuznetsov, Y. A. (1995). *Elements of Applied Bifurcation Theory*. Springer.

Kuznetsov, Y. A., and Levitin, V. V. (1996). CONTENT, a multiplatform continuation environment. Technical report, Centrum Wiskunde & Informatica.

Lanczos, C. (1950). An iteration method for the solution of the eigenvalue problem of linear differential and integral operators. *Journal of Research of the National Bureau of Standards*, **45**, 225–80.

Landau, L. D., and Lifshitz, I. M. (1987). *Fluid Mechanics*. Butterworth – Heinemann.

Lehoucq, R., Sorensen, D., and Yang, C. (1998). *ARPACK Users' Guide: Solution of Large-Scale Eigenvalue Problems with Implicitly Restarted Arnoldi Methods*. Software, Environments, Tools. Society for Industrial and Applied Mathematics.

Lehoucq, R. B., and Meerbergen, K. (1998). Using generalized Cayley transformations within an inexact rational Krylov sequence method. *SIAM Journal on Matrix Analysis and Applications*, **20**, 131–48.

Lenton, T. M., Held, H., Kriegler, E., et al. (2008). Tipping elements in the Earth's climate system. *Proceedings of the National Academy of Sciences of the United States of America*, **105**, 1786–93.

Liesen, J., and Strakoš, Z. (2013). *Krylov Subspace Methods: Principles and Analysis*. Oxford University Press.

Lippert, R., and Edelman, A. (2000). Nonlinear eigenvalue problems with orthogonality constraints. In Z. Bai, J. Demmel, J. Dongarra, A. Ruhe, and H. van der Vorst, editors, *Templates for the Solution of Algebraic Eigenvalue Problems: A Practical Guide* SIAM. https://netlib.org/utk/people/JackDongarra/etemplates/node343.html.

Loiseau, J.-C., Bucci, M. A., Cherubini, S., and Robinet, J.-C. (2019). Time-stepping and Krylov methods for large-scale instability problems. In A. Gelfgat, editor, *Computational Modelling of Bifurcations and Instabilities in Fluid Dynamics*, pages 33–73. Springer International.

Loisel, S., and Maxwell, P. (2018). Path-following method to determine the field of values of a matrix with high accuracy. *SIAM Journal on Matrix Analysis and Applications*, **39**, 1726–49.

Lorenz, E. N. (1963). Deterministic nonperiodic flow. *Journal of the Atmospheric Sciences*, **20**, 130–41.

Lust, K. (1997). *Numerical bifurcation analysis of periodic solutions of partial differential equations*. Ph.D. thesis, Katholieke Universiteit Leuven.

Lust, K., and Roose, D. (2000). Computation and bifurcation analysis of periodic solutions of large-scale systems. In E. Doedel and L. Tuckerman, editors, *Numerical Methods for Bifurcation Problems and Large-Scale Dynamical Systems*, volume 119 of Volumes in Mathematics and Its Applications, pages 265–301. Springer.

Ly, P.-M. T., Mitas, K. D. J., Thiele, U., and Gurevich, S. V. (2020). Two-dimensional patterns in dip coating – first steps on the continuation path. *Physica D: Nonlinear Phenomena*, **409**, 132485.

Mamun, C. K., and Tuckerman, L. S. (1995). Asymmetry and Hopf bifurcation in spherical Couette flow. *Physics of Fluids*, **7**, 80–91.

Marangoni, C. (1865). *Sull'espansione delle goccie d'un liquido galleggianti sulla superfice di altro liquido*. Fratelli Fusi.

Martins, J. R. R. A., Sturdza, P., and Alonso, J. J. (2003). The complex-step derivative approximation. *ACM Transactions on Mathematical Software*, **29**, 245–62.

Matsson, J. (2008). A student project on Rayleigh Bénard convection. *Paper presented at 2008 Annual Conference and Exposition, Pittsburgh, Pennsylvania*.

Mattheij, R. M. M., Rienstra, S. W., and Ten Thije o.g. Boonkkamp, J. H. M. (2005). *Partial Differential Equations: Modeling, Analysis, Computation (SIAM Monographs on Mathematical Modeling and Computation) (SIAM Models on Mathematical Modeling and Computation)*. Society for Industrial and Applied Mathematics.

Meerbergen, K. (1996). *Robust methods for the calculation of rightmost eigenvalues of nonsymmetric eigenvalue problems*. Ph.D. thesis, Katholieke Universiteit Leuven.

Meerbergen, K., and Spence, A. (2010). Inverse iteration for purely imaginary eigenvalues with application to the detection of Hopf bifurcations in large-scale problems. *SIAM Journal on Matrix Analysis and Applications*, **31**, 1982–99.

Mei, Z. (2000). *Numerical Bifurcation Analysis for Reaction–Diffusion Equations*. Springer, Berlin.

Meijerink, J., and van der Vorst, H. (1977). An iterative solution method for linear systems of which the coefficient matrix is a symmetric M-matrix. *Mathematics of Computation*, **31**, 148–62.

Molemaker, M. J., and Dijkstra, H. A. (2000). Multiple equilibria and stability of the North-Atlantic wind-driven ocean circulation. In E. Doedel and L. S. Tuckerman, editors, *Numerical Methods for Bifurcation Problems and Large-Scale Dynamical Systems*, volume 119 of The IMA Volumes in Mathematics and Its Applications, pages 35–65. Springer.

Morton, K., and Mayers, D. (2005). *Numerical Solution of Partial Differential Equations*. Cambridge University Press. Second edition.

Mullin, T., and Blohm, C. (2001). Bifurcation phenomena in a Taylor–Couette flow with asymmetric boundary conditions. *Physics of Fluids*, **13**, 136–40.

Mullin, T., Heise, M., and Pfister, G. (2017). Onset of cellular motion in Taylor–Couette flow. *Physical Review Fluids*, **2**, 2–7.

Mutabazi, I., Wesfreid, J. E., and Guyon, E. (2010). Dynamics of spatio-temporal cellular structures: Henri Bénard centenary review.

Net, M., Alonso, A., and Sánchez, J. (2003). From stationary to complex time-dependent flows in two-dimensional annular thermal convection at moderate Rayleigh numbers. *Physics of Fluids*, **15**, 1314–26.

Net, M., Garcia, F., and Sánchez, J. (2008). On the onset of low-Prandtl-number convection in rotating spherical shells: Non-slip boundary conditions. *Journal of Fluid Mechanics*, **601**, 317–37.

Nordström, J. (2022). Nonlinear and linearised primal and dual initial boundary value problems: When are they bounded? How are they connected? *Journal of Computational Physics*, **455**, 111001.

Notay, Y. (2012). Aggregation-based algebraic multigrid for convection–diffusion equations. *SIAM Journal on Scientific Computing*, **34**, A2288–A2316.

Paolucci, S. (1994). The differentially heated cavity. *Sadhana*, **19**, 619–47.

Parlett, B. N. (1980). *The Symmetric Eigenvalue Problem*. Prentice-Hall.

Parlett, B. N. (1990). Misconvergence in the Lanczos algorithm. In M. G. Cox and S. Hammarling, editors, *Reliable Numerical Computation*, Chapter 1. Clarendon Press.

Pawlowski, R. P., Shadid, J. N., Simonis, J. P., and Walker, H. F. (2006). Globalization techniques for Newton–Krylov methods and applications to the fully coupled solution of the Navier–Stokes equations. *SIAM Review*, **48**, 700–21.

Perko, L. (2013). *Differential Equations and Dynamical Systems*. Springer International.

Peyret, R., and Taylor, T. (1983). *Computational Methods for Fluid Flow*. Springer.

Pfister, G., Schmidt, H., Cliffe, K. A., and Mullin, T. (1988). Bifurcation phenomena in Taylor–Couette flow in a very short annulus. *Journal of Fluid Mechanics*, **191**, 1–18.

Platten, J. K., and Legros, J. C. (1984). *Convection in Liquids*. Springer International.

Puigjaner, D., Herrero, J., Simó, C., and Giralt, F. (2011). From steady solutions to chaotic flows in a Rayleigh–Bénard problem at moderate Rayleigh numbers. *Physica D*, **240**, 920–34.

Quarteroni, A., Sacco, R., and Saleri, F. (2007). *Numerical Mathematics*. Springer.

Rabinovich, M. I., Ezersky, A. B., and Weidman, P. D. (2000). *The Dynamics of Patterns*. World Scientific.

Rahmstorf, S. (2000). The thermohaline circulation: A system with dangerous thresholds? *Climatic Change*, **46**, 247–56.

Rayleigh, L. (1916). On convection currents in a horizontal layer of fluid, when the higher temperature is on the under side. *Philosophical Magazine*, **32**, 529–46.

Recktenwald, A., Lücke, M., and Müller, H. W. (1993). Taylor vortex formation in axial through-flow: Linear and weakly nonlinear analysis. *Physical Review E*, **48**, 4444–54.

Reisner, J., Mousseau, V., and Knoll, D. (2000). Application of the Newton–Krylov method to geophysical flows. *Monthly Weather Review*, **129**, 2404–15.

Reisner, J., Wyszogrodzki, A., Mousseau, V., and Knoll, D. (2003). An efficient physics based preconditioner for the fully implicit solution of small-scale thermally driven atmospheric flows. *Journal of Computational Physics*, **189**, 30–44.

Richtmyer, R., and Morton, K. (1967). *Difference Methods for Initial Value Problems*. Interscience.

Saad, Y. (1993). A flexible inner-outer preconditioned GMRES algorithm. *SIAM Journal on Scientific and Statistical Computing*, **14**, 461–9.

Saad, Y. (2003). *Iterative Methods for Sparse Linear Systems*. Society for Industrial and Applied Mathematics. Second edition.

Saad, Y. (2011). *Numerical Methods for Large Eigenvalue Problems*. Society for Industrial and Applied Mathematics.

Saad, Y., and Schultz, M. (1986). A generalized minimal residual algorithm for solving nonsymmetric linear systems. *SIAM Journal on Scientific and Statistical Computing*, **7**, 856–69.

Salinger, A. G., Lehoucq, R. B., Pawlowski, R. P., and Shadid, J. N. (2002). Computational bifurcation and stability studies of the 8:1 thermal cavity problem. *International Journal for Numerical Methods in Fluids*, **40**, 1059–73.

Sánchez, J., Marqués, F., and López, J. M. (2002). A continuation and bifurcation technique for Navier–Stokes flows. *Journal of Computational Physics*, **180**, 78–98.

Sánchez, J., Net, M., García-Archilla, B., and Simó, C. (2004). Newton–Krylov continuation of periodic orbits for Navier–Stokes flows. *Journal of Computational Physics*, **201**, 13–33.

Sánchez, J., Net, M., and Simó, C. (2010). Computation of invariant tori by Newton–Krylov methods in large-scale dissipative systems. *Physica D*, **239**, 123–33.

Sánchez, J., Net, M., and Vega, J. (2006). Amplitude equations close to a triple-(+1) bifurcation point of D_4-symmetric periodic orbits in $O(2)$-equivariant systems. *Discrete and Continuous Dynamical Systems-Series B*, **6**, 1357–80.

Sánchez Umbría, J., and Net, M. (2013). A parallel algorithm for the computation of invariant tori in large-scale dissipative systems. *Physica D: Nonlinear Phenomena*, **252**, 22–33.

Sánchez Umbría, J., and Net, M. (2019). Stationary flows and periodic dynamics of binary mixtures in tall laterally heated slots. In A. Gelfgat, editor, *Computational Modelling of Bifurcations and Instabilities in Fluid Dynamics*, pages 171–216. Springer International.

Saury, D., Benkhelifa, A., and Penot, F. (2012). Experimental determination of first bifurcations to unsteady natural convection in a differentially-heated cavity tilted from 0° to 180°. *Experimental Thermal and Fluid Science*, **38**, 74–84.

Seydel, R. (1994). *Practical Bifurcation and Stability Analysis: From Equilibrium to Chaos*. Springer.

Shampine, L. F. (2007). Accurate numerical derivatives in Matlab. *ACM Transactions on Mathematical Software*, **33**, 26–es.

Shroff, G., and Keller, H. (1993). Stabilization of unstable procedures: The recursive projection method. *SIAM Journal on Numerical Analysis*, **30**(4), 1099–120.

Silvester, D., and Wathen, A. (1994). Fast iterative solution of stabilised Stokes systems part II: Using general block preconditioners. *SIAM Journal on Numerical Analysis*, **31**, 1352–67.

Simó, C. (1990). Analytical and numerical computation of invariant manifolds. In C. Benest and C. Froeschlé, editors, *Modern Methods in Celestial Mechanics*, pages 285–330. Editions Frontières.
Also at www.maia.ub.es/dsg/2004/.

Sleijpen, G., and Fokkema, D. (1993). Bicgstab(l) for linear equations involving unsymmetric matrices with complex spectrum. *Electronic Transactions on Numerical Analysis*, pages 11–32.

Snyder, H. (1968). Stability of rotating Couette flow. II. Comparison with numerical results. *The Physics of Fluids*, **11**, 1599–605.

Sonneveld, P., and van Gijzen, M. B. (2008). Idr(s): A family of simple and fast algorithms for solving large nonsymmetric systems of linear equations. *SIAM Journal on Scientific Computing*, **31**, 1035–62.

Sorensen, D. (1992). Implicit application of polynomial filters in a k-step Arnoldi method. *SIAM Journal on Matrix Analysis and Applications*, **13**, 357–85.

Straughan, B. (2004). *The Energy Method, Stability and Nonlinear Convection; Second Edition*. Springer.

Strogatz, S. H. (1994). *Nonlinear Dynamics and Chaos: With Applications to Physics, Biology, Chemistry, and Engineering*. Perseus Books.

Taylor, G. I. (1923). Stability of a viscous liquid contained between two rotating cylinders. *Philosophical Transactions of the Royal Society A*, **223**, 289–343.

Thies, J., Röhrig-Zöllner, M., Overmars, N., et al. (2020). Phist: A pipelined, hybrid-parallel iterative solver toolkit. *ACM Transactions on Mathematical Software*, **46**, 1–26.

Thurner, S., Hanel, R., and Klimek, P. (2018). *Introduction to the Theory of Complex Systems*. Oxford University Press.

Tiesinga, G., Wubs, F., and Veldman, A. (2002). Bifurcation analysis of incompressible flow in a driven cavity by the Newton–Picard method. *Journal of Computational and Applied Mathematics*, **140**(1–2), 751–72.

Tinney, W., and Walker, J. (1967). Direct solutions of sparse network equations by optimally ordered triangular factorization. In *Proceedings of the IEEE 55*, pages 1801–9. Proceedings, Reading.

Trefethen, L. (2000). *Spectral Methods in MATLAB*. SIAM.

Tuckerman, L., and Barkley, D. (1988). Global bifurcation to travelling waves in axisymmetric convection. *Physical Review Letters*, **61**, 408–11.

Tuckerman, L. S., and Barkley, D. (2000). Bifurcation analysis for timesteppers. In E. Doedel and L. S. Tuckerman, editors, *Numerical Methods for Bifurcation Problems and Large-Scale Dynamical Systems*, volume 119 of IMA Volumes in Mathematics and Its Applications, pages 453–66. Springer.

Tuckerman, L. S., Langham, J., and Willis, A. (2019). Order-of-magnitude speedup for steady states and traveling waves via Stokes preconditioning in Channelflow and Openpipeflow. In A. Gelfgat, editor, *Computational Modelling of Bifurcations and Instabilities in Fluid Dynamics*, pages 3–31. Springer International.

Uecker, H. (2021). *Numerical Continuation and Bifurcation in Nonlinear PDEs*. SIAM.

van der Sluis, A., and van der Vorst, H. (1986). The rate of convergence of conjugate gradients. *Numerische Mathematik*, **48**, 543–60.

Van der Vaart, P. C. F., Schuttelaars, H. M., Calvete, D., and Dijkstra, H. A. (2002). Instability of time-dependent wind-driven ocean gyres. *Physics of Fluids*, **14**, 3601–15.

van der Vorst, H. (1992). Bi-CGSTAB: A fast and smoothly converging variant of Bi-CG for the solution of nonsymmetric linear systems. *SIAM Journal on Scientific and Statistical Computing*, **13**, 631–44.

van der Vorst, H., and Vuik, C. (1992). GMRESR: A family of nested GMRES methods. DOI: https://doi.org/10.1002/nla.1680010404.

van der Vorst, H. A. (2003). *Iterative Krylov Methods for Large Linear Systems*. Cambridge Monographs on Applied and Computational Mathematics. Cambridge University Press.

van Gijzen, M. B., Sleijpen, G. L. G., and Zemke, J.-P. M. (2015). Flexible and multi-shift induced dimension reduction algorithms for solving large sparse linear systems. *Numerical Linear Algebra with Applications*, **22**, 1–25.

van Veen, L., Kawahara, G., and Atsushi, M. (2011). On matrix-free computation of 2D unstable manifolds. *SIAM Journal on Scientific Computing*, **33**, 25–44.

Vanka, S. (1986). Block-implicit multigrid calculation of two-dimensional recirculating flows. *Computer Methods in Applied Mechanics and Engineering*, **59**, 29–48.

Varga, R. (1962). *Matrix Iterative Analysis*. Prentice Hall.

Veldman, A. (2010). Computational fluid dynamics. University of Groningen. Lecture Notes.

Verhulst, F. (2000). *Nonlinear Differential Equations and Dynamical Systems*. Springer International.

Verstappen, R., and Veldman, A. (1998). Spectro-consistent discretization of Navier–Stokes: A challenge to RANS and LES. *Journal of Engineering Mathematics*, **34**, 163–79.

Wathen, A., and Silvester, D. (1993). Fast iterative solution of stabilised Stokes systems. Part I: Using simple diagonal preconditioners. *SIAM Journal on Numerical Analysis*, **30**, 630–49.

Watkins, D. (2005). Product eigenvalue problems. *Siam Review – SIAM REV*, **47**, 3–40.

Webster, R. (2018). CLC in AMG solvers for saddle-point problems. *Numerical Linear Algebra with Applications*, **25**, e2142. e2142 nla.2142.

Wendl, M. C. (1999). General solution for the Couette flow profile. *Physical Review E*, **60**, 6192–4.

Wesseling, P. (1999). *Principles of Computational Fluid Dynamics*. Springer Series in Computational Mathematics. Springer.

Wilkinson, J. H. (1965). *The Algebraic Eigenvalue Problem*. Clarendon Press.

Wubs, F., Niet, A., and Dijkstra, H. (2006). The performance of implicit ocean models on B- and C-grids. *Journal of Computational Physics*, **211**, 210–28.

Wubs, F. W., and Thies, J. (2011). A robust two-level incomplete factorization for (Navier–)Stokes saddle point matrices. *SIAM Journal on Matrix Analysis and Applications*, **32**, 1475–99.

Xu, J., and Zikatanov, L. (2017). Algebraic multigrid methods. *Acta Numerica*, **26**, 591–721.

Young, D. (1971). *Iterative Solution of Large Linear Systems*. Academic Press.

Zikanov, O. (2010). *Essential Computational Fluid Dynamics*. Wiley.

Index